Springer Series in the Data Sciences

Springer Series in the Data Sciences focuses primarily on monographs and graduate level textbooks. The target audience includes students and researchers working in and across the fields of mathematics, theoretical computer science, and statistics. Data Analysis and Interpretation is a broad field encompassing some of the fastest-growing subjects in interdisciplinary statistics, mathematics and computer science. It encompasses a process of inspecting, cleaning, transforming, and modeling data with the goal of discovering useful information, suggesting conclusions, and supporting decision making. Data analysis has multiple facets and approaches, including diverse techniques under a variety of names, in different business, science, and social science domains. Springer Series in the Data Sciences addresses the needs of a broad spectrum of scientists and students who are utilizing quantitative methods in their daily research. The series is broad but structured, including topics within all core areas of the data sciences. The breadth of the series reflects the variation of scholarly projects currently underway in the field of machine learning.

More information about this series at http://www.springer.com/series/13852

Jeffrey C. Chen • Edward A. Rubin •
Gary J. Cornwall

Data Science for Public Policy

Jeffrey C. Chen [ID]
Bennett Institute for Public Policy
University of Cambridge
Cambridge, UK

Edward A. Rubin
Department of Economics
University of Oregon
Eugene, OR, USA

Gary J. Cornwall
Department of Commerce
Bureau of Economic Analysis
Suitland, MD, USA

ISSN 2365-5674   ISSN 2365-5682   (electronic)
Springer Series in the Data Sciences
ISBN 978-3-030-71354-6   ISBN 978-3-030-71352-2   (eBook)
https://doi.org/10.1007/978-3-030-71352-2

Mathematics Subject Classification: 98B82, 62-01, 62-04, 62-07, 62H11, 62H12, 62J05, 62J07, 68T50, 68-01, 91-01, 91C20, 68N01, 68N15, 62P20, 62P25, 62P12

This Springer imprint is published by the registered company Springer Nature Switzerland AG
The registered company address is: Gewerbestrasse 11, 6330 Cham, Switzerland

To our families:

Maya and Olivia

Michelle, Theo and Clara

Danielle and Claire

Preface

Public policy has long relied on the empirical traditions of econometrics and causal inference. When new regulations, laws, and new programs are created, policy analysts and social scientists draw on experimental and quasi-experimental methods to assess if these policy implementations have any effect. This empirical evidence is then added to a body of knowledge that identifies causes and effects, which in turn inform future policies that can improve people's lives. The feedback loop can take years (if not decades), but with the advent of data science, we are presented with the opportunity to increase the velocity and precision of decisions. While data science is integrated into many companies of private industry, its role in government is still evolving and being clarified.

Policy analysts strive to understand the measured effects of decisions, but analysts can conflate this measurement with predicting who will most benefit from a policy. Understanding an effect is the pursuit of knowledge, looking back into the annals of history to offer an explanation that satisfies the *whys*. In contrast, being able to anticipate what will happen to program participants is a skill that de-risks decisions, providing operational knowledge about *who* and *what*. These distinctions are not well known, yet government agencies are investing in their data science capabilities.

From working in public policy and government operations, we have had the opportunity to work on data science in this special context. When empowered by the government apparatus, data scientists can unlock the possibility of precision policy. Not only can data science inform policy makers and decision makers to arrive at on-point decisions, but they can develop machine learning applications that have transformative potential. Machine learning can surface patterns in complex datasets, enable precise targeting and prioritization use cases, and inform if not automate decisions. The possibilities and shortcomings have not been documented and illustrated for a public and social sector audience. This textbook introduces aspiring and veteran public servants to core concepts and aims to be the springboard for building data science capacity.

This book has been shaped by a number of brilliant and generous people. Dr. Kyle Bradbury, Dr. Ching-Ling Chen, Dr. Ying-Chih Chen, Christopher Eshleman, Dr. Tyrone W. A. Grandison, Aya Hamano, Annabel Jouard, Artem Kopelev, Dr. Jacob Model, Wade Petty, Alice Ramey, Robin Thottungal, and Levi Weible read many chapters of our text, suggesting ways to simplify complex ideas into a form that is appropriate for policy and strategy audiences. We are incredibly grateful for their contributions. Lastly, our formative experiences in government were enriched by dedicated experts, namely, Jeffrey Tyrens (New York City Mayor's Office of Operations), Jeffrey Roth (New York City Fire Department), Dr. Howard Friedman (United Nations Population Fund), Dr. Curt Tilmes (NASA— Goddard Space Flight Center), Dr. Dennis Fixler, and Dr. Benjamin Bridgman (U.S. Bureau of Economic Analysis).

Cambridge, UK   Jeffrey C. Chen

Eugene, USA   Edward A. Rubin

Suitland, USA   Gary J. Cornwall

Contents

Chapter 1

An Introduction

A campaign manager, an economist, and a data scientist walk into a bar after their candidate delivered a new stump speech. The campaign manager yells excitedly, "Did you hear that crowd? I have a great feeling about this. Everyone, next round's on me!"

The economist adds, "That new education initiative is sure to fill that tech jobs vacancy gap. I'm sure this'll also do wonders for our state's productivity numbers!"

Without looking up from her laptop, the data scientist sighs, then as the bearer of bad news informs the group, "I just plugged in the polling results that came out today and integrated it with my models. It looks like focusing on education would equate to a 70% chance of losing by more than five percentage points. We should instead focus on healthcare in this list of counties and voters with this social media profile, which would give us a good shot at a two-percentage point lead".

After a few moments of awkward silence, the campaign manager smoothly counters, "Well, at least we nailed our campaign music playlist!"

The data scientist mutters under her breath, "Well, about that. . . "

Data scientists are a relatively recent entrant into the ecosystem of public policy. Drawing upon an interdisciplinary skill set that joins scientific methods and algorithmic approaches, data scientists convert expansive, diverse, and timely data into useful insights and intelligent products. There are many aspects of data science that are shared with other fields. The social and natural sciences typically rely on descriptive analysis and causal inference to craft a narrative. In the policy space, social scientists will formulate normative theories that guide how institutions should act (e.g., education initiatives to close the job vacancy gap). Data science approaches problems by adding the edge of prediction. By leveraging all available information, good predictions can reliably anticipate what will happen next. By de-mystifying and de-risking the future, data scientists can develop *data products* that solve a user need through provision of forecasts, prioritization of resources, measuring effects of policies, among others.

In the public sector, data science can hold the potential of enabling precision policy, allowing governments and social sector to implement targeted programs to address society's biggest problems at scale. For example, to improve safety for firefighters after a tragic incident, Fire Department of New York (FDNY) developed a prediction algorithm known as FireCast to target inspections to fire traps before fires occur—a move from reactive to proactive. Political campaigns weave social media, survey, and demographic data to cluster the electorate into discrete segments and craft targeted messaging to each—a strategy to optimize resources. In many cases, data scientists helping campaigns also "nowcast" polls to help guide where resources should be invested. By leveraging data science to inform large decisions, and even micro-decisions on the ground, the spirit of a policy can be more accurately and precisely reflected in the everyday operations of delivering public services.

© Springer Nature Switzerland AG 2021
J. C. Chen et al., *Data Science for Public Policy*, Springer Series in the Data Sciences,
https://doi.org/10.1007/978-3-030-71352-2_1

1.1   Why we wrote this book

The authors of this book have led data science projects in over 40 domains, working for two city government administrations, two US presidential administrations, four research universities, large financial and e-commerce companies, consultancies, non-profits, and international development authorities, among others. Through our collective experiences, we found that the current state of data *anything* in the public policy is well developed in the space of evaluation and inference as it follows the traditions of the social sciences. But there are large gaps as well.

The tools of public policy focus on strategy and storytelling. Strategies focus on the long-term aim of public services, relying on retrospective evaluations to gage if programs are working as intended. Public policy and business schools teach correlative analysis and causal inference to extract the treatment effect of a policy. But there is a blind spot for data-driven tactics—the planned actions that help reach an ideal end state. Tactics maximize the chance of wins here, now, tomorrow, and the day after. Data science and prediction paradigms can enable timelier, smarter decisions to maximize success with the power of foresight. But to make use of data science requires a more forward-leaning posture that embraces experimentation. It also requires a focus on implementation—moving an idea to something real and tangible.

We hope that we can inspire the next generation of data scientists and help them see the distinctions between data-driven strategy and tactics. Seeing these differences, in turn, can help one imagine a future of public services that injects the intellectual adrenaline that fuels innovation. We have written this textbook for the public policy and business students who want to push beyond strategy decks and retrospective program evaluation and use data as a means of direct action. We have written this text for the government data analyst who wants to scale ideas beyond narrative using data to direct targeted interventions. We have written this for anyone who believes we can do more for society through data.

1.2   What we assume

Data science is a complex and sprawling subject area that can take years to master. For experts in the field, complex concepts are communicated in only a few pages in a peer-reviewed academic article. But even for those who are trained in basic econometrics, data science techniques can prove to be challenging to understand, as it requires viewing data problems from an entirely different perspective. For example, it might not be immediately evident why it would be acceptable to apply a non-parametric algorithm like the Random Forest algorithm for prediction problems when it does not have an intuitive linear interpretation. As we will see, the standards of inference and prediction are often conflated—they are not well defined in policy and add unnecessary normative bounds that suppress innovation. To add data science to one's repertoire requires an open mind.

Furthermore, the ethos of data scientists is often different than their social and natural science counterparts—there is a focus on *analyzing* data for a scientific purpose, but also *building* data-enabled applications. Therefore, implementation through programming is often necessary.

We lay out foundational concepts, doing so with the backdrop of public policy and strategy. Where possible, we tell the story of how data science and policy interact. For this book to be effective, we assume an understanding of basic probability theory and statistics with prior exposure to:

- *Summary statistics* such as mean, median, mode, standard deviation, and percentiles;
- *Data structures* such as tabular data in spreadsheets; and
- *Hypothesis testing* such as the T-test and the chi-square test.

We heavily rely on programming to bring to life practical data science applications. Prior experience in a spreadsheet software like Excel can be beneficial to understand data structures, and experience with a high-level programming such as Stata will ease the transition to a statistical programming language (namely, R).

1.3 How this book is structured

How we wrote this book. Through serving as advisors to both executives and senior officials, we have seen that there is much respect for the field of data science. Spending the first moments of a briefing in the technical weeds misses an opportunity to construct the architecture of an argument and gain trust in what we do. We thus take a different approach to introducing the technical machinery of data science. Each chapter begins with the big picture by placing a technical concept in the context of an experience or story from working in the public sector. These experiences easily generalize to any strategy environment and are intended to help the reader visualize how data science is relevant in these contexts. We then lay the technical foundation through a mixture of analogies, mathematical formulae, and illustrations. As data science is a discipline that emphasizes action, we likewise emphasize the role of writing code and provide well-tested coded.

At no point do we attempt to mathematically derive the full quantitative splendor of techniques. Other texts such as *Elements of Statistical Learning* and *Pattern Recognition and Machine Learning* already fill that role. We do, however, believe that the advantage of data science is implementation—*delivery is the strategy.* Where necessary, we illustrate key mathematics concepts through simple arithmetic examples. We have focused each chapter on knowing enough to enable thoughtful action so we help build momentum. For this reason, we dedicate significant portions of the text to extensive Do-It-Yourself (DIY) examples that walk through each step of implementing the code of a data science project. In many cases, the code can be applied to other problems. Most chapters have DIY sections that use real data to solve public problems while others interlace code to illustrate otherwise difficult concepts.

Data and code resources. We have assembled a wealth of datasets and coding resources on Github (https://github.com/DataScienceForPublicPolicy). On this page, there is an assortment of analyses and applications as well as novel datasets. For example, we include code repositories for building an interactive dashboard and the DIY for predicting where downed trees are likely located after a hurricane.

There is no better way to learn than doing. Thus, we have developed over 20 *Do It Yourself* (DIY) sections that illustrate the mechanics of data science projects and discuss how projects fit into the work of the public and social sectors. To download the full DIY repository, visit the following URL:

https://github.com/DataScienceForPublicPolicy/diys/archive/main.zip

and your web browser will download the full repository as a zip file. We recommend that you unzip the repository and place it in directory that you can easily access. For more about the DIYs, visit https://github.com/DataScienceForPublicPolicy/diys.

Roadmap. The next 15 chapters are arranged into four sections that cover issues that data scientists commonly encounter in the public sector.

As computer programming is core to data science, the first section (Chapters 2 through 5) introduces ways of working with data and code. We focus on building core programming skills in R—a commonly used statistical programming language. Chapter 2 lays a conceptual foundation of how programming works and how to install R. In Chapter 3, we introduce basic coding concepts such as types of objects, class, data structures among other elements of programming languages. Chapter 4 focuses on how to manipulate data so that it is useful for analysis and application. As the value of big data often lies in bringing information together from multiple datasets, Chapter 5 is dedicated to techniques for joining data together—a topic that is not often covered in policy or social science classes.

With the foundation laid, we turn our focus to using techniques that are already common in public policy. Chapter 6 focuses on visual analytics through exploratory data analysis—a way to formulate hypotheses and investigate the qualities of a dataset. Chapter 7 is our first step into proper statistical modeling, presenting the regression models for continuous outcomes such as revenue, traffic, crime, etc. Chapter 8 extends regression frameworks to classification problems (e.g., binary outcomes) and introduces regularized techniques that bridge traditional statistical models with machine learning approaches. Chapter 9 takes a step back from the analytical approaches to consider how a single model could be used for distinct types of applications, namely, descriptive analysis, causal inference, and prediction. The distinctions we draw in this chapter are key in understanding why social, natural, and data scientists have divergent views and ways of working.

The third section dives into supervised learning and unsupervised learning approaches—two branches of machine learning that support most data-driven use cases. Chapter 10 exclusively focuses on prediction, starting with simple methods like k-nearest neighbors, then advances to tree-based learning that is appropriate for the types of data used in the public sector. The performance gain of some of these techniques can be quite notable, thus mastering these methods can open new possibilities in policy environments. Chapter 11 explores unsupervised learning that brings structure to otherwise unlabeled data through clustering of observations. Chapters 12 and 13 cover how data science can help policy and strategy organizations make use of new types of data, namely, spatial data (e.g., rasters, vector files, mapping) and natural language processing, respectively.

The last section discusses the social and organizational issues that influence the success and acceptance of data science pursuits. Chapter 14 provides a survey of ethical considerations that are subject of vigorous debate, such as bias, fairness, transparency, and privacy. Chapter 15 lays out the concept of a *data product* that transforms data into a tool to address a user's needs. Lastly, Chapter 16 describes how to design and build data science teams, including typical roles, operating structures, and the hiring process.

Chapter 2

The Case for Programming

2.1 Doing visual analytics since the 1780s

Between 1786 and 1802, a Scottish engineer named William Playfair published *Commercial and Political Atlas*, a collection of data visualizations illustrating economic and political trends of the 18th century. The book introduced the first line and pie charts, skillfully designed and furnished with narrative to guide readers through the economic patterns of the day. While his inferences and conclusions were correlative and speculative, his work is often credited as the beginning of graphical statistics—gathering information in a form that easily conveys insight-packed patterns. For example, in Figure 2.1, England's imports and exports to and from Flanders are plotted for the period between 1700 and 1800 (Playfair 1801). It is striking how the charts resemble the modern line graphs that inundate our daily lives—a smooth pair of lines for each import and export move across the page, illustrating how trade balances increasingly favored England.

At the time when Playfair's work was published, an annual-level dataset containing variables for time, imports, and exports spanning an 100-year period would have been considered big data. Collecting such data would have required a significant time investment and coordination. Developing and perfecting an innovative format to communicate data was also no easy task. Indeed, it was the data science of its age and was the sort of high-browed novelty that drew the attention of royalty in Europe, particularly in France.

The time required to amass, rationalize, graph, and publish the data was expensive. Furthermore, the benefits from the insights could not have been habitually relied upon as information traveled slowly. The decision-making cadence was quite slower—data simply could not allow for real-time decisions. For example, the War of 1812 between the British and Americans ended on December 24, 1814, but the Americans' greatest victory of the war was won two weeks after the peace agreement was signed ("The Battle of New Orleans" 2010). Indeed, news traveled slowly in those days. Over time, the world has grown more interconnected and the speed of communication has greatly improved. The big data techniques of the industrial revolution became so commonplace that elements of it are covered in math curricula in grade schools.

With the advent of computer programming, the mastery required to draft a chart has been democratized in the form of software thanks to a new type of engineer—the software engineer. By leveraging computer programming, software engineers develop programs that allow even the most novice of analysts to apply complex algorithms. New generations of economists and quantitatively minded individuals can stand on Playfair's shoulders without years of apprenticeship using drafting tools. Each analytical task is programmed to imitate an expert's abilities: data are transformed by a computer program following a series of well-planned, pre-defined steps, returning a specific result. There are inordinate number of micro-decisions that determine the success and utility of a graph—how thick to make a line, what color should the palette be, etc. Engineers can set sensible defaults that make those decisions for the user, which reduces the intellectual overhead. For most who have not had the opportunity to program, interaction with data is usually through *What You See Is What You Get* (WYSIWYG) applications—you can directly manipulate visual elements on the screen into the desired form. The output is exactly what is seen on screen. This format is ideal for quickly and tangibly

© Springer Nature Switzerland AG 2021
J. C. Chen et al., *Data Science for Public Policy*, Springer Series in the Data Sciences,
https://doi.org/10.1007/978-3-030-71352-2_2

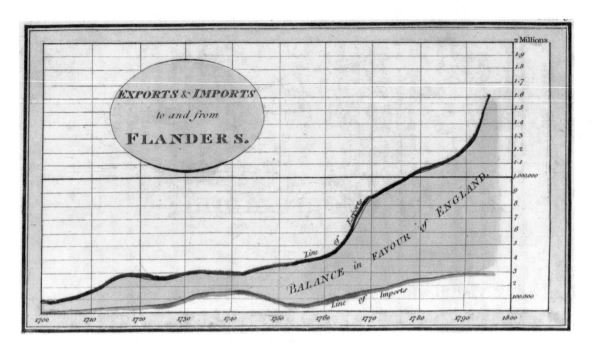

Figure 2.1: Plate 9 from Playfair's Commercial and Political Atlas, 1801.

exploring new data, grabbing the data bull by the horns and showing it who's boss.

Why would anyone need something more than this?

Wrangling that bull is a manual process—every step of the process involves some clicking, typing, and dragging. Suppose you need to calculate the mean for each of 100 variables in a spreadsheet using a WYSIWYG interface. The process might resemble the following:

1. Identify a suitable row to place the calculations.
2. Type in the command "average" in a cell.
3. Navigate the data to the relevant range values on which to calculate the average.
4. Navigate or type in the location of the range within the "average" function. Press enter.
5. Copy and paste this command across 100 columns.
6. Sometimes, the command might be accidentally applied to too few (e.g., 92, 93, 94 columns) or too many columns (e.g., 103, 104, 106 columns)—some edits may be in order.

The process is only a half-dozen actions if you know your way around the spreadsheet, but many more without the know-how of productivity-enhancing hacks. Alternatively, this whole process can also be expressed with a single command in a programming language (or two commands if one also accounts for the line of code to import data). The colMeans function in the R statistical programming language, for example, can calculate the mean for as many columns as the computer's memory can hold. The np.sum function in the Python programming language's numpy library can achieve the same result. The difference between one and six actions to calculating a number is not particularly large—especially if the task is done only once.

What if the same problem were scaled to an arbitrarily large number of spreadsheets, say 1,000 or 10,000 or 1,000,000? Achieving the objective becomes burdensome by the sheer scale of the problem. What if the task also were expanded to processing billions of records, producing forecasts, and writing simple descriptive summaries of the data? The manual spreadsheet edits would be beyond arduous, let alone the amount of review required to pinpoint errors. Checking for errors would then compound the time required to complete the task.

The advantage of programming becomes far clearer. By programming, repetitive tasks can be delegated to a computer. Complicated tasks only have to be defined once, then applied to any number of other instances.

When applied to policy, programming has the potential of transforming the art of policy into a science of precision policy. A data scientist can process an inordinate amount of data to train an algorithm to predict the seemingly unpredictable, like whether someone will recidivate, hit economic troubles, or need mental health care. Programming also allows huge sums of data to be processed to make even more profound causal analyses.

In this chapter, we take the first step toward programming by introducing conceptual fundamentals: how does programming work, where do we *get* a programming language, how to interact with code, and how to justify its use. In subsequent chapters, we build upon these ideas to develop programs that load, transform, and apply data in impactful ways.

2.2 How does programming work?

Programming languages allow users to communicate with a computer's Central Processing Unit (CPU) in a way that makes sense to both humans and computers. The CPU is the brain of a computer, comprised of a large number of switches. The computer can carry out basic computing functions like math and input/output operations by flipping switches on and off in a certain patterns and sequences. In a way, CPUs allow computers to think. CPUs do not speak human language, however. Instead, they rely on instructions communicated in *machine language*. For a programmer to pass instructions to the CPU, she needs the help of a programming language to translate between human commands and machine language.

It seems abstract at first, but we regularly use intermediary devises to get the job done. For example, to drive a car from point A to point B, a driver needs to regulate the speed using both gas and brake pedals, as well as guide the vehicle through the streets through a series of well-timed, coordinated rotations of the steering wheel. The user actions are translated into signals that a car understands. For example, when a driver presses down on a gas pedal, the throttle control receives the force and translates it into an amount of fuel that is injected into the engine. The result: *acceleration*. The system that controls acceleration is a layer that abstracts the acceleration process into something intuitive for humans. Otherwise, the driver would need to squeeze a syringe of fuel into the engine and hope for the best.

There are different *types* of programming languages for different types of problems. Some will abstract the interaction with the CPU to high-level functions (e.g., running an entire algorithm) while others are useful for low-level functions that are in the computational weeds. We can group programming languages into three broad groups:

- *Compiled languages* rely on a *compiler* program to translate source code into binary machine code, which on its own is self-contained and executable. After translation, the resulting binary code is self contained and can be executed without any reliance on the source code. In data science, a common compiled language is C. As the code is compiled into binary once the source code is completed, the performance tends to be faster.

- *Interpreted languages* rely on an *interpreter* program to execute source code. Unlike compiled languages, interpreted languages are fully reliant on the interpreter program, meaning that each time a user wants to take an action, the source code needs to be re-run and interpreted through the program. The interactivity of web browsers is made possible by an interpreted scripting language known as JavaScript. For advanced statistical research, the R is one of the interpreted languages of choice for statisticians, social scientists, and data scientists.

- *Pseudocode languages* (or p-code) are hybrids between compiled and interpreted languages. The source code is compiled into bytecode—a compact instruction set that is designed for specialized, efficient interpreters when executed. A common example of a p-code language is Python.

What do programming languages mean for data science? Data science tends to have the greatest payoff when learning patterns at a large scale. To achieve broad impact, programming is a must. Two languages have proven themselves to be the workhorses of data science: R and Python.

R is a common language among statisticians, epidemiologists, social scientists, and natural scientists as it is designed for mathematical computation. The language is optimized specifically for all things math and statistics, ranging from matrix algebra to more complex optimization. The implications are that the language is quite extensible. In fact, thousands of *code libraries*—sets of functions that work together—extend R's usefulness from visualizations to machine learning. The language does not naturally lend itself to creating standalone web applications, but recent advancements have vastly improved R's ability to support production-grade use cases.

Python is a general-purpose scripting language and a favorite of software engineering teams. Not only is it used for deploying web applications such as websites and Application Programming Interfaces (APIs), but it has been extended to mathematical and statistical computation through an extensive set of code libraries. A key feature of Python is its simplicity and readability.

Which one language should a beginner choose first? The decision rests on the tasks you need to accomplish. For settings where data informs strategy—identifying which levers can be pulled to affect change, learning R is a sound choice. It is more than able to conduct data analyses, build sophisticated prediction models, visualize insights, and develop interactive tools. For cases where data-driven applications need to be scaled to a large number of users—like streaming media, social media ads, or financial services—starting with Python makes more sense.

Of course data scientist can start with one language and then learn another. In fact, knowing more than one language makes data scientists more versatile and marketable. There are other languages that complement these core languages. Structured Query Language (SQL), for example, is the most common language for managing databases and extracting data. JavaScript is common for high-end custom visualizations. Scala is a language used for writing jobs for real-time data frameworks like Spark.

In this text, we introduce data science through R. Let's get the software up and running.

2.3   Setting up R and RStudio

R is an open-source language developed and maintained by a large community of high-powered contributors who make available the core program and the thousands of libraries that extend its functionality through the Comprehensive R Archive Network. As it is open source, it is free to use. In fact, many of the most influential programming languages in use today are open source as *everyone* can benefit from *everyone else's* code contributions.

2.3.1   Installing R

To install R, visit the CRAN website:

- Visit https://cran.r-project.org/.
- Select "Download R" for your computer's operating system or OS (Windows, Linux, or Mac).
- Download the latest **.pkg** file that your OS can support.
- Open and install the **.downloaded** file by following the instructions.

Upon installing R, open the software. The first window that appears is a nondescript window known as the *console*, which is the gateway to R (see Figure 2.2[1]). R code can be directly entered into the console, which then is interpreted into machine language so that the CPU can perform the desired task. While entering code directly into console can achieve a desired result, it is a short-lived victory as the code disappears into the ether—it is not saved. It is best practice to write code into the *script editor* where it can be saved. The script editor is a feature that is available in the toolbar above the console.

The R GUI (graphical user interface) is the bare bone basic designed for when R was strictly for math and statistics. An Integrated Development Environment (IDE) known as RStudio can be installed to better manage the data science workflow.

---

[1]While this screen capture is from a Mac computer, the layout is quite similar on a Windows machine.

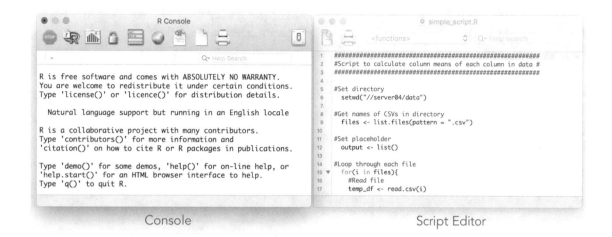

Figure 2.2: The Base R console (left) and the script editor (right).

2.3.2 Installing RStudio

Using the basic windows in R can feel like driving a car with only a seat and a steering wheel—it lacks the additional features to make driving a fully intuitive, integrated experience. RStudio adds an interactive development environment that sits on top of R to make the programming process more comfortable and intuitive. Note, however, that RStudio is not R.

One useful feature in RStudio is code highlighting, which helps the coder see which parts of the code contain functions, variables, comments, among others. The highlighting also identifies when code is incomplete or erroneously written which helps prevent bugs from entering the code. Another useful feature is how RStudio handles graphics. If a piece of code outputs hundreds of graphs, R will render hundreds of separate windows. RStudio, in contrast, will capture each of the windows produced by the script and place them into a gallery of plots, thereby keeping the workflow as clean and manageable as possible. The number of productivity-enhancing features continues to grow since the introduction of RStudio in 2011.

One key thing to remember: *Since RStudio is an IDE, it requires R, but R does not require RStudio.*

To install RStudio, follow these steps:

- Visit the RStudio website https://www.rstudio.com.
- Navigate to the Products > RStudio > RStudio Desktop > Download.
- In the Download section, select the *Installer* that corresponds to your computer's operating system (e.g., Windows 32-bit, Windows 64-bit, Linux or Mac, etc.).
- Open and follow the installation wizard.

If R has not yet been installed, make sure to do so before opening RStudio. During initialization, the program will look for R and ask if the detected version is correct. Upon installing both programs, open RStudio by clicking on its icon on your Computer.

When you open RStudio for the first time, the interface is laid out as four panes, as seen in Figure 2.3. Like the basic R windows R, RStudio contains a *script editor* in which code can be written, edited, and saved to disk. In addition to smart highlighting the different parts of a script such as functions and variable names, the user can work on multiple scripts simultaneously.

All R code is executed in the *console*, including the code written in the script editor. The console pane is also shared with the computer's command line terminal, which is quite convenient for more complex projects that require access to deeper functionality.

The *environment pane* provides an inventory of the data, functions, and other *objects* that are being used in the current session of R. In some statistical analysis software like Stata, only one dataset can be managed

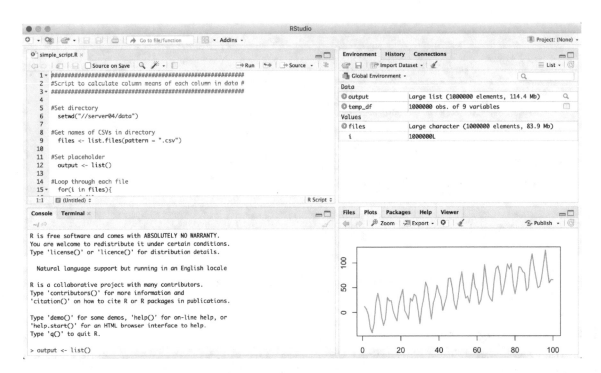

Figure 2.3: A view of RStudio being used to write this chapter: the script editor (top left), environment pane (top right), plots (bottom right), and console (bottom left).

at a time. As you can see, R allows for multiple data objects to be used at the same time, so as long there is memory available. The environment tab also shares the pane with the *History* tab that keeps track of all commands that have been executed in the current session. If there is a database (e.g., OBDC) or computing framework (e.g., Spark) being used in the current session, the *Connections* tab makes the process relatively painless.

Lastly, all graphs and figures are rendered in the *plots* pane. One of the nifty features of the plots pane is it stores the history of the current session's plots, making it easy to toggle and review a string of graphs. The pane is also home to

- the *Files* tab that shows all files and folders in the current directory;
- the *Packages* tab from which one can install and manage code libraries;
- the *Help* tab that users can search coding reference materials; and
- the *Viewer* tab where interactive web-based visualizations are rendered.

For the most part, the layout of RStudio panes is fairly consistent with new tabs added and styling differences from one version to the next.

2.3.3   DIY: Running your first code snippet

DIYs are designed to be hands-on. To follow along, download the DIYs repository from Github (https://github.com/DataScienceForPublicPolicy/diys). The R Markdown file for this example is labeled diy-ch02-running-snippet.Rmd.

In this first DIY, let's demystify the idea of *running code*. When writing code, it is important to make sure the code is easily understood—your assumptions and logic are clear and produce the intended result. This clarify ensures that a program that becomes relied upon in a policy process can be used by *anyone*. The code snippet below involves two steps. The first step is to simulate a random normal distribution with $n = 10000$ observations, storing the values in x. Then we plot that result as a histogram with 100 light blue

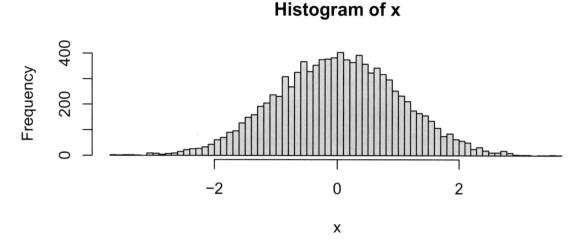

Figure 2.4: Histogram for a simulated normal distribution.

bins. The lines that begin with a *hash* contain comments that describe what will happen in the lines that follow. Comments help the writers and readers of code to better understand the steps in a program. Copy the code from "# Simulate 10,000..." through "hist(x..." and paste this into a new script in **RStudio**.

```
# Simulate 10,000 values drawn from a normal distribution
x <- rnorm(10000)

# Plot graph
hist(x, breaks = 100, col = "lightblue")
```

You can run this code snippet using one of two options:

1. *The Run button*: Highlight all lines, then click on the Run button at the top of the script editor.
2. *Hot keys*: Highlight all lines (Control + A), then press Control + Enter.

Both options send the highlighted code in a script to the console where it is interpreted (run), but the Hot Key approach is far faster, removing the need for the mouse. The result should look similar to the histogram in Figure 2.4.[2] Granted, it is a simple example, but it is just the tip of a massive iceberg of data science possibilities.

2.4   Making the case for open-source software

Most people who work with data rely on spreadsheet software such as Excel and Open Office to conduct accounting and statistical tasks. For the better part of the desktop computing era, spreadsheets were all that was needed to use data, and thus for many, the idea of writing code is quite daunting and seemingly unnecessary. But to create the next generation of public services, as society grows larger and more complex, we need to go beyond what has been the norm.

As a data scientist in the social and public sectors, you may need to persuade others that open-source software is critical to your organization's future. You will need to convince stakeholders why coding matters and how it can be a force multiplier. We recommend focusing on what *can* be achieved through coding, balanced with practical considerations such as *cost, security, and customer support*.

---

[2]The histogram will look similar but not exactly the same, as the $n = 10000$ observations are randomly drawn. In subsequent chapters, we show a simple way to ensure that the same simulation can be replicated on any computer.

Enhanced Capabilities. While spreadsheet software can accommodate data up to approximately one million records, its ability to be flexible and responsive is quite limited with larger datasets. Furthermore, it lacks many of the basic tooling and algorithms used for sophisticated use cases such as risk management, micro-targeting, interactive visualizations, among others. R, in contrast, can handle as much as data as memory available on your computer. It also has thousands of *libraries* that are similar to plug-ins to vastly expand its capabilities, ranging from applying artificial intelligence techniques to genetics to producing real-time forecasts for stock markets. R does not replace spreadsheet software, but rather opens a new world of possibilities to take organizations to the next level.

Cost. The cost of new software is often a concern for organizations with lean budgets. Fortunately, open-source software such as R and Python are free to use and accessible by anyone. Since there is no cost barrier, this pair of programming languages have been adopted by a large, vibrant global community of practitioners and experts who contribute, maintain, and extend the functionality of the software.

Security. In an age of heightened cyberrisks, Chief Information Officers (CIOs) and Chief Information Security Officers (CISOs) will need to be vigilant to safeguard their IT systems. Part of their security strategy is to rely on software from trusted vendors who maintain their own proprietary software. This strategy also implies that open-source code that is openly exposed and maintained has more cyberrisks. Ironically, according to the NIST National Vulnerabilities Database that records all reported vulnerabilities in software, there have not been any reported vulnerabilities with R, but many more for proprietary software. Since a large community relies on and contributes to the software, there are far more people scrutinizing the source code than just a single company's development team. Indeed, it is a different model for viewing software security, but as IT paradigms change, these open-source approaches offer new benefits that were not previously possible through proprietary software alone.

Customer support. Like life insurance, some IT offices need peace of mind when buying a commercial software—it comes with customer support if help is needed. Open-source software, in contrast, does not offer customer support. Because it is a different paradigm of working, troubleshooting code has become much simpler with online forums like *Stack Overflow* in which users post questions and a global audience offers possible solutions. However, if worse comes to worse, there are a variety of enterprise-grade open-source software available for R and Python. RStudio offers an enhanced commercial version of R with software support and companies like Continuum maintain special builds of Python that are enhanced and offer technical support as well.

Human capital. An overlooked consideration is the human capital perspective. New generations of analysts and programmers *expect* to use R and Python in their jobs. It is what is taught in university and it is what they want to use to do their work. In some organizations, staff are unproductive and leave when forced to use antiquated software without tooling to streamline their day-to-day tasks. Advocate for open source. At the end of the day, work is about your team and the team must have the right tools in order to do the job with pride and do it well.

Chapter 3

Elements of Programming

3.1 Data are everywhere

Data are everywhere we look. They are not necessarily numeric values stored in a spreadsheet but can be embedded in everyday media, imagery, and text. Stock market performance, for example, is rarely communicated to the public in the form of a spreadsheet (perhaps to investors). Rather it is communicated in the form of concise statements. Let's take the example of a CNBC online article (Imbert 2017) about stock market performance:

> *The Dow Jones industrial average erased earlier losses today to end 56.97 points higher at 21,865.37...*

An investor or an economist can interpret a single statement and draw upon their past experience to make sense of it. In a concise line, we can infer there is some volatility in a specific index but at the end of the day, the market ended on a high note. Reviewing news articles and reports one at a time is effective in small batches. But with high-frequency news cycles and continuous content generation, the volume of information may be too much for any one person or group of people to analyze in a timely and accurate fashion.

Computers can help, but they need some guidance. Programmers exercise their creative fibers to layout rules and procedures in the form of software. The software is comprised of, among other things, *functions* that dictate how the computer should transform a raw piece of data using clearly defined procedures in order to derive value from it. The data, however, needs to follow basic conventions. The *functions* within a program are built with a certain *data type* in mind. Addition and subtraction functions only work if given a numeric value. A function could also be built to review a sentence for useful information. For example, a simple program could boil the CNBC article into a few morsels of information:

- The "Dow Jones industrial average" references the name of a specific stock market index. A function could check if the name "Dow Jones" is in the sentence, then store the information.
- The values of "56.97" and "21,865.37" are numbers (with units). If given a pair of numbers, the function could reason that the smaller of the two numbers is likely a change in index and the larger number is the index volume.
- The direction of growth and volatility is embedded in the phrases "erased", "earlier losses", and "higher".

Once information is extracted, the data are only useful if they are stored. Often times, data are assigned to *objects* that are useful building blocks to *structure, manipulate,* and *use* the data. Then, the objects can be stored to some type of file. These steps are some of the common tasks of programming—transforming data into the right form.

Data structured from a sentence.

© Springer Nature Switzerland AG 2021
J. C. Chen et al., *Data Science for Public Policy*, Springer Series in the Data Sciences,
https://doi.org/10.1007/978-3-030-71352-2_3

source	market	change	volume	date	description
CNBC.com	Dow Jones	56.97	21865	2017/08/29	erased, earlier losses, higher

Modern data science is only made possible through programming. It is through programming that computers analyze satellite imagery or guess topics in textual information. As a first step toward algorithmic splendor, this chapter presents fundamentals of programming with an eye toward data. We cover the basic skills that allow data scientists to store data, manipulate different types of information, and load and executive code.[1]

3.2 Data types

In R, data are encoded (often implicitly) as one of several types of data, e.g., `numeric`, `logical`, `character`, `factor`, or `date`. In this section, we describe the properties of each.

3.2.1 `numeric`

As you may have guessed, *numeric* data are numbers. The numeric type is R's catch-all for all numbers, encompassing both integer and continuous numbers. Below are four examples of numeric data in R.

```
4
56.4
4/9
pi
```

3.2.2 `character`

Words, sentences, letters, among other sequence of letters and numbers are often encoded as `character` data (also known as *strings*) in datasets of linguistic text (e.g., court transcripts, novels, text messages, or government audit reports). We R that a value is character by wrapping the value in quotation marks (e.g., `"Text"`). Both single quotes (') and double quotes (") work.

```
"text"
"Text is fun!"
"Let's have an example with an interior quotation mark."
```

3.2.3 `logical`

If we want to test whether something is *true* or *false*, then we have stumbled into the world of logic. Because programming and data science frequently rely on *logical* data (or *Booleans*), they get their own data type. In R, logical data are `TRUE` or `FALSE`, which can be abbreviated as `T` or `F`. As is always the case in R: *the letters' cases matter*, meaning you must use `TRUE` and cannot use `true`. Below are the only four values that logical values can assume:

```
TRUE
T
FALSE
F
```

But how do we produce logical values? Often times, a relational operator is used to test whether a statement is `TRUE` or `FALSE`. For instance:

```
# Is 3 less than 4?
3 < 4
```

---

[1]An abridged, hands-on version of this chapter is available in the DIY repository. To follow along, download the DIYs repository from Github (https://github.com/DataScienceForPublicPolicy/diys). The R Markdown file for this DIY is `diy-ch03-basics.Rmd`.

```
# Is 3 greater than 4?
3 > 4

# Is 3 equal to 4?
3 == 4
```

In combination with a logical operator, we can evaluate two or more relational statements.

```
# Is 3 less than 4 AND greater than 1?
(3 < 4) & (3 > 1)

# Is 3 less than 4 OR greater than 1?
(3 < 4) | (3 > 1)
```

There are a variety of relational operators similar to the examples above. Logical operators will come in handy for filtering and subsetting data in later chapters. Table 3.1 presents some of R's most common logical operators.

Table 3.1: Common relational and logical operators.

Operator	Example	Translation
>	x > y	x is greater than y.
<	x < y	x is less than y.
>=	x >= y	x is greater than or equal to y.
<=	x <= y	x is less than or equal to y.
==	x == y	x is equal to y.
&	x > y & x > z	Both x > y and x > z are true.
|	x > y | x > z	Either x > y or x > z are true.

These logical operators are particularly common for `numeric` data, but they can be used for any type of data. For instance, if we execute each of these statements, we will notice that R is indeed case sensitive: `"a"` is not the same as `"A"`, and, in fact, `"a"` precedes `"A"` in R's ordering of letters. Numbers precede letters.

```
# Is "A" equal to "a"? (Is 'R' case sensitive?)
"A" == "a"

# Is "a" less than "A"?
"a" < "A"

# Is 3 less than "a"?
3 < "a"

# Is numeric 3 equal to character 3?
3 == "3"
```

Finally, it is worth noting that you can transform `TRUE` and `FALSE` to binary numeric (1 and 0) by multiplying them by the number 1. This is particularly useful for creating *dummy variables* quickly.

```
TRUE * 1
1 * FALSE
```

3.2.4 `factor`

Factors are a special data type in R, because they allow you to take one type of variable (e.g., `numeric` or `character`) and add a few more layers of information to make it interpretable. We can add labels to data so that values "Sun" and "Mon" are labeled "Sunday" and "Monday". Also, we have the option to indicate the order of that data, meaning Sunday precedes Monday, etc. You can manipulate these labels, the ordering, and a number of other factor options through the functions `factor` and `as.factor`.

In the age of big data, we may want to eliminate redundant information. Factors can help. For example, a dataset containing 50 million people's annual check-up covering 10-year period might have 500 million name records—450 million of which are repeats. Factors can encode each name as a unique identifying number, thereby removing the need to store redundant information and greatly reduces file sizes. The benefits of factors are thus quite notable.

We will more fully explore `factor` data in the graphing section.

3.2.5 `date`

While there are many ways to think about dates—`character` data like "January" or `numeric` data like 1—dates are so common (and oddly challenging to work with) that they get their own type of data, namely, the `date` type. We will dive more into working with dates in the next chapter.

3.2.6 The `class` function

Inferring the type of data by simply looking at the data can be misleading. The question of type is particularly important when you read data into R from another source—we will not always know whether R read a variable as `numeric`, `character`, `logical`, or perhaps `factor`. The `class` function works wonders.

A Quick aside on functions: This is the first time in this book that we officially use function. We will cover functions in much greater depth throughout the book, but let's first cover a few of the basics. Functions take **arguments** (data that are fed or passed to a function) and produce **output**. For example, the class function takes one argument like the number `56.97` and outputs a character that describes the class of whatever argument we inputted into the function—`"numeric"` in our case. Functions help us complete tasks and analyses that may be too difficult or time-consuming to code by ourselves. Functions also excel at repetitive tasks—you can easily tell R to do something a million times so you do not have to copy and paste a million times. You can also write your own functions in R—but that will come later.

Let's see this `class` function in action.

```
# Numeric
class(56.97)
class(57)
class(57L)
class(1/3)

# Character
Class("Dow Jones")

# Logical
class(TRUE)
class(F)
class(3 > 2)
class(1 * TRUE)

# Factor
class(factor(3))
```

```
# Date
Sys.Date()
class(Sys.Date())
as.Date("2015-12-31")
class(as.Date("2015-12-31"))

# Fun
class(class)
```

One more note on functions: Many functions in R can take many arguments, and sometimes arguments are optional. To keep track of which values you are assigning to which arguments, it is often best to use the argument's name. In the `class` examples that we just covered, we implicitly assigned the various values that we fed to `class` to its sole argument `x`. We could have also used the argument's name.

```
# Without the argument's name
class(TRUE)

# With the argument's name
class(x = TRUE)
```

Both options produce the same output, but the second example—in which we explicitly use the argument's name—makes it a bit clearer what we are doing. We admit it does not truly matter for `class`, since there is only the possibility of one argument. However, we will soon meet some functions where explicitly naming arguments helps you and your collaborators.

3.3   Objects in R

Simply entering data into R's console is not enough. The console runs commands blindly, but the output evaporates into the digital ether as it is not automatically captured. We need a way to store data in our computer's memory. To be able to access and re-access the data that we store, we give each piece of data a name. In R parlance, we *assign* the values (the data) to objects stored in memory (the names).

R's assignment operator `<-` exists solely for the task: assigning a value (on the right-hand side) to the name (on the left-hand side). If you are reading code aloud,[2] `object <- value` is typically read "*object* gets *value*".

```
a <- 3
string_ex <- "Some funny/clever words."
some.logic <- TRUE
```

As you can see, we have a freedom in the names that can be used for the objects. However, we can mitigate the heartache of troubleshooting erroneous code by using common variable name conventions. Variable names should avoid (1) a number as the first character in the name (e.g., `"1variable"` is not a good idea) as this will not register as an object, (2) any spaces in the name (e.g., `"variable 1"`), (3) a name of an existing function.

Once we assign a value to an object, we can access that object's value and perform operations on that value. For instance, we can assign a value of three to object `a`, as above, and then we can (1) check the value held in `a` by executing `a` in the console, (2) check the class of `a`, and (3) multiply `a` by 3:

```
# Assign value 3 to object named 'a'
a <- 3

# (1) Check a's value
a
```

---

[2]Yes, it does happen.

```
# (2) Check a's class
class(a)

# (3) Multiply a by 3
a * 3
```

3.4   R's object classes

Just as data in R come in different *types*, objects in R come in different *classes*, e.g., a vector, a matrix, a data.frame, or a list. Each of the object classes presents data in a slightly different structure, making it better for some uses and worse for others. Here, we review four of the most common data structures now and will discuss others as they arise throughout the book.

3.4.1   vector

vector objects are one of the simplest and most used classes of objects in R. Specifically, (in R) a vector is a one-dimensional "list" of values. The values in a vector can be logical, numeric, factor, or character, but they must all be of the same type. If the objects are not all of the same type, R will *coerce* them to the most conducive type.

The simplest way to create a vector is with the concatenate function: c. To create a vector of values, you feed the c function your values, each separated by a comma (,). For instance, to make a vector out of the numbers 1, 2, and 3, we write c(1, 2, 3). And to create a vector of the characters "apple" and "banana", we write c("apple", "banana").

```
# Vector of numbers
c(1, 2, 3)

# Vector of characters
c("apple", "banana")
```

So what happens when we create a vector of a number and a character?

```
# Mixed vector
c(1, "banana")
```

R *coerces everything to character*. This result may not be what you expected—you may have expected R to allow a vector to contain objects of different types. At some point, the folks who developed R made this decision—and millions of other little decisions—that led to the current implementation of R. As you get deeper into R, you will find that there is a logic to these decisions (e.g., why R tells you it is TRUE that "apple pie" < "banana cream pie"), but R will still occasionally surprise you. And that is part of learning to program in R. So don't be afraid to dive into the rabbit holes that surprise you. You may lose a few hours in the short run, but in the long run you'll become an excellent data scientist.[3]

3.4.2   matrix

The next class of objects is R's matrix class. In R, matrices are very similar to the matrices taught in mathematics courses: they have rows and columns filled with data. However, in addition to filling your matrices with numbers, as you did in your math courses, in R you can also fill your matrix with other types of data—e.g., logical, character, factor—just as you can with vectors.

You generally will need to feed two arguments to R's matrix function:

1. The data that you want placed into your matrix, as a vector.

---

[3] And R Wizard.

2. The number of rows (`nrow`) in your matrix.

For example, let's imagine we want a matrix with 2 rows and 3 columns that contains the integers one through six. Then we need to feed `matrix` (1) a vector containing the integers one through six, and (2) `nrow = 2`, telling R we want two rows.

Aside on creating numeric sequences: As we covered before, you can easily create short numeric vectors using the `c` function, e.g., `c(1, 2, 3)`. But this construction quickly gets tedious for longer vectors—even creating a vector of the integers between 1 and 100 would be a nightmare with only the `c` function. Lucky for us, R's makers included a quick and easy fix: `a:b` generates a numeric vector containing the integers from `a` to `b` (inclusive of `a` and `b`). For example, `1:100` will create a vector with the integers 1 to 100.[4]

```
# Old school
c(1, 2, 3, 4, 5, 6)

# New school
1:6
```

Using our new sequence knowledge, we can create our 2×3 matrix with the integers one through six[5].

```
matrix(data = 1:6, nrow = 2)
```

And there we have it. Notice that because the `matrix` function takes two arguments, we explicitly named the arguments[6] and we separated the two arguments' assignments with a comma. Also notice that by default, R filled the matrix *by its columns* (top to bottom and then left to right). If you would rather fill your matrix *by its rows*, you can add the optional argument `byrow = TRUE` to the `matrix` function:

```
matrix(data = 1:6, nrow = 2, byrow = TRUE)
```

As we mentioned earlier, you can create matrices filled with essentially any type of data.

```
# Character
matrix(data = c("Six", "words", "can", "be", "a", "lot."), ncol = 3, byrow = T)

# Logical
matrix(data = c(T, T, F, F, F, T), ncol = 3)
```

Finally, notice that you can force R to fill in a `matrix` by specifying both the number of rows (`nrow`) *and* the number of columns (`ncol`) for a `matrix`.

```
matrix(data = 1, nrow = 3, ncol = 2)
```

3.4.3 data.frame

As a budding data scientist, one day you will wake up and find yourself surrounded by `data.frame`s. The `data.frame` is the basic object for data entry and analysis in R—much like a table of data or a spreadsheet, but *better*. Just like a `matrix`, a `data.frame` has rows and columns. Unlike a `matrix`, a `data.frame` can have different types of data in each of its columns. In addition, you will name each of the columns in your `data.frame`. In general, you define a `data.frame` using the following conventions:

```
data.frame(
  column1 = values...,
  column2 = values...,
  ...
  columnN = values
)
```

---

[4]For more complicated tasks, there's also the sequence function `seq`, which takes the starting value for the sequence (`from`), the stopping value for the sequence (`to`), and the *distance* between each of the items in the sequence (`by`).

[5]Two rows and three columns.

[6]Not necessary but highly helpful.

More concretely, let's create a two-column `data.frame`. The first column will be called `some_letters` and will contain... some letters. The second column with be called `some_numbers`...

```
data.frame(
    some_letters = c("a", "A", "b", "B"),
    some_numbers = 1:4
)
```

One of the main requirements for a `data.frame` is that R needs to be able to fill each of the `data.frame`'s columns. In practice, this requirement means one of two things:

1. The values you specify for each column have the same number of elements (as they do above).

2. The numbers of elements you specify for each column are multiples of each other.

For an example of the second option:

```
data.frame(
    some_letters = c("a", "A", "b", "B"),
    some_numbers = 1:2
)
```

R repeats the shorter vector (the second column in this example) until the two columns are the same length. If the numbers of elements are not multiples (e.g., 4 and 3), R will refuse to construct the `data.frame`, and you will get an error. For example,

```
data.frame(
    some_letters = c("a", "A", "b", "B"),
    some_numbers = 1:3
)
```

```
Error in data.frame(some_letters = c("a", "A", "b", "B"), some_numbers = 1:3) :
  arguments imply differing number of rows: 4, 3
```

Notice that R's error message gives us a hint to what the problem is: `arguments imply differing number of rows: 4, 3`. Generally, error messages in R will contain a hint of the problem, but it takes some experience to understand what R is really trying to say (e.g., here we need to know the rules about column lengths in a `data.frame`). As a result, error messages can seem a bit cryptic while you are in the early stages of learning R.

Aside: In the `data.frame` examples above, we wrote `data.frame(` and then went to the next line to define the columns. R does not stop at the end of the line, because it "knows" that the `data.frame` definition is unfinished. You don't have to define a `data.frame` over multiple lines[7], but using multiple lines tends to make your code more legible and easier to troubleshoot.[8]

3.4.4   `list`

R's `list` class of objects is exactly what its name implies—it is a list of things. Similar to the `vector`, a `list` has only one dimension—its length. Unlike the `vector`, a `list` can contain objects of many different types. The most incredible feature of `list` is its elements can be *anything*—it is remarkably flexible. We are not resigned to simple objects and formats like `1` or `T` or `"Some text"`. A single `list` can have `vectors`, `data.frames`, and even other `lists` as its elements.

The conventions for creating a `list` mirror what we've discussed so far—especially with the `vector`-creating c function: Simply pass the objects of interest to the `list` as the arguments in the `list` function:

```
list(
    "apples",
```

---

[7]You can do it in one line: `data.frame(some_letters = c("a", "A", "b", "B"), some_numbers = 1:3)`.

[8]In longer code examples in later chapters, we condense the code layout.

```
    c(T, F, T),
    matrix(data = 1:4, nrow = 2),
    data.frame(var1 = 9:10, var2 = 11:12)
  )
```

Here we have a (very mixed) four-element `list` with individual elements:

1. a one-element `character` vector,

2. a three-element `logical` vector,

3. a 2×2 `matrix`, and

4. a two-variable `data.frame`.

As with most objects in R—as we have just illustrated with the columns of the `data.frame`—each individual element of the `list` can be named. Let's re-create the `list`, naming its elements with original, creative names.

```
list(
  el_1 = "apples",
  el_2 = c(T, F, T),
  el_3 = matrix(data = 1:4, nrow = 2),
  el_4 = data.frame(var1 = 9:10, var2 = 11:12)
)
```

For complex programming projects, `lists` are indeed the workhorse of efficient processing.

3.4.5 The class function, v2

Previously we used the `class` function to learn the data types of various objects (e.g., `class(3.1415)`). The `class` function can also be used to learn about an object's class. For example,

```
# Check the class of a matrix
class(matrix(data = 1:4, nrow = 2))
```

```
## [1] "matrix" "array"
```

```
# Check the class of a data.frame
class(data.frame(a = 1:3, b = 1:6))
```

```
## [1] "data.frame"
```

```
# Check the class of a list
class(list(a = 1:3, b = 1:6))
```

```
## [1] "list"
```

However, if we check the class of a `vector`, R will not indicate it is a `"vector"`. Rather, R will return the type of vector it is—numeric, `logical`, `character`, etc.

```
# Numeric vector class
class(c(1, 2, 3))
```

```
## [1] "numeric"
```

```
# Character vector class
class(c("a", "b", "c"))
```

```
## [1] "character"
```

```
# Logical vector class
class(c(T, F, T))
```

```
## [1] "logical"
```

```
# Coercion to character
class(c(T, 3, "Hi"))
```

```
## [1] "character"
```

The takeaway? R "thinks" about different types of vectors as different classes of objects.

In addition, each object class also has its own function to check whether an object belongs to the class. For example, `is.matrix` can check whether an object is a `matrix`. Type "`is.`" into the `RStudio` console and hit the tab key—you will get a glimpse of all of these `is.*` class functions.

3.4.6 More classes

R makes use of many other object classes. Many of the object classes are specific to an application (e.g., geographic data like rasters or shapefile) or packages (e.g., when you run a regression with the function `lm`, the function produces a special `lm` object).

3.5 Packages

3.5.1 Base R and the need to extend functionality

R is furnished with a large set of core functions. You can perform mathematical operations (addition `+`, division `/`), conduct statistical summaries (`mean`, `sd`), run a regression (e.g., `lm`), plot a histogram (`hist`), and even make a map (`plot`). This original set of functions that shipped with R is called *base* R—the basic setup when you open the box. One will eventually find themselves in a situation that requires more complex and specialized functionality that is not possible by any function in base R.

Some users have the adventurous itch to program specialized functions, which may sound daunting. But fear not, there is a whole universe of functions awaiting your use on the Comprehensive R Archive Network (CRAN), as well as on Github. Specifically, functions are available as families of functions, formally called *packages*.

For example, imagine that a colleague sends you an Excel spreadsheet that you then need to load into R. Herein lies a common problem: *base R does not have a function to read directly from a spreadsheet*. But, as its name implies the `readxl` package on CRAN provides exactly the functionality you need.

3.5.2 Installing packages

A large community of R users and developers contribute and maintain a vast archive of packages.[9] These packages span computationally intensive tasks like reading satellite imagery to more visual tasks like creating interactive maps for websites. It is easy to tap into these vast resources, and to get started with a package like `readxl`, we only need two inputs: (1) an internet connection and (2) R's `install.packages` function. We can install the `readxl` package by executing the following line in the R console. To break it down, the `install.packages` function requires the name of a known package on CRAN in the form of a string value.

```
install.packages("readxl")
```

For some packages, R will also install package dependencies—other packages on which the requested package relies. This is one of the beautiful things about R: it is an extensible language that builds on existing functionality. Entire families of packages are dependent on common functions, ensuring that a common logic is employed in making the magic happen.

There are sources (e.g., METACRAN) that allow you to search CRAN for packages, but, as with many R-related tasks, Google and asking friends and colleagues are going to be your best bets.

---

[9]Visit CRAN for a full list https://cran.r-project.org/web/packages/available_packages_by_name.html.

3.5.3 Loading packages

Like installing a new air conditioning unit, the unit is only useful when it is plugged in and turned on. R packages follow the same logic. Installing a package simply means that you now have access to the package on your machine. To take advantage of the functionality, packages need to be loaded into the R environment during each session.

To load a package, use the `library` function in combination with the package's name. Note that whereas `install.packages` expects the package name in string format, the `library` function can accommodate both strings (e.g., "readxl") or the package as a name (e.g., `readxl`) as it is now an installed resource.

```
library(readxl)
```

The `library` function will only load one package at a time, meaning that every package will need to be called individually. If we needed to load two installed packages—`readxl` and a package for text manipulation known as `tidytext`—we would need to do the following:

```
library(readxl)
library(tidytext)
```

For highly complex projects requiring many packages, this can be cumbersome. In the next section, we present a solution to this problem.

When a package is loaded, it is only temporarily held in the computer's memory. Whenever R and RStudio are closed or restarted, all loaded packages are removed from memory. The point: *you will need to load the packages again, but you do not need to install them again.*

3.5.4 Package management and `pacman`

While installing, loading, and updating packages in R is fairly quick and easy, the `pacman` package further streamlines and simplifies these process (and many related processes). What if we needed to load these three packages? The manual option requires one to embrace the tedium:

```
library(readxl)
library(tidytext)
library(ggplot2)
...
```

And the tedium does not end on your computer. When the code is shared with a collaborator, they may not have all of the packages installed, meaning the script will *crash*, requiring them to backtrack through the error log and install all of the missing packages. This scenario happens all too often, but it need not be the case.

Meet: `pacman`.

Upon installing `pacman`, the mental overhead of keeping track of all packages is quickly minimized:

```
library(pacman)
p_load(readxl, tidytext, ggplot2)
```

The `p_load` function within the `pacman` package makes package management easy. Not only does `p_load` install requested but missing packages, it loads them. We can also directly call the `p_load` function using the double colon "`::`" operator, which is shorthand for calling a specific function in a package.

```
pacman::p_load(readxl, tidytext, ggplot2)
```

Including this line of code at the top of each R script only requires collaborators to have `pacman` installed and the package does all of the heavy lifting of keeping packages up-to-date and in sync.

Commonly used packages. There are thousands of packages available in R—the language is truly extensible. But one thing to note is that not all packages are of the same high quality—some are relatively new

Table 3.2: Recommended packages.

Package	Description
haven	Read Excel files.
readr	Read multiple types of standard files such as CSV.
readxl	Read Excel files.
rjson	Read JSON files
XML	Read XML files.
data.table	Extension of data frames for fast data processing.
dplyr	Tools for efficiently manipulating data.
lubridate	Date manipulation tools.
reshape2	Simple functions for fast data reshaping.
stringr	String manipulation functions.
tidyr	Basic tools to "tidy" one's data.
fastlink	Probabilistic record linkage tools.
openssl	Toolkit for encryption.
ggplot2	Data visualization suite.
gridExtra	Utilities to layout sets of graphs.
glmnet	Elastic net, LASSO, and Ridge regressions.
hdm	Regularized regressions for causal inference.
caret	Model building tools for classification and regression.
e1071	Support vector machines.
kknn	K-nearest neighbors.
ranger	Fast implementation of the Random Forest algorithm
rpart	Classification and Regression Trees (CART) algorithm.
rpart.plot	Visualization tools for CART models.
cluster	Tools for clustering data.
dendextend	Tools for visualizing dendrograms.
sf	Spatial data processing tools.
tidytext	Text mining tools.

and still have their fair share of bugs. We recommend the following 27 packages that have largely proved tried and true. Using `pacman`, install these packages, as they will re-appear again and again throughout this book (Table 3.2).

3.6 Data input/output

When Captain Jean Luc Picard says "Make it so, Number Two", he is giving the order to execute a command that he specified prior to his statement. All the necessary information needed by the starship's crew to act is embedded in that order, ensuring that a precise output is delivered according to Picard's input. Our interaction with computers is no different. Computer programs cannot guess what we intend to do—not yet. Computers are designed to interact and respond to humans given specific rules and protocols, and importing data is no different. This is the fundamental idea that underlies data *Input/Output* or I/O.

Suppose you need to analyze an Excel file of stock market data. In order to load data, we need three pieces of information: (1) where to look for a file (*directory*), (2) how to load the file (*function*), and (3) the name of file (*file*).

3.6.1 Directories

R can import any kind of file, but it first needs to know where to look. Directory paths are a special kind of string that dictate how R should traverse folders on your computer to get to a file. Suppose the Marvel superhero Black Panther wanted to pull fuel cost data for his jet vehicle that is stored in a directory on an Apple computer. The path may look something like /Users/BlackPanther/Tech_and_Travel/Bills/. On a Windows machine, the path may look like C:\\Users\\BlackPanther\\Tech_and_Travel\\Bills\\.

In order to open files in these directories, R expects the paths to follow a specific format. Each subfolder in a directory path is denoted with a forward slash. Back slashes can be used, but only on Windows machines. Furthermore, there must be two backslashes for every slash. Table 3.3 illustrates valid and invalid directory paths. Note that the erroneous path that begins with \Users conflates the combination \U with special characters rather than as strings representing paths.[10]

While it might not seem to matter, we recommend one additional best practice: *avoid adding spaces to paths* like Black Panther. Spaces can be challenging to work with in some programming languages, whereas underscores (_) and hyphens (-) are reasonable substitutes. Keep this best practice in mind when creating new folders.

Table 3.3: Examples of valid and invalid directory paths.

Path	About
\Users\BlackPanther\Tech and Travel\Bills\	Erroneous path
/Users/Black Panther/Tech and Travel/Bills/	Acceptable path on Mac but could be better
/Users/Black_Panther/Tech_and_Travel/Bills/	Acceptable path on Mac and Windows Machines
C:\\Users\\BlackPanther\\Tech_and_Travel\\Bills\\	Acceptable path in Windows

Let's now put paths to work, as shown in the code block below. R is always "pointed" at a directory, but *which one is it?* The getwd function will retrieve the path of the *working directory* (or *WD*) and print it to console. But *what if that directory is not the right one?* Perhaps the data are in a different folder. We can pass a directory path to the setwd function to "set the working directory". And as a final check, *does the directory contain the files of interest?* Using dir, R will return a list of files and folders in the current directory.

```
# Check path
getwd()

# Set directory path
main_dir <- "your-path-goes-here"
setwd(main_dir)

# Get list of files in current directory
dir()
```

A few best practices. The setwd should only be called once and only once, and it is a standard practice to set your directory at the top of your script. This keeps the code clean and flexible, forcing all data processing steps to adhere to a defined folder structure. But this also has practical implications for how folders are set up. All files required for a project should be well organized in the same folder, even if it means using multiple subfolders.

This may seem like a draconian set of practices, but there are reasons for the madness. Suppose you need to move a code project from your desktop computer to a server directory, which requires every single setwd to be updated. As you might imagine, updating the directory path in one command at the top of the script is faster and more accurate than updating it in 100 lines of code. Setting the working directory to the top level

---

[10]The use of these special characters is covered in Chapter 4.

of a project folder also allows you to take advantage of *relative paths*—calling on subfolders of the working directory.

3.6.2 Load functions

How exactly do we load a dataset into R? The answer depends on where the data file is located (the path) and the data storage format. It turns out there are hundreds of storage formats, but there are a few that are commonly used in the public sector, namely, Comma Separated Values (`csv`), JavaScript Object Notation (`json`), Excel (`xls`, `xlsx`), Extensible Markup Language (`xml`), and Text (`txt`).[11] For each format, there is a specific function or set of functions that load the data into the R environment.

For example, to open `csv` files "`fuel_bills.csv`", "`electricity_bills.csv`", and "`insurance_bills.csv`" located in directory "`/Users/Black Panther/Tech and Travel/Bills/`", we can use the `read.csv` function (Base R)[12]. The simplest approach is to concatenate the directory and file name into one string and pass it to load function, which then assigns the data to a data frame `bills`.

```
# Read with directory path and file in one string
fuel_bills <- read.csv("/Users/Black Panther/Tech and Travel/Bills/fuel_bills.csv")
elec_bills <- read.csv("/Users/Black Panther/Tech and Travel/Bills/electricity_bills.csv")
ins_bills <- read.csv("/Users/Black Panther/Tech and Travel/Bills/insurance_bills.csv")
```

The first approach, while straight forward, is not transferrable. If the Black Panther shares his `Tech and Travel` folder with Tony Stark, the load function will not be able to find the `csv` files as the path will likely be different. Then, Mr. Stark would need to manually change the paths for each bill file—not an efficient use of superhero time. Instead, consider the approach below that requires the user to change only the `path` for the top-level directory, which then is set as master directory. Each file in the `Bills` folder is located using a relative path.

We cover the nuances of importing files in more detail in Chapter 4.

```
# CHANGE master path
path <- "/Users/Tony/Tech and Travel/"

# Set working directory
setwd(path)

# Import
fuel_bills <- read.csv("Bills/fuel_bills.csv")
elec_bills <- read.csv("Bills/electricity_bills.csv")
ins_bills <- read.csv("Bills/insurance_bills.csv")
```

3.6.3 Datasets

As you may recall from Chap. 1, we have made available a number of datasets through a Github repository.[13] Let's set up the data folder so you can access the data.

In the code chunk below, modify the directory in `setwd` to a path on your computer. By executing the download and unzip steps, R will save a copy of the `data-sets` repository from Github and unzip it into the directory of your choice. The new folder will be labeled `data-sets-master`.

```
# Set directory (modify this string)
setwd("//directory/of/your/choice")
```

---

[11]There are other formats that are more subject area specific such as NetCDF (`nd`) for earth science, but to keep the list concise, we focus on the ones listed in the text.

[12]For more efficient data import of large datasets, consider the `read_csv` function in the `readr` package.

[13]Visit https://github.com/DataScienceForPublicPolicy/data-sets.

```
# Download repository
link <- "https://github.com/DataScienceForPublicPolicy/data-sets/archive/master.zip"
download.file(url = link,destfile = "dspp.zip")

# Unzip file
unzip(zipfile = "dspp.zip")
```

The `data` folder in `data-sets-master` contains all datasets that have been assembled in this book.

3.7 Finding help

3.7.1 Help function

Developers interested in driving adoption of their packages understand that ease of use and clear documentation win the day. Thus, each `R` function is equipped with documentation to illustrate its syntax and use. As a first line of support, you can access any function's help file using `?<function_name>` in the console, provided that the package has been loaded.

For instance, suppose you are interested in importing a spreadsheet using the `read_xlsx` from the `readxl` package, but you would like to have a better understanding on how to use it. Assuming that `readxl` has been loaded, simply execute the following in the console `?read_xlsx`. If the package has not already been loaded, include the package's name, double colons (`::`), and the function's name, e.g., `?readxl::read_xlsx`. Recall that the double colon trick essentially is short hand for obtaining a specific function from a library without importing the entire library.

3.7.2 Google and online communities

One of the best aspects of `R` is the vibrant and (generally) helpful online community. By simply searching "*R*", you will likely discover that you are not the first to face the challenge at hand. To illustrate the point:

- *Leading zeros* often are required to convert a number to a certain format. One of the zipcodes for Somerville, MA is `02144`, which sometimes is erroneously converted into a number `2144`. To find a solution to this formatting issue, simply search "*R leading zeros*".
- *JSON* or JavaScript Object Notation is a data format that is increasingly used for web data. It requires a package to read and import. Search "*R import JSON*" to find a solution.

It quickly becomes apparent that Stack Overflow is the premier resource for programming and Cross Validated for statistics and machine learning. More often than not, these online communities will have an answer and `R` coders hunting for an answer will copy and paste code without checking what it does. User beware, the answer might not be to your question, may not be the most efficient solution, or could be wrong. Nonetheless, these online resources are a great starting point.

3.8 Beyond this chapter

3.8.1 Best practices

As you may have noticed, many of the code snippets in this chapter contained comments preceded by a hash (`#`).[14] The hash tells `R` to ignore whatever follows the hash *on the same line as the* `#`. Thus, the hash allows you to make notes—or *comments*—to yourself, your collaborators, or anyone else who may have to read through your code. These comments document what was the thought process behind the code so that you or anyone else can read and understand the `R` script. For example:

---

[14]Also called a *pound sign* or *number sign* or *octothorpe* in the olden days before Twitter.

```
# Make sure the packages are installed and loaded.
p_load(readxl, dplyr)

# Navigate to the data directory to load the data.
setwd("/Users/Someone/OnlyData")

# Load the xlsx file (named 'fake_data.xlsx')
fake_df <- read_xlsx("fake_data.xlsx")
```

In the example above, each comment concisely describes what the code will do in the next line and why we want to do it. Often the *why* is the most important part: with some experience, you will be able to figure out what each line of an R script does, but it is much more difficult to figure out *why* the author wrote the 641 lines of code in the middle of a 1200-line R script.

Another popular use of comments in R—and in many other languages—is to flag questions, issues, or unfinished items. The idea here is to start the comment with a quick flag, for instance, # TODO. This flag helps you find the unfinished parts of your document—or the parts that you want your collaborators to finish for you. For instance, we might start a document

```
# Make sure the packages are installed and loaded
# FIXME: There are better packages for large datasets.
p_load(readxl, dplyr)

# Navigate to the data directory to load the data
setwd("/Users/Someone/OnlyData")

# Load the xlsx file (named 'fake_data.xlsx')
fake_df <- read_xlsx("fake_data.xlsx")

# TODO: Clean the dataset. Check out 'stringr' and 'data.table'
```

You can also use commenting (#) to tell R to ignore lines of code. This usage of the hash is called *commenting out* your code, as it allows you to turn off a line of code, e.g.,

```
# Standard line of code
two_plus_two <- 2 + 2

# Commented out line of code
# Two_times_two <- 2 * 2
```

In general, commenting lines of code out is most helpful for temporary changes and/or testing code. If you really do not need a line, just delete it. It is worth saying that the style and amount of commenting, flagging, and commenting-out are highly subjective. If you find them burdensome, then you probably should cut back a bit or try a different strategy. That said, it's pretty rare for 0 comments to be an optimal number of comments in an R script.

3.8.2   Further study

In-depth coverage of R. The ecosystem of educational materials for programming has significantly evolved and matured in recent memory. For a more in-depth text on programming in R for absolute beginners, consider reading *The Art of R Programming: A Tour of Statistical Software Design* by Norman Matloff, which covers data structures, mathematical simulations, and many other topics. We do recognize that some readers are transitioning from other social science software. A Rosetta Stone can be helpful to facilitate the transition:

- *R for Stata Users* by Joseph M. Hilbe and Robert A. Muenchen.

- *R for SAS and SPSS Users* by Robert A. Muenchen.

- *R for Excel Users: An Introduction to R for Excel Analysts* by John L Taveras.

Programming requires one to build *muscle memory*—there is no better way of learning than doing. The `swirl` package (install by executing `pacman::p_load(swirl)`) is a free interactive environment that teaches `R` programming through in-console exercises. Our students from previous data science courses have relied on this package to build up their core skills with great success. For more information, visit: https://swirlstats.com.

Beyond `R`. Aspiring data scientists will no doubt encounter many programming languages throughout their career in the field. While one could start with `R`, she will eventually encounter `Python`—another core data science language. A basic command of this ubiquitous language will prove useful for adapting to different tech stacks and requirements. But why learn more than one language?

Teams tend to be more efficient when using a common language, removing the need to code switch. New joiners to the team may need to conform to the team's standard language. In addition, the use cases differ between languages. While `R` is more than sufficient for analysis and small to medium-scale data applications, `Python` is well suited for building scalable software projects that have a more demanding technology requirement. As developing software in `Python` follows different workflows and development paradigms, we recommend building Python skills from the ground up using *Learn Python The Hard Way* by Zed A. Shaw. Originally a web book, the content has since evolved to include written content and videos to walk through how Python works.

While some data can be stored in files in a folder (e.g., CSVs, JSONs, Excel), some data is too large and too complex to sit in a single file. Instead, data is stored on a relational database. Being able to *query* a database—or extract data from a data storage system—is among the most foundational skills required of data scientists or data analysts. Structured Query Language (SQL) is a universal standard for interacting with relational databases. For a gentle introduction, we recommend *Practical SQL: A Beginner's Guide to Storytelling with Data* by Anthony DeBarros, which walks through how to create a database, use "Create-Read-Update and Delete" (CRUD) operations, and basic analysis using SQL.

3.9 DIY: Loading solar energy data from the web

DIYs are designed to be hands-on. To follow along, download the DIYs repository from Github (https://github.com/DataScienceForPublicPolicy/diys). The R Markdown file for this example is labeled `diy-ch03-load-data.Rmd`.

For this DIY, we will import and examine data about solar energy—a natural resource that is playing an increasingly important role in fulfilling energy demand. In the period between 2008 and 2017, annual net generation of solar energy produced by utility-scale facilities had grown by 61.3 times, from 864 thousand megawatt hours to 52,958 thousand megawatt hours (Energy Information Administration 2019). At the same time, solar also became more economical: photovoltaic module costs fell from \$3.37/peak watt to \$0.48/peak watt—an 86% reduction in cost.

The increased affordability of solar, among other advanced technologies, opens the possibility for constructing buildings that are hyper energy efficient. For example, the Net Zero Energy Residential Test Facility is a house that produces as much energy as it uses over the course of a year. Engineered and constructed by the National Institute of Standards and Technology (NIST) in Gaithersburg, MD, the lab home was designed to be approximately 60 percent more energy efficient than typical homes. In order to achieve net zero energy, the lab home needs to produce an energy surplus and overcome the energy consumption of a simulated family of four. In fact, within its first year, the facility produced an energy surplus that is enough to power an electric car for 1400 miles (National Institute of Standards and Technology 2016).

The test bed also generates an extraordinary amount of data that help engineers study how to make more energy-efficient homes and may one day inform building standards and policies. We tap into one slice of the net zero house's photovoltaic time series data, which is made available publicly at https://s3.amazonaws.com/nist-netzero/2015-data-files/PV-hour.csv. This dataset contains hourly estimates of solar energy production and exposure on the Net Zero home's solar panels.

Suppose we would like to know how much energy is produced from the solar arrays and how variable is that energy production over the course of a year. If information on solar energy production is known for the Net Zero home, decision makers and home owners can decide if investing in solar would be cost-effective.[15] With only a few lines of code, we can import and summarize the information. To start, we load the `pacman` package and use `p_load` to load the `readr` package.

```
# Load libraries
pacman::p_load(readr)
```

With the basic functions loaded, `read_csv` is used to import a comma separated values (CSV) file stored at the given `url`, which is then loaded into memory as a data frame labeled `solar`. Notice that the path to the data file is not on your computer. R is able to directly download and read files from the internet.

Once the data are loaded, we check the dimensions of the data frame using `dim`: it has n = *8,737* with *32* columns. As we are primarily interested in the total amount of sunlight shining on the solar arrays at any given hour (kWh), we focus in on the variable `PV_PVInsolationHArray` in data frame `solar` and review the first five observations.

```
# Check dimensions
dim(solar)
```

```
## [1] 8760    32
```

```
# Extract solar array variable
solar_array <-
solar$PV_PVInsolationHArray

# Retrieve 5 observations from top of dataset
head(solar_array, 5)
```

```
## [1] 0.05740833 0.06028630 0.05881078 0.05886550 0.05928754
```

To answer our questions, we can obtain a concise `summary` to gage the hourly variability. The photovoltaic arrays are exposed to fairly small amounts of energy for the majority of hours, as indicated by the small median relative to the mean, but there are occasional periods with intense energy exposure.

```
summary(solar_array)
```

```
##      Min.  1st Qu.   Median     Mean  3rd Qu.     Max.
##   0.00000  0.05585  0.33865  9.45263 14.84311 54.84352
```

This can be more concisely summarized as the coefficient of variation ($CV = \frac{\text{standard deviation}}{\text{mean}}$)—the ratio of the standard deviation and the mean. Values of the CV that exceed $CV = 1$ indicate greater dispersion. We compute the `mean` and standard deviation `sd`. The resulting CV indicates that one standard deviation is 1.5 times as wide as the mean, suggesting that the hourly solar energy generation can be quite variable.

```
# Calculate mean
a <- mean(solar_array)

# Calculate standard deviation
b <- sd(solar_array)

# Calculate coefficient of variation
b / a
```

```
## [1] 1.537536
```

---

[15]There are a many factors that should inform these decisions. Here, we aim to illustrate the basics of loading and extracting insight from scientific data.

With only a few lines of code, we have shown how programming can make data analysis an efficient process. In the context of policy, programming removes the tedium of manually working with spreadsheets, allowing an analysis to be delivered using only a few decisive keystrokes.

Chapter 4

Transforming Data

Data processing can occupy between 50% and 80% of a data scientist's time (Lohr 2014), yet this formative step in a data science project is overshadowed by the end product. More often than not, stakeholders will request data and expect it to be delivered quickly and effortlessly without realizing the substantial effort that is put into producing a *useful* dataset.

Processing raw data into analysis-ready form involves a set of techniques and transformations. The specific steps, however, are dependent on the subject area and use case. The US GDP, for example, involves hundreds of economists and statisticians. Data are downloaded from published survey data and administrative records, inspected for outliers and missing values, joined together, reshaped into a useful form, then used as inputs for estimating the state of the economy. Satellite imagery collected by NASA and NOAA are downloaded in batches of "granules" that need to be rectified, stitched together, and otherwise processed into a ready-for-use imagery. In short, there is always substantial data processing behind any published dataset. We tend to only see the end product and not the underlying machinery.

Where do we start? Data scientists often conceptualize data processing in terms of Extract-Transform-Load (ETL)—a common flow for readying data (Figure 4.1).[1] The first step acknowledges that data comes from *somewhere*, often times a directory of files or a database. The *Extract* step simply means to download or move data to a *staging area* where the raw copy of the data is stored in its native format. Data is then *imported*—sometimes referred to as *parsed*—from the raw data. The *transform* step is a feat of data wizardry, pushing the data through steps to (not necessarily in this order):

- *clean* and *convert* into the appropriate data types;
- *extract* from text if not already structured as a tabular dataset;
- *feature engineering* or create new variables;
- *reshape* into a useful data structure;
- *join* with other datasets; and
- *aggregate* into summary metrics.

In larger organizations, entire teams of *data engineers* are the data refinery that converts raw, untamed data into usable information. Their refined outputs are *loaded* into a database (e.g., a relational databases), data warehouse, or other useful format (e.g., CSVs, Rdas) for storage. It is only at this point that data is made available for analysis. Interestingly, computational research pays little attention to the role of the ETL process or data processing, in general, although it may be the source of major performance gains.

In this textbook, we dedicate two chapters to illustrate workflows for transforming data into data science-ready form. In this chapter, we begin with a broad overview of the mechanics of data processing. First, we focus on processing at the data record level, walking through how to load raw data, screen it for peculiarities,

---

[1]The term "ETL" technically refers to a specific paradigm for big data processing. However, the concept is relevant for many data applications. An alternative is Extract-Load-Transform (ELT) in which the "staging" step is removed and places the onus of transformation on downstream users.

© Springer Nature Switzerland AG 2021
J. C. Chen et al., *Data Science for Public Policy*, Springer Series in the Data Sciences,
https://doi.org/10.1007/978-3-030-71352-2_4

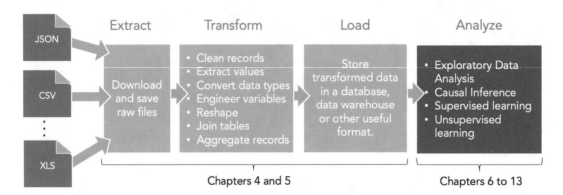

Figure 4.1: The flow of an Extract, Transform, and Load (ETL) process.

and transform data values into useful, analysis-ready information. Second, we take a step back to focus on the macroscopic structure of data—how to sculpt the structure of a dataset (rows and columns) into an appropriate format. We round off the chapter by introducing efficiency augmenting programming paradigms, namely, control structures (e.g., loops, if statements) that reduce repetitive coding and user-written functions that allow scripts to be generalized and re-used.

4.1   Importing and assembling data

The first step of any data project is to *load* a dataset. New programmers will find the *input-output* (I/O) steps at times confusing. Sometimes, analysts can find themselves investing substantial effort to find *a* single feasible method to open a file using code. As a result, analysts resort to precarious data practices. In particular, we have observed that:

- There are unnecessary rows before the header of the file. The analyst will manually delete those rows in a spreadsheet software, which fails to preserve the edit trail.
- The number of columns differs between two files are to be appended. The analyst will either delete columns or add empty columns to ensure the right number of columns exist, which fails to preserve the edit trail.
- Column names are not as desired. The analyst will manually edit header names.
- Manual edits overwrite the original file. The consequence of an erroneous edit is that there is no hope of restoring the raw data.
- Manual edits are learned and shared as an oral history rather than a codified procedure.

The first steps in data processing should be relatively painless without manual intervention. If processing steps are not captured in the code, the transformations will also not be reproducible, allowing analyst-specific judgments to go untraced and documented. Furthermore, manual editing fails to capitalize on the efficiency gains of well-documented code. In short, manual edits in the import step are common habits that undermine the success of a data science project.

In this section, we illustrate how to load and assemble a simple dataset from data files that each have quirks. As mapped in Figure 4.2, our goal is to assemble a gas price dataset for two US ports (from Energy Information Administration (2017)). We show how to load a CSV of gas prices for the US Gulf Coast and two worksheets from an Excel workbook for NY Harbor gas prices. The columns in each dataset need to be renamed for consistency, then the datasets are joined together into a daily time series with three columns: `date`, `gulf.price`, and `ny.price`. We then save the processed data as a CSV and calculate a correlation coefficient between the two gas price variables.

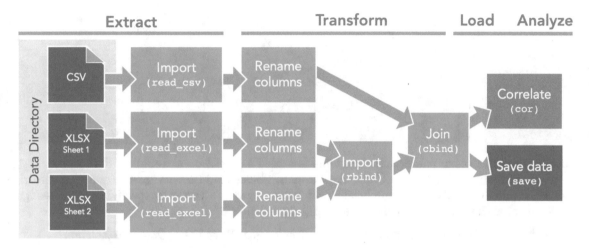

Figure 4.2: A simple flow for extracting and assembling a simple dataset.

4.1.1 Loading files

Data can be stored in many formats such as extensible markup language (XML), comma separated values (CSV), and Excel files. Some fields have their own formats, such as NetCDF for earth science and .wav files for audio. Some data formats require special software to load while others simply need time and patience to crack the code. Fortunately, R is equipped to read virtually all data formats. In Table 4.1, we provide a recommended set of functions that are easy to use and flexible.

Table 4.1: Recommended functions to load an assortment of files.

Function	Package	Description
read_excel	readxl	Load Excel files into a data frame. Note that with excel files, generally need to specify which sheet to load using the *sheet* argument and may need to skip a number of rows at the beginning of worksheet to indicate the header row using *skip*.
read_csv	readr	Load any comma separated file into data frame format.
read_delim	readr	Load delimited file using any delimiter such as tab delimited ("\") using the *sep* argument.
readLines	base R	Read a text file, line by line. This is helpful for working with free form text.
xmlToDataFrame	XML	Read XML into a data frame if the structure is simple and flat. Note that XML files tend to have a more complex hierarchical structure.
xmlToList	XML	If the XML is complicated, read each element as a list.
fromJSON	rjson	Read JSON into list. If JSON is a flat, non-hierarchical file, then the resulting object can be simply converted into a data frame.
read_dta	haven	Load Stata data file.
read_sas	haven	Load SAS data files.
read_spss	haven	Load SPSS data files.
load	base R	Load saved R dataset that can contain multiple objects. Note that this does not need to be assigned to another object.
readRDS	base R	Read individually saved R object.

Loading a CSV. Before loading a file, consider opening files in a code editor to take a peek at the structure.[2] Otherwise, blindly loading a file can also give clues as well. Let's start with the CSV `doe_usgulf.csv` using the `read_csv` function (`readr` package), assign the data to the object `gulf`, then inspect the first few rows using `head`.

```
# Load package
pacman::p_load(readr)

# Read file
gulf <- read_csv("data/doe_usgulf.csv")

# Preview first three rows
head(gulf, 3)
```

Because we used the `readr` package, the `gulf` file is loaded as a *tibble*—a variant of data frames designed for more efficient data manipulation. Unlike data frames, tibbles read headers "as-is", retaining the original capitalization and punctuation. Column headers can be retrieved as a vector by specifying `colnames(gulf)`. When doing so, we find that the second column of `gulf` is labeled:

"U.S. Gulf Coast Conventional Gasoline Regular Spot Price FOB (Dollars per Gallon)"

Long variable names are unwieldy. It is good practice to keep names short, in the same case (e.g., lower case) and without spaces and punctuation with the exception of periods. Furthermore, the column name should be intuitive and meaningful. As the column names are quite long, we overwrite them with new concise labels. Notice that we did not need to open the CSV in Excel to manually edit the column names!

```
colnames(gulf) <- c("date", "gulf.price")
```

Loading an Excel file. Excel files are designed for readability and analytical convenience in a WYSIWYG interface. However, reading an Excel file in R can be a challenge without the proper tooling. The specialized styles applied to the data are not intended to be machine readable. Also, an Excel file can hold more than one table (e.g., worksheets)—like multiple datasets in one file. Have no fear, `read_excel` (`readxl` package) makes the process of reading individual worksheets simple.

Below, we blindly load the first worksheet (`sheet = 1`) of our Excel file `doe_ny.xlsx`, then use `head` to examine its contents. Unlike our CSV, our first attempt requires a minor tweak: *The headers are located in the second row of the sheet, which has the implication that the first row needs to be skipped.*

```
# Load package
pacman::p_load(readxl)

# Read Excel
attempt1 <- read_excel("data/doe_ny.xlsx", sheet = 1)

# Preview first three rows
head(attempt1, 3)
```

At this point, some analysts would be tempted to manually delete the first row. Resist the temptation. Most load functions are furnished with a `skip` argument that allows the user to specify the number of rows to skip. Below, we load the first and second sheets of `doe_ny.xlsx`, skipping the first row in each, then rename the headers.

```
# Load first two sheets
sheet1 <- read_excel("data/doe_ny.xlsx", sheet = 1, skip = 1)
sheet2 <- read_excel("data/doe_ny.xlsx", sheet = 2, skip = 1)

# Rename columns
```

---

[2]Keep in mind that some files may be too large to comfortably open in a code editor.

```
colnames(sheet1) <- c("date", "ny.price")
colnames(sheet2) <- c("date", "ny.price")
```

Appending data. If we examine the date range (`range`) for each *sheet1* and *sheet2*, the pair of worksheets contain data from the same time series of spot prices, but cover different periods of time. Naturally, these two tables should be reunited to yield one complete data frame. Adding the rows of one dataset to the bottom of another similar dataset is known as *appending*. The two can be combined into a complete data frame by appending one data frame to the other using the `rbind` function.

```
ny <- rbind(sheet1, sheet2)
```

The above scenario is the ideal: two data frames with the same number of columns and the same column names. But what if one data frame has at least one more column than the other? The `rbind.fill` function (`plyr` package) appends the two data frames, filling any additional non-matching columns with `NA` values. Note that if two columns in two different data frames represent the same concept but have different names, `rbind.fill` does not know to make the connection. It is of the utmost importance that headers are exactly labeled as `R` interprets column names literally.

Pro Tip. In the example below, we used the double colon operator "`::`" to specifically call the `rbind.fill` function within `plyr`. There is good reason for this: *many packages use the same function names*. Both the `plyr` and `dplyr` packages, for example, have a `summarise` function. The double colon operator is useful for disambiguating between packages.

```
ny <- plyr::rbind.fill(sheet1, sheet2)
```

Combining the datasets by row. The two data frames `gulf` and `ny` contain the same number of rows and the same dates. If two datasets are in the *same exact order* and have the *same number of rows*, we can use the `cbind` function to combine the datasets.

```
# Combine data
prices <- cbind(gulf, ny)
```

The above result has duplicate information: the `date` column appears twice. To ensure the data is neatly organized, we keep all columns from the `gulf` dataset and only the second column from the `ny` dataset. The resulting data frame contains three columns containing daily spot prices from two sources. As we will see in Chapter 5, two data frames can be combined in a more sophisticated way by *matching* on common variables. For now, `cbind` is sufficient to complete this processing task.

```
# Combine data
prices <- cbind(gulf, ny = ny[, 2])
```

Save the data. Processed datasets should be saved for future use. There are many different file formats in use, some more `R` specific while others are more general data formats. In general, we recommend using formats that ensure for maximum accessibility for fellow data scientists, regardless of software.

For data science teams that primarily use `R`, consider `Rda`—it allows one or more data objects from your current session to be saved, retaining their names, data types, and formats. `Rda` files also substantially compress the size of the file to save on disk space yet make it possible to open a file quickly. While convenient for `R` users, data scientists using other languages can find it challenging to access your work.

For wider distribution, comma separated values (`CSV`) and JavaScript Object Notation (`JSON`) are universal formats that are used by all data scientists. Unlike `Rda`, only one data object can be saved per file. Furthermore, these formats do not retain the data types, requiring additional documentation to be included to help others to translate each variable.

Nonetheless, `CSV` and `JSON` files reign supreme. When saving these files, here are two tips of the trade:

- To save a `CSV` while maintaining the data frame's shape, specify `row.names = FALSE` when using the `write.csv` function. Otherwise, the `CSV` will be written with an additional nameless column for row number.

- To save a JSON, a data frame needs to be converted into a JSON object using the rjson package, then saved to file using the write function.

```
# Save Rda file
save(prices, file = "gas_prices.Rda")

# Save CSV
write.csv(prices, "gas_prices.csv", row.names = FALSE)

# Save JSON
# Convert to JSON
json_out <- rjson::toJSON(prices)

# Export
write(json_out, "gas_prices.json")
```

4.2   Manipulating values

After loading a dataset, it will invariably require some cleaning and processing. More often than not, working with raw data means working with string values. In fact, dates are often assumed to be text characters upon loading. It is easy to imagine that necessary data are embedded in a sentence. For example, below is a vector of four string values:

```
budget <- c("Captain's Log, Stardate 1511.8. I have $10.20 for a big galactic mac.",
            "The ensign has $1,20 in her pocket.",
            "The ExO has $0.25 left after paying for overpriced warp core fuel.",
            "Chief medical officer is the high roller with $53,13.")
```

What if we need to calculate the total amount of pocket change available to buy galactic big macs? All four elements contain dollar values that can be extracted using *regular expressions* or *regex*—a series of characters that describe a regularly occurring text pattern.

To retrieve the dollar amounts embedded in the text, we first need to replace commas with a period using str_replace_all(), assigning the result to a new object new. Note that in some regions, such as Europe, commas are used as decimals rather than periods.

```
pacman::p_load(stringr)
new <- str_replace_all(budget, ",", "\\.")
```

Second, the dollar amounts need to be defined as a string pattern: a dollar sign followed by one to two digits, followed by a period, then another two digits ("\\$\\d{1,2}\\.\\d{2}"). Using the str_extract_all function in the **stringr** package, we extract all matching substrings, specifying simplify = TRUE to return a vector:

```
funds <- str_extract_all(new, "\\$\\d{1,2}\\.\\d{2}", simplify = TRUE)
print(funds)
```

```
##      [,1]
## [1,] "$10.20"
## [2,] "$1.20"
## [3,] "$0.25"
## [4,] "$53.13"
```

Third, we should replace dollar sign with blank, then strip out any leading white space using str_trim().

```
funds <- str_replace_all(funds, "\\$","")
funds <- str_trim(funds)
print(funds)
```

```
## [1] "10.20" "1.20"  "0.25"  "53.13"
```

Lastly, we convert the character vector to numeric values using `as.numeric`, then sum the vector.

```
money <- as.numeric(funds)
print(paste0("Total galactic big mac funds = $", sum(money)))
```

```
## [1] "Total galactic big mac funds = $64.78"
```

In steps one through three, you will have noticed that the characters `"$"`, `"."`, and `"d"` were preceded by double backslash. These are known as *escaped characters* as the double backslash preceding the characters changes their meanings. In step two, a sequence of unusual characters (" `\\$\\d{1,2}\\.\\d{2}` ") was used to find the "`$x.xx`" pattern, which can be broken into specific commands:

- `\\$` is a dollar sign.
- `\\d{1,2}` is a series of numerical characters that is between one to two digits long.
- `\\.` is a period.
- `\\d{2}` is a series of numerical characters that is exactly two digits long.

Mastering *regex* is a productivity multiplier, opening the possibility of ultra-precise text replacement, extraction, and other manipulation. There are plenty of scenarios where raw data is not quality controlled and mass errors plague the usefulness of the data. An analyst may spend days if not weeks or months cleaning data by hand (or rather through find and replace). With regex, haphazard cleaning is no longer an issue. To make the most of regex requires a command of both *text manipulation functions* that are designed to interpret regex as well as *regex* itself.

4.2.1 Text manipulation functions

Find and replace functionality are common in word processing and spreadsheet software, but are not particularly efficient with complex string patterns. Text manipulation functions in R go beyond find and replace operations making it possible to find, extract, and edit string patterns. In Table 4.2, we list two sets of text manipulation functions. The first set contains core functions that are furnished with `Base R`—these closely resemble string manipulation functions implemented in most programming languages. While they are powerful, they at times may require some additional trial and error to use. Alternatively, the `stringr` package simplifies text manipulation and syntax, and in some case extends functionality. For the most part, we will use `stringr` in this text.

Table 4.2: Recommended text manipulation functions.

Description	Base R	stringr
Returns either the index position of a matched string or the string containing the matched portion.	grep	str_which
Returns a logical vector indicating if a matched string was found.	grepl	str_detect
Searches for a specified pattern and replaces with user-specified substring.	gsub	str_replace_all
Remove matched pattern from string.	gsub	str_remove_all
Returns the first position of matched pattern in a string.	regexpr	str_locate
Returns the position of all matched patterns in a string.	gregexpr	str_locate_all
Extract substring based on matched pattern. Note that regmatches is used in conjunction with regexpr or gregexpr.	regmatches	str_extract_all
Splits strings into a list of values based on a delimiter.	strsplit	str_split

Description	Base R	stringr
Extract substring based on start and end positions	substr	str_sub
Trim whitespace on either end of string (excessive spaces)	trimws	str_trim
Returns number of characters in string	nchar	str_length
Returns the number of matched patterns		str_count
Convert to upper case	toupper	str_to_upper
Convert to lower case	tolower	str_to_lower
Convert to title case		str_to_title
Pad string (e.g., add leading zeros to string)	sprintf	str_pad

Complex "find" operations are possible with stringr functions. To illustrate, below we have four sentences that indicate when four US laws were signed.

```
laws <- c(". Dodd-Frank Act was signed into federal law on July 21, 2010.",
          "Glass-Steagall Act was signed into federal law by FDR on June 16, 1933",
          "Hatch Act went into effect on August 2, 1939",
          "Sarbanes-Oxley Act was signed into law on July 30, 2002")
```

Suppose we need to find acts that are named for two congressmen. The str_detect() function produces a Boolean vector if a regex pattern [A-z]{1,}-[A-z]{1,} is matched for each element in the string vector laws. This is particularly useful if we would like to identify and subset elements that match a text pattern.

```
str_detect(laws, "[A-z]{1,}-[A-z]{1,}")
```

```
## [1]  TRUE  TRUE FALSE  TRUE
```

Otherwise, we can extract the specific senator names using str_extract_all, returning a vector of string values.

```
str_extract_all(laws, "[A-z]{1,}-[A-z]{1,}", simplify = TRUE)
```

```
##       [,1]
## [1,] "Dodd-Frank"
## [2,] "Glass-Steagall"
## [3,] ""
## [4,] "Sarbanes-Oxley"
```

4.2.2   Regular Expressions (RegEx)

The secret ingredient of the text manipulation functions is the regex expressions that dictate a text pattern. With only a few characters, even the most complex pattern can be concisely described. There are a few basic building blocks of regex that make text search possible. There are *character classes* that define specific types of character (e.g., numbers, letters, punctuation) quantifies that indicate how long a pattern should be, escape characters that allow certain characters to hold dual meanings, among others. But first, there are a few basic characters that make both simple and complex pattern searches possible and knowledge of these cleverly designed operators may go a long way:

(1) Alternatives (e.g., "OR" searches) can be surfaced by using a pipe "|". For example, a string search for "Bob or Moe" can be represented as "Bob|Moe".

(2) The extent of a search is denoted by parentheses (). For example, a string search for "Jenny" or an alternative spelling like Jenny would be represented as "Jenn(y|i)".

(3) A search for one specific character should be placed between square brackets [].

(4) The length of a match is specified using curly brackets {}.

In New York City, the famed avenue *Broadway* may be written and abbreviated in a number of ways. The vector `streets` contains a few instances of spellings of Broadway mixed in with other streets that start with the letter B.

```
# A sampling of street names
streets <- c("Bruckner Blvd", "Bowery", "Broadway", "Bway", "Bdway",
        "Broad Street", "Bridge Street", "B'way")

# Search for two specific options
str_detect(streets, "Broadway|Bdway")

# Search for two variations of Broadway
str_detect(streets, "B(road|')way")

# Search for cases where either d or apostrophe are between B and way
str_detect(streets, "B[d']way")
```

Escaped Characters. Quite a few single characters hold a special meaning in addition to the literal meaning. To disambiguate their meaning, a backslash precedes these characters to denote the alternative meaning. A few include

- \n: new line.
- \r: carriage return.
- \t: tab.
- \': single quote when in a string enclosed in single quotes (`'Nay, I can\'t'`).
- \": double quote when in a string enclosed in double quotes (`"I have a \"guy\"."`).

In other cases, double backslashes should be used:

- \\.: period. Otherwise, un-escaped periods indicate searches for *any* single character.
- \\$: dollar sign. A dollar sign without backslashes indicates to find patterns at the end of a string.

Character Classes. A *character class* or *character set* is used to identify specific characters within a string. How would one represent "12.301.1034" or "?!?!?!"? One or more of the following character classes can do the job:

- [:punct:]: is a catch all for any and all punctuation such as periods, commas, semicolons, etc. To find a specific punctuation, simply enclose the characters between two brackets. For example, to find only commas and carrots, use "[<>,]".
- [:alpha:]: Alphabetic characters such as a, b, c, etc. With other languages including R, searches for any letter combinations are denoted as [a-z] for lower case, [A-Z] for upper case, and [A-z] for mixed case.
- [:digit:]: Numerical values. With other languages including R, it is commonly written as \\d or [0-9]. For any non-digit, write \\D.
- [:alnum:]: Alphanumeric characters (mix of letters and numbers). With other languages including R, it is indicated using to as [0-9A-Za-z] or \\w. For any non-alphanumeric character, use \\W.
- [:space:]: Spaces such as tabs, carriage returns, etc. For any white space, use \\s. For any non-whitespace character, use \\S.
- [:graph:]: Human readable characters including [:alnum:] and [:punct:].
- \\b: Used to denote "whole words". \\b should be placed before and after a regex pattern. For example, \\b\\w{10}\\b indicates a 10 letter word.

These are only a few of the most important character classes. It is worth noting that many of the R specific character classes may differ from other data science programming languages.

Quantifiers.

Each character class on its own indicates a search for any character of that class. For example, a search for the pattern [[:alpha]] will return any element with an alphabetic character—it is not very specific. In practice, most character searches will involve a search for more than just one character. To indicate such a search, regex relies on *quantifiers* to indicate the length of patterns. For example, a search for a year between the year 1980 and 2000 will require exactly four digits, but a search for the speed of a gust of wind will likely vary between 1 and 3 digits. The following six quantifiers provide flexibility and specificity to accomplish targeted search tasks:

- {n}: match pattern n times for a preceding character class. For example, "\\d{4}" looks for a four-digit number.

- {n, m}: match pattern at least n times and not more than n times for a preceding character class. For example, \\d{1,4} looks for a number between one and four digits.

- {n, }: match at least n times for a preceding character class. For example, \\d{4,} searches for a number that has at least four digits.

- *: Wildcard, or match at least 0 times.

- +: Match at least once.

- ?: Match at most once.

In the example below, quantifiers are used to extract specific number patterns with a high degree of accuracy.

```
big_dates <- c("Octavian became Augustus on 16 Jan 27 BCE",
              "In the year 2000, a computer bug was expected to topple society.",
              "In the 5400000000 years, our sun will become a red dwarf.")


# Return a 9 digit number
str_extract(big_dates, "\\d{9}")


# Return a 4 digit substring that is flanked by empty value at either end
str_extract(big_dates, "\\b\\d{4}\\b")


# Match a date that follows 16 January 27 BCE
str_extract(big_dates, "\\d{2}\\s\\w{3}\\s\\d{2}\\s\\w{3}")
```

Positions. Regex also builds in functionality to search for patterns based on location of a substring, such as at the start or end of a string. There are quite a few other position matching patterns, but the following two are the main workhorses:

- $: Search at the end of a string.

- ^: Start of string when placed at the beginning of a regex pattern.

To demonstrate these patterns, we'll apply str_extract to three headlines from the BBC.

```
headlines <- c("May to deliver speech on Brexit",
              "Pound falls with May's comments",
              "May: Brexit plans to be laid out in new year")


# Find elements that contain May at the beginning of the string
str_extract(headlines, "^May")
```

```
# Find elements that contain Brexit at the beginning of the string
str_extract(headlines, "Brexit$")
```

Together, these character combinations make it possible to accomplish complex string manipulation tasks.

4.2.3 DIY: Working with PII

> *DIYs are designed to be hands-on. To follow along, download the DIYs repository from Github (*https://github.com/DataScienceForPublicPolicy/diys*). The R Markdown file for this example is labeled* `diy-ch04-pii.Rmd`*.*

In an increasingly digital world, data privacy is a sensitive issue that has taken center stage. At the center of it is the safeguarding of personally identifiable information (PII). Legislation in the European Union, namely, the General Data Protection Regulation or GDPR, requires companies to protect the personal data of European Union (EU) citizens associated with transactions conducted in the EU (Burgess 2019). The US Census Bureau, which administers the decennial census, must apply disclosure avoidance practices in order so that individuals cannot be identified (Abowd 2018). Thus, anonymization has become a common task when processing and using sensitive PII data.

Redaction. For example, the first element in the vector below contains hypothetical PII and sensitive information—John's social security number and balance in his savings account are shown. When presented with many lines of sensitive information, one could review each sentence and manually redact sensitive information, but given thousands if not millions of pieces of sensitive information, this is simply not feasible.

```
statement <- c("John Doe (SSN: 012-34-5678) has $2303 in savings in his account.",
               "Georgette Smith  (SSN: 000-99-0000) owes $323 to the IRS.",
               "Alexander Doesmith (SSN: 098-76-5432) was fined $14321 for overdue books.")
```

Using a combination of regex and `stringr` we can redact sensitive information with placeholders. To remove the SSN, we need a regex pattern that captures a pattern with three digits (\\d{3}), a hyphen, two digits (\\d{2}), a hyphen, then four digits (\\d{4}), or when combined: \\d{3}-\\d{2}-\\d{4}. The matched pattern is then replaced with XXXXX.

```
pacman::p_load(stringr)
new_statement <- str_replace(statement,"\\d{3}-\\d{2}-\\d{4}", "XXXXX")
print(new_statement)
```

```
## [1] "John Doe (SSN: XXXXX) has $2303 in savings in his account."
## [2] "Georgette Smith  (SSN: XXXXX) owes $323 to the IRS."
## [3] "Alexander Doesmith (SSN: XXXXX) was fined $14321 for overdue books."
```

Next, we replace the dollar value by matching a string that starts with the dollar sign (\\$) followed by at least one digit (\\d{1,}). And finally, John Doe's first and last name are replaced by looking for two substrings that each have at least one uppercase letter with an unspecified length ([A-z]{1,} [A-z]{1,}) and are found at the beginning of the string (^). The resulting sentence has little to no information about the individual in question.

```
# Find a replace dollar amount
new_statement <- str_replace(new_statement,"\\$(\\d{1,})", "XXXXX")

# Find and replace first and last name
new_statement <- str_replace(new_statement,"^[A-z]{1,} [A-z]{1,}", "XXXXX")

print(new_statement)
```

```
## [1] "XXXXX (SSN: XXXXX) has XXXXX in savings in his account."
## [2] "XXXXX  (SSN: XXXXX) owes XXXXX to the IRS."
## [3] "XXXXX (SSN: XXXXX) was fined XXXXX for overdue books."
```

Extraction. Alternatively, we can extract identifying information and create a structured dataset. Using the same regex statements, we can apply `str_extract` to create a three variable data frame containing person name, SSN, and money—all done with minimal effort.

```
ref_table <- data.frame(name = str_extract(statement,"^[A-z]{1,} [A-z]{1,}"),
                        ssn = str_extract(statement,"\\d{3}-\\d{2}-\\d{4}"),
                        money = str_extract(statement,"\\$(\\d{1,})"))
print(ref_table)
```

```
##                   name         ssn   money
## 1            John Doe 012-34-5678  $2303
## 2     Georgette Smith 000-99-0000   $323
## 3 Alexander Doesmith 098-76-5432 $14321
```

In either case, it is advisable to conduct randomized spot checks to determine if the regex accurately identifies the desired patterns.

4.2.4   Working with dates

Most programming languages are unable to "auto-magically" recognize a date variable, requiring coders to specify the format using a set of date symbols. The `lubridate` package significantly lowers the bar for R users, making the process of working and manipulating dates more seamless and less prone to error.

```
pacman::p_load(lubridate)
```

Upon loading lubridate, we can convert string and numeric objects with values that represent dates into date objects. The `as_date` function is the easiest to use, automatically detecting the date format and assumes that the date is recorded in UTC (see `d0`). In case `as_date` is unable to detect the date, a user can define the date format using functions such as `mdy` and `ymd`. Both are able to accommodate both string and numeric formats. Once data are converted into date objects, it is easy to derive new time-related information. To calculate duration between two dates is as simple as subtraction.

```
# Convert to date format
d0 <- as_date("2010-08-20")
d1 <- mdy("01/20/2010")
d2 <- mdy_hm("01/20/2010 00:00 AM")
d3 <- ymd(20100101)

# Date calculations
d1 - d3
```

Furthermore, we can extract parts of a date from a date object. Rather than extracting parts of a date using regex statements, the `lubridate` package makes it possible to use simplified functions to extract each `year`, `month`, `quarter`, `day`, `hour`, `minute`, and `second`. For example, we can extract the `year` and `quarter` and concatenate into a string using `paste0`.

```
# Extract time values
y1 <- year(d0)
q1 <- quarter(d0)

# Concatenate
paste0(y1, "Q", q1)
```

In base R, we can alternatively use the `format` function to specify an output format using date symbols. The examples below illustrate three different versions of the same month-year combination.

```
# YYYY-MM format
format(d0, "%Y-%m")
```

```
# Month Name - YYYY format
format(d0, "%B %Y")
```

```
# Month Abbreviation - YY format
format(d0, "%b-%y")
```

Either option gets the job done, but the choice depends on one's willingness to become familiar with date symbols.

4.3 The structure of data

4.3.1 Matrix or data frame?

Last chapter, we introduced matrices and data frames as options for storing multivariate data. Matrices are vectors of the same variable type, but in two dimensions ($n \times m$ dimensions). Data frames are a generalization of matrices that allow for each column of data to hold different data types as well as refer to individual columns by a user-specified name.

When should each matrices and data frames be used? From a pure logistical perspective, data frames are more flexible with respect to its ability to store multiple data types. Some code libraries are built specifically for matrices and others for data frames. Ultimately, data frames are generally a safe bet. They can be subsetted, sorted, reshaped, collapsed, and merged, retaining the variable types (Table 4.3).

Table 4.3: Matrices versus Data Frames.

	Matrix	Data Frame
Pros	Memory efficient. Good for advanced mathematical operations.	Store mixed types of data types. Allows user to refer to columns by an explicit name.
Cons	Able to store one data type at a time—leads to slightly more work required to manage multiple matrices. Columns can only be referred to by index number.	Not as memory efficient.

To illustrate data in action, we will draw from a dataset containing all reported thefts, burglaries, and robberies in Chicago between 2014 and 2018 (Chicago Police Department 2018). Provided as an `Rda` file, the file contains a data frame `crimes` with $n = 428,250$ with 17 variables.

```
load("data/chicago_crime.Rda")
```

4.3.2 Array indexes

We can think of data frames and matrices as data organized along a grid—essentially a spreadsheet. Each cell in a data frame can be located using a coordinate set of *array indexes*, or a number that indicate the location along a row or column. The first, fifth, and millionth rows have index values of 1, 5, and 1000000. The second, ninth, and one-thousandth columns have values of 2, 9, and 1000. This is similar to other languages like FORTRAN, MATLAB, and Julia. In contrast, array indexes in programming languages such as Python, Perl, C, Java, and Ruby start with 0—often a point of confusion when transitioning between languages.

Why do array indexes matter? Understanding this is the gateway to subsetting data, or slicing and dicing as one pleases. To extract a specific part of the data frame requires one to refer to the array indexes in a specific syntax:

```
crimes[7, 6]
```

where `crimes` is the name of a data frame. 7 refers to the 7th row and 6 is the 6th column—both of which are sandwiched in square brackets and separated by a comma. Executing the above state will extract the value in the seventh row and sixth column.

What if we want to extract an entire row? Simply specify the row index of interest and leave the column blank returns the desired result. The output is a single row data frame describing a robbery that occurred on "March 18, 2015 at 11:55 PM".

```
crimes[7, ]
```

To *extract multiple records by row index* depends on whether the request is sequential or piecemeal. Below, the first line extracts a range of rows from the 2nd through 4th rows in `crimes`, whereas the second extracts two non-overlapping ranges that are specified as a numeric vector.

```
crimes[2:4, ] # Apply the index range to extract rows
crimes[c(1:2, 10:11), ] # Specific indexes
```

The same notation can be used *to extract all street blocks that had a crime reported* by typing the number 4 after the comma in square brackets. The number 4 is the column index that contains `block` names. Rather than returning a data frame, the result is a vector containing the partially redacted street block.

```
crimes[,4]
```

In addition, data frames provide a few additional options for extracting the `block` column.

```
crimes[, "block"] # Extract column with"block" label
crimes[["block"]]
# List syntax to extract column from data frame
crimes$block
# Compact version of data column manipulation
```

To *extract two or more columns* follows a familiar pattern, making use of either a range of column indexes or a vector of column names.

```
crimes[, 3:4] # Extract multiple columns
crimes[, c("date", "block")] # Multiple column labels
```

4.3.3 Subsetting

Extracting specific *subsets* is perhaps one of the most essential skills of data processing as it allows one to focus on specific aspects of the data. In addition to calling specific array indexes to subset data, we can extract specific rows based on user-specified criteria, such as "crimes that happened in 2015" or "robberies that involved a handgun". Here, we describe three approaches using logical vectors, the `which` command, and the `filter` command in the `dplyr` library. The first two options are possible using the pre-built functionality of R.

The **logical vector** approach first creates a Boolean vector that describes some logical comparison, then passing the logical vector to the data frame to subset.

```
# Logical vector
group1 <- crimes$year == 2015

# Select based on vector
Out <- crimes[group1, ]
```

While this is a perfectly valid way of subsetting data, the **which** approach offers speed gains. The logical vector is passed to the `which` function to retrieve the row index of each `TRUE` value, thereby reducing the number of elements that need to be evaluated in the subset. The relevant row indexes are then passed to the data frame, returning a result far faster than the first option.

```
# Logical vector
Group1 <- which(crimes$year == 2015)

# Select based on vector
Out <- crimes[group1, ]
```

Lastly, and perhaps more convenient, is the `filter` function in the `dplyr` package. Not only is the syntax concise, the processing time is fast. The command only requires the data frame to be specified once and variables can be specified without the dollar sign notation.

```
pacman::p_load(dplyr)
out <- filter(crimes, year == 2015)
```

The `dplyr` allows for *piping* of operations, meaning that multiple steps can be combined into one function call. The pipe operator `%>%` allows the output of one step to be input into another, thereby creating a pipeline. For example, the statement below is the same as the above.

```
out <- crimes %>% filter(year == 2015)
```

To get comfortable, let's walk through a few different scenarios.

- ***Example 1***: Find all crimes that were armed. To do this, we use `str_detect` function from `stringr` to search for the term `ARMED` from the `description` field in `crimes`, then use the result to subset using `filter`.
- ***Example 2***: Find all cases that where the `primary.type` of crime was either a `THEFT` or `BURGLARY`. So that the statement can be one compact line, we use the `%in%` operator to search if any of the elements in the vector `c("THEFT", "BURGLARY")` are found in `primary.type`.
- ***Example 3***: Extract rows that meet two criteria is as simple as using logical operators. The pipe operator `"|"` should be used to represent *OR*, whereas `"&"` represents *AND*. To find all cases where an `arrest` occurred after the `year` 2016 :

```
# Example 1
out <- filter(crimes, str_detect(crimes$description, "ARMED"))

# Example 2
out <- filter(crimes, primary.type %in% c("THEFT", "BURGLARY"))

# Example 3
out <- filter(crimes, arrest == TRUE & year > 2016)
```

4.3.4 Sorting and re-ordering

At times, data frames need to be re-arranged to make it easier to navigate the data.

Re-arranging rows and columns. For one thing, columns and rows can be re-arranged or even removed by listing column indexes in a desired order. By passing a vector of column indexes to a data frame, we can not only re-order columns but also remove them. In the example below, we first re-order columns, moving columns 10 through 17 to the beginning of the dataset and moving one through nine to the end. Then, we create a new dataset `crime_new` dropping columns 10 and 13 by placing a negative (-) before the column indexes.

```
# Re-order
crimes <- crimes[, c(10:17, 1:9)]
```

```
# Drop columns
crime_new <- crimes[, -c(10, 13)]
```

It is more likely that we would organize our data calling specific columns by name. The standard approach
is to pass a string vector of column names, requiring each name to be placed in quotations. A more concise
approach using the `select` function from `dplyr` lists the names of desired columns without quotations.
Below, we illustrate how to create a data frame of geographic coordinates.

```
# Base R
crimes_geo <- crimes[, c("latitude", "longitude")]

# dplyr approach
crimes_geo <- select(crimes, latitude, longitude)
```

To *sort a dataset* can be done using `order`, which returns a vector of row indices in ascending order based
on the variable of interest. To sort descending, add we can either `"-"` before the variable if it is a numeric
or specify `decreasing = T`. The `order` function should not be confused with its cousin the `sort` function,
which is designed to order and return a vector of values.

```
# Base R - sort ascending
crimes[order(crimes$latitude, crimes$longitude), ]

# Base R - sort descending (2 options)
crimes[order(crimes$latitude, crimes$longitude, decreasing = TRUE), ]
crimes[order(-crimes$latitude, -crimes$longitude), ]
```

The base R options are somewhat clunky, requiring repetition of the name of the data frame. In contrast, the
`arrange` function in the `dplyr` package reduces redundancy. Similar to `order`, the `arrange` function sorts
ascending by default.

```
# dplyr
crimes <- arrange(crimes, latitude, longitude)
crimes <- arrange(crimes, desc(latitude), desc(longitude))
```

4.3.5   Aggregating data

Seeing the pattern in the data can be challenging when it is too granular. Aggregating data (or "rolling up")
into coarser units shows the landscape of the forest rather than the shape of an individual tree. It is also a
strategy for anonymizing data by stratum or groups of individuals while communicating insight.

Here, we introduce the `summarise` and `group_by` functions from `dplyr`. The former `summarise` is used to
calculate summary metrics on a set of variables, such as the count of crimes or the average latitude of
crimes. The `group_by` indicates along which variables should aggregation occur. Together, a highly granular
dataset can be converted into a more compact summary set, whether a time series of activity over time or
a cross-sectional snapshot.

Suppose we would like to see the total number of each type of crime for each ward of Chicago. We can
easily send the `crimes` dataset into a pipeline like the one below. A `group_by` statement indicates that all
subsequent calculations will be applied for each combination of type of crime (`primary.type`) and ward of
the city (`ward`). The data is then summarized by calculating two variables: the number of crimes `num.crimes`
using the count function `n` and the `arrest.rate` by taking the `mean` of the `arrest` variable. This process
is a new data frame with only $n = 151$ and $k = 4$. It is clear how this can be used—turning raw data into
aggregate, communicable insights.

```
crimes %>%
  group_by(primary.type, ward) %>%
  summarise(num.crimes = n(),
```

```
                    arrest.rate = 100 * mean(arrest))
```

Any variable can be used in the `group_by` statement, separating each dimension by a comma. The `summarise` function is quite flexible encompassing a large number of summary statistics:

- *Central tendency*: mean, median
- *Dispersion*: sd, IQR, mad
- *Range*: min, max, quantile
- *Position*: first, last, nth,
- *Count*: n, n_distinct
- *Logical*: any, all

Constructing Time Series. Are there more arrests over time? Do they follow a predictable seasonal pattern? Individual event records often need to be rolled up into a *time series*—data that is aggregate by some time unit in order to see patterns over time. Suppose we would like to visualize the arrest rates in Chicago—the proportion of crimes that resulted in an arrest. To learn if arrest rates have evolved over time, we process the data following two simple steps: (1) *Create quarterly time variables* using a combination of `mdy_hms` and `quarter` functions from `lubridate`. The resulting time variable will be used in `group_by`. (2) *Summarize the data* into a new data frame. When we `plot` the resulting 20-observation data frame `crimes_qtr`, we can see that arrest rates have fallen over time (Figure 4.3).

```
# Convert to date form and extract the quarter
crimes$date1 <- mdy_hms(crimes$date)
crimes$qtr <- quarter(crimes$date1, with_year = TRUE)

# Summarized data
crimes_qtr <- crimes  %>%
group_by(qtr) %>%
summarise(arrest.rate = 100 * mean(arrest))

# Plot a line chart
plot(crimes_qtr$arrest.rate,
      type = "b", col = "blue", cex.lab = 0.9,
      ylab = "Arrest Rate (%)",
      xlab = "Time (Quarters)")
```

4.3.6 Reshaping data

The shape and structure of a dataset usually go hand-in-hand with how it will be used. Data are generally stored in either *long* form or *wide* form as shown in Figure 4.4. Wide form stores each observation (unit of analysis) as a row and its attributes (variables) as a set of columns. As we look across a row, we see all information that describes that observation. In our example, each row is a Chicago Police Ward, each quarter is a separate variable, and arrest rates in each cell. This format is well suited for visual inspection and comparison between observations. Wide form, however, is not conducive for statistical analysis.

Long form, in contrast, spreads an observation's attributes over multiple rows—there are as many rows as *observations* × *attributes*. Most transaction and event logs (e.g., financial records, call log) are recorded as long form with an observation ID, a time stamp, and columns for each metrics. Panel datasets are a special case of long form that tracks observations over time. Table (A) is a panel dataset in which each row is a different ward in Chicago (e.g., 10, 11, 12) in a quarter (e.g., 2013.1, 2014.1).

Data will likely need to be *reshaped* between long and wide forms. For those who have familiarity with spreadsheet software, reshaping data essentially is the same as working with a pivot table. In R, we rely on a package called `reshape2` to reduce the process to a pair of functions: `melt` and `dcast`. The former `melt` converts data into long form, whereas the latter `dcast` converts data into wide form.

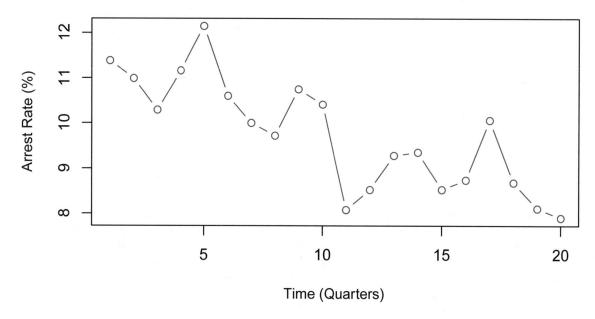

Figure 4.3: Time series of crime over time in Chicago.

(A) Long		
qtr	ward	arrest.rate
2014.1	10	14.96
2014.1	11	11.5
2014.1	12	12.2
2014.2	10	15.23
2014.2	11	9.12
2014.2	12	14.19
2014.3	10	9.41
2014.3	11	9.76
2014.3	12	13.25
.	.	.
.	.	.
.	.	.

(B) Wide			
qtr	ward 10	ward 11	ward 12
2014.1	14.96	11.5	12.2
2014.2	15.23	9.12	14.19
2014.3	9.41	9.76	13.25
.	.	.	.
.	.	.	.
.	.	.	.

Figure 4.4: Data in long form (A) and wide form (B). Two cells are color-coded to illustrate how the same value is positioned in each format.

```
pacman::p_load(reshape2)
```

Long to Wide. Suppose we summarized the crimes data into a quarterly dataset of arrest rates by type of crime and ward. This data would be returned in long form.

```
crimes_long <- crimes   %>%
          group_by(qtr, ward, primary.type) %>%
```

```
summarise(arrest.rate = 100 * mean(arrest),
          num.crimes = n())
```

The `dcast` function converts data from long form to wide form and expects a data frame, a formula, and at least one value variable. Let's walk through the example below that produces a new data frame where each row is a type of crime in a ward and each column is the arrest rate for each time period.

- The formula dictates the desired structure. In the formula `ward + primary.type ~ period`, for example, variables to the left of the tilde `~` are treated as row identifiers and variables to the right are dimensions that will be spread across the columns. Multiple dimensions can be specified, each separated by an addition symbol `+`.
- The parameter `value.var` indicates the variable that will fill the cells of the new data structure, which in this case is the `arrest.rate`.

The resulting data frame has $n = 151$ rows and $k = 22$ columns.

```
crimes_wide <- dcast(crimes_long,
                     ward + primary.type ~ qtr,
                     value.var = "arrest.rate")
```

Suppose we would like to reshape both the `arrest.rate` and the `num.crimes`. This requires two modifications to the above code: (1) We will need to install and load the `data.table` package, which converts data frames into a class of objects for memory efficient and speedy data manipulation. From that package, we will need to use `setDT` to convert our data frame into the appropriate format. (2) Specify a string vector of variable names for the `value.var`.

The resulting data frame has $n = 150$ rows and $k = 42$ columns.

```
pacman::p_load(data.table)
crimes_widetoo <- dcast(setDT(crimes_long),
                        ward + primary.type ~ qtr,
                        value.var = c("arrest.rate", "num.crimes"))
```

Wide to Long. It is just as simple to roll the data back into long form using the `melt` function. The example below re-converts the `crimes_wide` data frame into long form. Remember that all but two variables contain `arrest.rate` for each quarter. Moving from wide to long means that all the arrest.rate columns will be consolidated into two variables: column headers will be captured as a new time variable and column values will be arrest rates. The function expects

- a data frame (`crimes_wide`).
- `id.vars`: a vector of variable names or column indexes that indicate row identifiers.
- `variable.name`: a name of a new variable to which the wide column names will be stored.
- `value.var`: a name of a new variable to store the values.

```
crimes_longtoo <-   melt(crimes_wide,
                         id.vars = c("ward", "primary.type"),
                         variable.name = "qtr",
                         value.name = "arrest.rate")
```

4.4 Control structures

Data scraped from websites and media sources are becoming increasingly more common in the digital age. Investment companies regularly use news reports to detect sentiment toward companies in order to predict the direction of stock prices and earnings reports. Economists use scraped data from e-commerce sites to measure inflation. Scraping is quite repetitive. The same set of pre-defined actions are applied to many different webpages on a website. The format of that webpage dictates the types of actions. One could manually target websites and choose the code to be applied, but a far more scalable solution is to take advantage of control structures.

Given some variable, a control structure determines how a program will treat that variable given pre-determined conditions and parameters. The webpage formats of a New York Times article and a Fox News article, for example, are fundamentally different, thus a scalable web scraping program should code that are fitting to each type in order to get the desired data. Likewise, a script designed to extract debits and credits from a financial statement may look at the header of a table, then decide how to handle each type of value. In this section, we will cover two commonly used control structures that make programming far more efficient: if statements and loops.

4.4.1   If statement

If statements evaluate a logical statement, then execute a script based on whether the evaluated statement is true or false. If the statement is TRUE, then the code block is executed. Below, we have written some logic to evaluate if a manager has spent over his allocated budget for the year, returning a printed statement if spending exceeded acceptable tolerance of $400.

```
  spending <- 450
  if(spending > 400){
# If statement true, run script between curly brackets
print("You're over budget. Cut back.")
  }
```

```
## [1] "You're over budget. Cut back."
```

In cases where there are two or more choices, if...else statements would be appropriate. In addition to the if() statement, an else statement can be included to handle cases where the logical statement is FALSE.

```
  spending <- 399
  if(spending >= 400){
# If statement true, run script goes here
print("You're over budget. Cut back.")
  } else {
# Else, run script goes here
print("You're within budget, but watch it.")
  }
```

The complexity of these statements can accommodate more complex logic:

```
  spending <- 500
  if(spending >= 450){
    print("High roller, eh? Is your debt supported by a wealthy donor?")
  } else if(spending >= 400){
    print("You're over budget. Cut back.")
  } else if(spending >= 50 && spending < 400){
    print("You're within budget. Good job.")
  } else if(spending < 50 &){
    print("You might be under-reporting your spending.")
  }
```

If statements work is designed to evaluate *one logical statement at a time* and requires a *loop* to evaluate for each element of a vector—a control structure we cover in the next subsection. For simple if statements, we can rely on the ifelse function that evaluates a logical condition for each element in a data frame or vector, then returns a vector with desired transformations.

```
# Vector of spending for multiple managers
spending <- c(50, 200, 500)
```

```
# Ifelse returning overbudget flags
out <- ifelse(spending >= 400, "Over-budget", "Within Budget")
```

4.4.2 For-loops

Many programming tasks apply the same code over and over. Rather than repeating tasks manually, loops can reduce redundant tasks by applying the same function over a range of elements, objects, etc. It is an efficiency augmenting functionality that is central to program and allows computers to perform tasks in a semi-automated way. Conceptually, for loops can be likened to an assembly line in a car factory. In order to build a car, a series of well-defined, well-timed processes need to be coordinated in a serial fashion. To build 500 cars, the process needs to be executed 500 times. For-loops are essentially the same: Given a well-defined set of tasks, the manufacturing process runs 500 times.

Let's take the following toy example in which we print to console a set of numbers. The code block essentially says "print values for the range of 1 through 5", where i is an *index value*. When executing the statement, R will push the first value in the sequence of 1:5 into the index (in this case, it's the number 1), then the code block in between the curly brackets ({}) will be executed, treating i as if it's the number 1. Upon executing the code without error, R will advance to the next value in the sequence and repeat the process until all values in the sequence have been completed.

```
for(i in 1:5){
    print(paste0("Car #", i, " is done."))
}
```

```
## [1] "Car #1 is done."
## [1] "Car #2 is done."
## [1] "Car #3 is done."
## [1] "Car #4 is done."
## [1] "Car #5 is done."
```

We can do the same for a vector or list of values. In the example below, the vector news contains six terms. Using a for-loop, we can print out each word and append the number of characters to a placeholder vector word_length.

```
# Vector of words
news <- c("The","Dow","Is","Up","By","400pts")
word_length <- c()

  for(i in news){
    print(i)
    word_length <- c(word_length, nchar(i))
  }
```

```
## [1] "The"
## [1] "Dow"
## [1] "Is"
## [1] "Up"
## [1] "By"
## [1] "400pts"
```

A few things to keep in mind when using loops. First, loops are convenient to use but there are often times more efficient functions that handle *vectorized* problems—or a function that is designed to be applied to all elements of a vector. For example, while the word_length example uses a loop to calculate the length of each word in the news vector, the nchar function can actually be applied to all words in the news vector without a loop. Second, all objects created in a loop are stored in the programming environment. While this means that all objects will be available after the loop is complete, it also means each subsequent iteration's results will overwrite the previous. To save the result of each iteration requires one to explicitly indicate how results should be stored.

If a calculation needs to be applied to each of one million files and the results need to be saved, then for-loops are a good option. *A common paradigm* involves the following:

1. Create empty placeholder object (e.g., a vector, matrix, or data frame) to store results;
2. Initialize loop; then
3. Append outputs to placeholder at the end of each loop iteration.

In the example below, we simulate 10,000 observations of a Random Walk, which is a time series in which the best forecast of today's value is yesterday's value plus an error term u_t:

$$\hat{y}_t = y_{t-1} + u_t$$

Many real-world phenomena can be described as random walks, such as exchange rates and stock prices. For statistical research, it is often necessary to simulate a random walk process in order to show how a new algorithm performs. Below, we simulate a random walk in which we create a placeholder data frame `df` in iteration 1. In each subsequent iteration, we add the previous period's value with a random growth selected from `alpha`, then append the result to `df`. To simulate 10,000 observations requires roughly 7 seconds to complete. What if we wanted to reduce the compute time? In the age of big data, it might be necessary to run millions of simulations, thus 7 seconds may actually be an eternity.

```r
# Set placeholder data frame with n rows
n <- 10000
alpha <- c(-1, 1)

# Simulate Random Walk
start <- Sys.time()
  for(i in 1:n){
    if(i == 1){
      df <- data.frame(iter = i,
                       y = sample(alpha, 1))
    } else {
      df <- rbind(df,
                  data.frame(iter = i,
                             y = df$y[i-1] + sample(alpha, 1)))
    }
  }

# Timer stop
Sys.time() - start
```

```
## Time difference of 14.13205 secs
```

If we know the total space required for the simulation, we can set up `df` with a pre-specified number of rows and overwrite individual cells when needed. Such a simple change vastly reduces the processing time to less than 1.5 seconds—only one-quarter of the time! *Why?* Part of the answer lies in the functional differences between appending to data frames and overwriting cells in a data frame. When we use `rbind`, R is forced to overwrite the entire data frame with an additional row. As objects become large, the write time compounds. In contrast, overwriting individual cells only replaces specific cells, removing unnecessary read-write time.

```r
# Set placeholder data frame with n rows
n <-10000
df <-
data.frame(iteration =1:n,
                 y = numeric(n))
alpha <- c(-1, 1)
```

```
# Simulate Random Walk
start <- Sys.time()
  for(i in 1:n){
    if(i == 1){
      df$y[i] <- sample(alpha, 1)
    } else {
       df$y[i] <- df$y[i-1] + sample(alpha, 1)
    }
 }

# Timer stop
Sys.time() - start
```

```
## Time difference of 1.079904 secs
```

Even more surprising, there are simpler functions that can simulate a random walk in far less time and with less code. In fact, a simple combination of the `sample` and `cumsum` functions can randomly draw $n = 10000$ values of u_t and calculate a vector of elements with cumulative sums (effectively a random walk), all in less than 0.01 seconds. This is all to show that loops are easy to use, but be sure to plan out your loops carefully to maximize efficiency and it is worth conducting some due diligence to identify possible alternatives.

```
# Simulate Random Walk in one line
start <- Sys.time()
out <- cumsum(sample(alpha, n, replace = T))
Sys.time() - start
```

```
## Time difference of 0.00270009 secs
```

4.4.3   While

Whereas for-loops require a range of values through which to iterate, `while` statements continue to be executed until a condition is no longer true. We can imagine scenarios in which the number of required iterations is an unknown but a pre-defined criteria is known. Optimization problems with a systems of equations, for example, require that multiple candidate solutions be tested in order to find a solution that minimizes an error function—while loops are great for automating searches.

A few key ingredients are required. The `while` statement requires a logical statement (e.g., $x > 2$)—the *goal*. So as long x is greater than 2, the statement will continue to run. There also needs to be an *update step* in the script that ensures that values that are to be evaluated are tuned toward satisfying the goal so that eventually x is no longer greater than 2. Otherwise, the while loop will continue for infinitum, also known as an *infinite loop*.

To put this into perspective, we construct a while loop to search for a value of x that minimizes the formula $y = x^{\frac{3}{2}} - 8x + 100$. Before the loop, we set initial value of $x = -100$, set *step* = 1 to control the fineness of the search interval for x, create a countervariable i to keep track of the number of iterations, and a placeholder vector `vals` to store values of y. The while loop starts by evaluating if one of two conditions is true: if there are less than two iterations (`i < 2`) or if we have found a minimum value (if current value of y in iteration i is less than the previous iteration). If either is true, then the script within the curly braces is executed, otherwise the loop terminates. Notice that *step* is added to x in order to advance the loop to the next set of conditions to be evaluated.

```
# Set initial value of x and i
x <- -100
step <- 1
i <- 0
vals <- c()
```

```
# Find a minimum of the function
# Where previous value is greater than current
while(i < 2 || vals[i-1] >= vals[i]){
   vals <- c(vals, x^2 -  8*x + 100)
   i <- length(vals)
   x <- x + step
}
```

While these loops are useful for searching for solutions, the problem needs to be well defined and well bounded.

4.5   Functions

Data manipulation tasks are often repeated for many different projects and it is not uncommon for two or more scripts to contain the same exact steps. Rather than tediously modifying scripts, try to live by the following philosophy: once, then never again. The idea is that if code is well designed and documented, it can serve as re-usable tools that can be re-applied to similar problems. This is the basis of *user-written functions*: a coder can define some set of standard required inputs on which a set of steps can be applied to produce a standard output, then their work can be shared with many others who face similar problems

A function is constructed as follows. Using `function`, a set of parameters are specified upfront as the key ingredients, then a set of coded procedures are executed using those same parameters. In the example below, we construct a function `meanToo` that replicates the functionality of the `mean` function. `meanToo` requires a data frame `df` and a variable name in string format `var.name`. Within the curly brackets, we insert code that uses the input parameters as placeholders. The mean is calculated as the sum of `var.name` in data frame `df` divided by the number of rows in `df`, then assigns the result to the object `temp`. These calculations are temporary and only executed with a *local environment*—they are wiped once the function finishes running. However, the `temp` object can be extracted by passing it to `return`. All of the above steps are assigned to the `meanToo` object that is treated like any other function.

It is good practice to include documentation about what the function does and how to use it. At a minimum, include what the function is, the arguments, and the output. Typically in the culture of open-source software, code should be written with others in mind: keep it transparent and re-usable.

```
meanToo <- function(df, var.name, ...){
  #
  ## Calculate mean of a variable in a data frame
  #
  # Args:
  #    df = data frame
  #    var.name = variable name (character)
  #
  # Returns:
  # A numeric value

  # Code goes here
 temp <- sum(df[[var.name]])/nrow(df)

  # Return desired result
  return(temp)
}
```

To use the function, first run the function code above, then fill in the input parameters. Any script can be converted into a function for general use. While it does require some time upfront to properly write a usable function, getting into the habit is worth the investment for efficiency and sharing your technical perspective.

```
meanToo(data, "x1")
```

4.6 Beyond this chapter

4.6.1 Best practices

Writing just any piece of code is easy. The real challenge is to ensure that it is re-usable, functional, and adopted.

Data processing is not like a vampiric feeding frenzy of the sort that might be seen on Buffy the Vampire slayer (the TV show): one simply does not arbitrarily chomp at and slice data without any thought as to what comes next. It requires discipline and finesse to acquire a raw piece of information and extract the empirical substance. It is the foundational step of any successful data science project.

Imagine that you have conducted a three-month analytics project, rushing to obtain to a result. If asked to provide backup, documentation, and a process, it may very well require an additional few months to decipher your passionate but hasty explorations. Where did this data come from? Why did I take a dot product in this equation? What happened to the raw file? Who has Bob mentioned in this email? Which directors have seen the report? Ultimately, the data scientist is in control and should be empowered to conduct data processing in a structured, scientific manner. Here are five tips that can bulletproof data processing:

- *Define a file structure and stick to it.* All work product needs to be chronicled in a coherent structure. At a minimum, create a folder for *scripts* to store all code, another for *raw* data, a folder for *outputs* from data processing, and another for *documentation*. Choose any file structure, but make sure it is consistent and always followed. Name scripts and data in a logical, intuitive format. And perhaps more importantly, create a *Read Me* file—a simple text document that explains the contents of the project and keeps a log of versions.

- *Adopt and stick to a coding style.* Each file, data field, and function should follow a style so that the logic is cohesive, readable, and traceable. The Google `R` Style Guide (Google 2009), for example, sets easy to follow guidelines. According to the style guide, variable names should be labeled as lower-case characters with a `"."` as a separator: `variable.name`. Functions should be named in proper case without spaces: `NewFunction` or `MyFunction`. Regardless of the style guide, file names should be meaningful and the only punctuation used should be `"_"` for spaces and `"."` to accompany the file type: `gdp_2014_10_02.json`. But most importantly, the code needs to make common sense. Be vigilant with your code style as it may be the difference between a successful adopted project and a failed project.

- *Version up.* Each time code is written, modified, amended, or updated, a new version should be saved. Cloud-based services such as Github make version control simple and also allow groups of coders to work together. If storing work product on a local server or computer, simply save the file with a new version number as well as making an entry into a log or Read Me file. The log file can be as simple as a text file with time stamps indicating the date of the edit, the file name, and the nature of the edit.

- *Think modularly with code.* One of the greatest human inventions is the assembly line. Each step in the process is discrete, requiring workers to specialize in one set of well-defined tasks. As a worker becomes specialized, production speed increases as they refine their ability to accomplish a specific task. If any step in the process needs to be improved, revised, or replaced, all other steps in the process are not disrupted—only one step needs to be addressed. This standardization of processes is the only way to scale. Programming and data processing are the same—the goal is to produce well-written code that can reliably accomplish and re-accomplish the same tasks using newly available data. In essence, one needs to construct code that lays the pipeline for data to flow from loading, to cleansing and standardization of values, re-structuring into usable form, and storing in an accessible format. Each module of the process should be able to be swapped out like an interchangeable piece if need be.

- *Think about data in layers.* All data should ideally stem from a single "source of truth" – one harmonized set of data should serve as the basis for all applications, whether its supporting inspectors in the field or informing high-level policy making. It is thus not uncommon to have data stored in varying degrees of processing. NASA earth science satellite imagery, for example, is stored in at least six different levels of varying amounts of processing. Level 0 data are raw satellite instrument data at full resolution. Level 1B contained reconstructed, calibrated, and processed data that serve as the core source data for downstream applications. Whereas Level 4 are outputs from specific models and analyses. Before commencing data processing, first consider what end users want to accomplish, then process the dataset into a common, standard form that can serve as a lowest common denominator. It is from that source dataset that all use case specific derived data should come.

It will undoubtedly require some time to first familiarize oneself with the steps of data processing before coding projects become tidy engineering feats. Nonetheless, these best practices are worthy ideals to which to strive.

4.6.2   Further study

Indeed, data processing is an expansive discipline, extending well beyond the capabilities of R alone. In fact, R relies on other programming languages, software and hardware infrastructure to extend its functionality. Here, we briefly touch upon other topics that are worth studying, such as data processing packages, parallel, computation, databases, and business intelligence.

Data processing with R. Preparing data will always occupy a significant amount of a data scientist's time—it is unavoidable. However, faster one can process data, the more time there will be to experiment and develop high-impact use cases. This, of course, requires substantial practice to develop programming muscle memory.

While base R can handle much of a data scientist's processing requirements, there are more efficient data processing packages, often times built on top of faster languages (e.g., C). In this textbook, for example, we rely on the `tidyverse` suite of packages that emphasize readability and extensibility—there are many neat tricks of the trade. For in-depth treatment, consider reading *R for Data Science: Import, tidy, transform, visualize, and model data* (Wickham and Grolemund 2017). The textbook provides many useful examples and illustrates the effective ways to mold data.

The `data.table` package is an alternative that has much of the same functionality as `dplyr`. The package's syntax is less readable and more concise. Relative to `dplyr`, it offers speed and memory efficiency gains, especially with large datasets. To get started with `data.table`, visit the package's reference site.[3].

Parallel computation. What if it were possible to split a large processing job across multiple machines or nodes? This would reduce the time needed to work with large datasets. This is the simple idea behind parallel computation: spread the load across multiple workers to shorten the time needed to accomplish a task. Early implementations of parallel processing required significant information technology experience; However, in today's day and age, open-source technologies have lowered the bar to entry. In fact, it is possible to even take advantage of the multiple cores on a laptop's processor to hasten a processing job. The `doParallel` and `parallel` packages allow programmers to exert control over how many cores, processors, or machines will be used to split a job, then facilitate the coordination task of managing the multiple parallel tasks. For a gentle introduction, review the vignette for the `doParallel` package.[4] There are also parallel computing resources for virtually any use case available on CRAN.[5] Among the most popular is *Apache Spark (Spark* for short), a framework designed to handle real-time streaming and large-scale data on server infrastructure. In fact, Spark has become an industry standard for heavy data processing and modeling. For a gentle introduction to Spark, consider reading *Mastering Spark with R* (Luraschi, Kuo, and Ruiz 2019).[6]

Advanced topics in R. Many data scientists will exhaust the functionality of current packages and find themselves writing a custom package for specific use cases. Building a package is accessible to anyone. To

---

[3]See the `data.table` reference site at https://rdatatable.gitlab.io/data.table
[4]See https://cran.r-project.org/web/packages/doParallel/
[5]See https://cran.r-project.org/web/views/HighPerformanceComputing.html
[6]Also available at https://therinspark.com/

get started requires the `devtools` and `roxygen2` packages. However, to design a well-functioning, cohesive package requires an understanding of software design, object-oriented programming , and abstraction of functionality from a more efficient language for use in `R`. For more technically advanced readers, consider reading *Advanced R* (Wickham 2017).

Databases. While some data projects can be built around a single file (e.g., CSVs and Excel files), other projects have more complex requirements. For instance, some datasets are only useful if updated regularly over time (e.g., COVID-19 new case counts, patient records). Other datasets are too large to be stored in a single file and need to be subdivided for efficient storage (e.g., one day's worth of satellite imagery captured across the world). To create and maintain a gold copy a large, ever-changing dataset, consider setting up a database.

Databases store and catalog data in a well-structured electronic format. Databases can be hosted on a laptop computer, but often times large data requires databases to be hosted on a powerful server or cloud environment built with advanced computation hardware. Relying a Database Manage System (DBMS)—a software to manage and interact with the data—users can perform CRUD operations, namely, *Create* (add a new record), *Read* (read or load a file), *Update* (modify an existing record), and *Delete* (remove a record). These four operations not only enable data analyses and exploration, but also serve as the foundation on which technologically advanced use cases are built (e.g., APIs, prediction engines, etc.). For tabular data that is typical in policy settings, data scientists will query and extract data using Structured Query Language (SQL)—a special language for CRUD operations. A command of basic database operations will prove invaluable throughout a career as a data scientist.

For a conceptual overview of database design, consider *Database Design for Mere Mortals: A Hands-On Guide to Relational Database Design* (Hernandez 2013). For a general introduction to SQL for a variety of DBMS, consider *SQL Queries for Mere Mortals: A Hands-On Guide to Data Manipulation in SQL* (Viescas 2018).

Businesses have sales, revenue, cost, and customer engagement data. Public sector agencies also have "business" data on their operations, such as complaints, phone calls, work orders, among others. A natural evolution from databases is *Business Intelligence* (BI), or analysis of business data. In both cases, executives and managers can use BI tools (e.g., Tableau, QlikView) to summarize and drill down on key operational metrics to monitor the current state of affairs. Imagine a case in which a manager clicks on a bar in a bar graph, then all metrics are subset to data that are contained within that bar's stratum. On a technical level, this simple click commands the BI tool to query and aggregate data rapidly, which in turn requires data to be pre-processed and stored in an analysis-ready format. For this reason, BI tools are built on *data warehouses*—a type of databasing system that is optimized for analysis and querying for *business intelligence* (BI) applications. Raw data are pushed through an Extract-Transform-Load (ETL) process to clean, blend, reshape, and aggregate data.[7] For a conceptual overview of BI applications and data warehousing, consider reading the following articles:

- Chaudhuri, Surajit and Umeshwar Dayal. 1997. "An overview of data warehousing and OLAP technology". *SIGMOD Rec.* 26, (1): 65–74. DOI: https://doi.org/10.1145/248603.248616
- Chen, Hsinchun, Chiang, Roger H. L. and Veda C. Storey. 2012. "Business Intelligence and Analytics: From Big Data to Big Impact". *MIS Quarterly* 36 (4): 1165–1188.

Data scientists will no doubt encounter BI in their work. It is an important part of the broader data ecosystem, but itself is not data science. Whereas BI is focused on monitoring and reporting (i.e., describing the past), data science focuses on inference and making predictions for the future. Nonetheless, both disciplines can benefit from data warehousing and ETL.

---

[7]While we borrowed the term *ETL* to describe data processing earlier in the chapter, ETL has its origins in BI and data warehousing applications.

Chapter 5

Record Linkage

5.1   Edward Kennedy, Bill de Blasio, and Bayerische Motoren Werke

During the summer of 2004, U.S. Senator Edward "Ted" Kennedy told the Senate Judiciary Committee that he was stopped at the airport five times in March and April of that year (Goo 2004). When attempting to check in at the airport, airlines staff would refuse to issue a ticket. When pressed as to why, staff were unable to offer a justification despite the senator's 42 years of flying through East Coast airports. Eventually, supervisors would override the hold-ups. It was later revealed by the Transportation Security Administration (TSA) that the senator's name was on the *no-fly* list—a list maintained by the United States federal government's Terrorist Screening Center (TSC) that indicates who is disallowed from boarding commercial aircraft to and from airports in the United States (Burns 2012). At the time, airlines were given the responsibility of stopping individuals from boarding planes if a match was found, but as the underlying data used to create the list was classified and sensitive, airlines were given limited identifying information such as the names of persons of interest. It so happened that the name "Edward Kennedy" was on the list—a generic alias used by terrorist organizations. The name is also exceedingly common. A LinkedIn search conducted in mid-2019 found 8,023 people who go by the name "Edward Kennedy", thus it should not be a surprise that one of the most prominent Congressmen in the US was flagged as a possible no-fly list suspect when screened only on first and last name. The senator's experience is a classic case of mistaken identity—when an entity is incorrectly identified due to imprecise or insufficient information.

It is easy to misidentify a person based on their name and even easier to miss if an individual maintains an abundance of identities. The 109th Mayor of New York City Bill de Blasio, for example, was not always known as *Bill de Blasio*. In fact, he has changed his name twice and has had three legal names, including *Warren Wilhelm Jr.*, *Warren de Blasio-Wilhelm* and *Bill de Blasio* (Smith 2013). His current name only first appeared in 1990 as a working name and was made official through a court petition in 2001. In addition, his close network addresses him as *Billy* (Wolper 2012). While his names are well covered in the press, it may be challenging to piece together his history if his other aliases are not already linked to the same entity.

Name changes are quite common throughout society. Some people may choose to change part of or their entire name at major life events such as marriage, or more clandestine encounters with witness protection. Nicknames (e.g., Bob = Robert, Dick = Richard, Jen = Jennifer) and stage names (e.g., George Michael = Georgios Kyriacos Panayiotou, Stevie Wonder = Stevland Hardaway Judkins) are common place. Organizations regularly update their brands by changing their logo and how they refer to themselves (e.g., Bayerische Motoren Werke = BMW, Klynveld Peat Marwick Goerdeler = KPMG, Defense Advanced Research Projects Agency = DARPA). Nonetheless, reconciling entities is a fundamental part of any data project, enabling profound insights: it allows *record linkage* to be conducted on two or more datasets so they can be matched, thereby enriching the story about any single entity. If done well, record linkage can help answer questions that are central to operating organizations such as

© Springer Nature Switzerland AG 2021   61
J. C. Chen et al., *Data Science for Public Policy*, Springer Series in the Data Sciences,
https://doi.org/10.1007/978-3-030-71352-2_5

- Two companies just merged and the executives need a new company-wide marketing strategy. What is the overlap in the customer base? Are the customers similar? Can services be cross-sold, and to whom?
- Suspected terrorists are up to no good. Can knowing all their aliases prevent them from getting on a plane?
- A jail is reviewing its logs. How many people recidivate more than once per year? What policy treatments minimize recidivism?
- In a genetics lab, can two nucleotide sequences be matched?
- A policy analyst gains access to social services records from multiple agencies. What can be learned about service users when examining all of their touches across the social services system?
- Two polling firms have two different polling universes. Do their surveys cover the same communities? Do their results tell the same story?

Despite the potential gains of record linkage, we cannot forget that linkage can easily be botched if not done so with careful planning and deliberation. In this chapter, we lay out principles of linking data efficiently and accurately. We begin by describing basic concepts of data linkage, then illustrate the mechanics of deterministic and probabilistic record linkage. The chapter closes with an example that links two real-world international sanctions lists.

5.2   How does record linkage work?

Flavors of record linkage. Record linkage aims to solve the following problem: *given an entity i, find all ways to uniquely identify i in any set of information*. We should be able to link an entity's records across two or more datasets and enrich our insights at an individual level. If done well, security agencies should be able to accurately identify which Edward Kennedy is a terrorism suspect while sparing innocent bystanders. If done well, an e-commerce company should be able to have a full picture of purchasing activity from multiple vendor databases.

The matching process begins with *pre-processing data* in which typographical formats are standardized (e.g., punctuation, capitalization, spacing) and the concepts captured in each field are conformed (e.g., date information is parsed into its month-day-year components). Simple changes to identifying fields like first name, last name, and birth date improve the chance of character-by-character matches or *exact matches*. However, some identifiers have more complicated spelling and recording errors that require approximate matching through *comparison functions*. Names like "Culbert" and "Colbert" have a higher chance of being the same than the name "Cuthbert". Comparison functions cast a wider net in making connections. At this stage, records can be *de-duplicated* using the cleaned identifying fields.

Beyond the preparatory steps, two record linkage strategies dominate the discipline. The simpler, more common approach is *deterministic record linkage*, in which a set of unique identifiers are used to match records character-by-character. A set of rules typically dictates which fields are matched and in what order, chasing both common and edge cases. A deterministic strategy could first search for exact matches among identifying fields (e.g., full name), followed by combination of identifiers with partial information (e.g., first initial, last name, and date of birth). Despite their conceptual simplicity and ease of implementation, deterministic methods have a number of shortcomings. Developing matching rules can be an artisanal affair, requiring a substantial time investment to adjust to each additional dataset. Furthermore, an exact match can be overly restrictive given spelling errors, thereby leading to many missed matching opportunities. Lastly, while some identifying fields possess greater identification power than others, deterministic methods ignore their truthiness, giving all fields equal weight regardless of their quality.

A sophisticated, more computationally intensive alternative is *probabilistic record linkage* focused on calculating the probability that any two record pairs are a match. Given two datasets with n_A and n_B records, comparison functions evaluate all combinations of potential matches ($n_A \times n_B$) based on identifying fields and calculate a probability that a given pair of records are a match. A threshold is tuned to separate likely matches from non-matches. However, $n_A \times n_B$ possibilities can be an awfully large number of matches, especially in the age of big data. To *reduce the search space* and simplify the linkage process, matching can employ *blocking* that stratifies data, then matches stratum of records to reduce the number of candidate pairs, then matches within each stratum using fields with more precise identifying information. Imagine being tasked with linking two mailing lists collected by two companies reflecting consumer activity in the past month. From a normative lens, an individual residing in New York probably should not be matched with a record from Montana, thus matching should be contained to matching states, which in turn reduces the number of potential matches that are considered. While probabilistic record linkage seems more technically complicated, it also requires far less human overhead.

Relational Models. An analyst conducting deterministic record linkage has a list of 1000 people A and another list of 3000 people B. The output is a list of $n = 3000000$ people—there are more records than in either of the input lists! *What happened?* It is no coincidence that the table contains exactly the product of 1000×3000, suggesting a failure to de-duplicate the data. In fact, inadequately de-duplicated data can actually *increase* the number of duplicates during the joining process and run the risk of a *Cartesian product* in which each record in table A is matched with every row in B and vice versa. Joins that result in Cartesian products are also a form of *many-to-many* (m:m) matches that rarely have any practical use (see Figure 5.1). Furthermore, when working with large datasets, simple joins become unnecessarily time-consuming as the computer matches every record, then outputs an inordinate number of duplicates.

When matching data, it is worth thinking about the relational model, or how two sets of data should relate and combine with one another. For the most part, analysts think of matching relationships in terms of *one-to-one* (1:1) and *one-to-many* (1:m) matches. A *one-to-one* match occurs when any record in A can be joined with at most one and only one record from B. For example, if two social welfare providers wanted to better understand what their users need, they may match unique identifiers (e.g., social security numbers) from their respective user lists and take a look at the one-to-one relationships (overlaps) as well as the unmatched records. A one-to-one match would also apply when matching a passenger on a flight to the no-fly list.

A *one-to-many* match occurs when one record in X is related to many records in Y. It is often applied in the context of relating an entity to its history over time. The suspect of a crime, for example, may have their DNA tested and matched with blood samples collected from multiple unsolved cases. A query for a frequent international traveler passing through border control will turn up a history of multiple trips. A tax filer should have a one-to-many match with multiple previous tax returns.

Record linkage requires a surprising degree of clarity in one's analytical objectives—without knowing precisely how you will use the data, a match may produce misleading results.

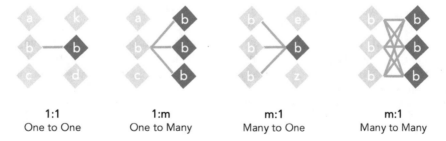

1:1	1:m	m:1	m:1
One to One	One to Many	Many to One	Many to Many

Figure 5.1: Illustration of how the uniqueness of an identifier affects the number of possible matches for each record.

5.3 Pre-processing the data

Humans can spot and mentally adjust for imperfections in text. Computers, on the other hand, need to be explicitly programmed to do so. Text standardization techniques, many of which covered in the previous chapter, play a critical role in efficient, scalable, and accurate record linkage. The strategy needs to be robust, able to harmonize identifiers not only in your dataset, but in any other dataset. At a minimum, basic text

Table 5.1: A pair of tables that contain information on individual filers (filers table) and tax filing data (filings data).

Filers Table				Filings Table				
first.name	mi	last.name	dob	first.name	last.name	birth.year	filing.year	income
Arthur	C.	Duran	9/2/1956	Arthur	duran	1956	2010	73000
alex	NA	Coyle	10/12/1965	Arthur	Duran	1956	2011	75400
Guy	N.	Ferdinand	5/2/1930	Gerak	Jaffa	1946	2010	102142
Franz	O.	Li	7/23/1985	JET	FERDINAND	1945	2010	231442
Jet	X.	Ferdinand	3/8/1945	ALEX	COYLE	1965	2010	66030
Richard	E.	Hawking	10/10/1951	Alex	Coyle	1965	2011	42123
richard	e	___Hawking	10/10/1951	Alex	coyle	1965	2012	54285
Stephen	T.	Colbert	5/13/1964	stephen	colbert	1964	2010	320140
Stephen	T.	Culbert	5/13/1964	stephan	colbert	1964	2012	520120
Gerak	G.	Jaffa	1/6/1946	Richerd	HAWKING	1951	2010	84200

standardization will consider typographical formatting and conforming of measurement concepts as will be described below.

Let's read in data on hypothetical tax `filers` ("filers table") with their tax `filings` history ("filings table"), a sample of which is presented in Table 5.1. For the remainder of this chapter, we will illustrate the matching process using this pair of hypothetical datasets.

```
# Load readr package
pacman::p_load(readr)

# Load person and filings tables
filers <- read_csv("tables/ch5_tab_person.csv")
filing <- read_csv("tables/ch5_tab_filings.csv")
```

Character standardization. A cursory scan of the *filers* table finds a number of typographical errors. Our pre-processing objective should be to ensure that names that are visibly the same are represented the same. Text characters can be standardized by stripping spaces that come before and after strings known as white space (`str_trim`), removing unnecessary punctuation and spaces (`str_remove_all` with a regex pattern `[[:punct:][:space:]]`), then converting strings to lower case (`str_to_lower`). As the procedure will be applied and re-applied, it is a good idea to write a re-usable function that strings together these three `stringr` package functions:

```
textStandard <- function(string_val){
  #
  # Basic text standardization
  # Args:
  #   string_val = a string or string vector
  # Returns:
  #   Transformed, trimmed, lower case strings
  #

  require(stringr)

  output <- string_val %>%
          str_remove_all("[[:punct:][:space:]]") %>%
          str_to_lower() %>%
          str_trim()
```

Table 5.2: Standardized text.

Filers Table		Filings Table	
FN	LN	FN	LN
arthur	duran	arthur	duran
alex	coyle	arthur	duran
guy	ferdinand	gerak	jaffa
franz	li	jet	ferdinand
jet	ferdinand	alex	coyle
richard	hawking	alex	coyle
richard	hawking	alex	coyle
stephen	colbert	stephen	colbert
stephen	culbert	stephan	colbert
gerak	jaffa	richerd	hawking

```
}
```

`textStandard` is then applied to the `first.name`, `middle.initial`, and `last.name` fields using a for-loop to minimize code redundancy. The transformed text are stored as new fields with the prefix "cleaned" to preserve the provenance of the data.

```
# Standardize first three columns in filers table
person_identifiers <- colnames(filers)[1:3]
for(i in person_identifiers){
filers[[paste0("clean.", i)]] <- textStandard(filers[[i]])
}
```

```
# Standardize first two columns in filings table
filing_identifiers <- colnames(filing)[1:2]
for(i in filing_identifiers){
filing[[paste0("clean.", i)]] <- textStandard(filing[[i]])
}
```

While this simple cleansing step can reconcile names like "Richard E. Hawking" versus "richard e ___Hawking" to "richard e hawking" as seen in Table 5.2, it is not effective for variations in spelling like "Stephen Colbert" versus "Stephan Colbert". Later this chapter, we will introduce fuzzy matching techniques that overcome this obstacle.

Conforming concepts. Some fields contain similar information with differing degrees of precision. Geographic information such as US zip codes can be reported with five digits or the more specific nine-digit format; birth date information can contain only the year while other fields include the month-day-year; and names of places may be reported in full or abbreviated. In any of these cases, the data need to be conformed to reflect the same concept with the same degree of precision, typically using the lowest common denominator.

For example, the `filers` table contains birth date information (month, day, and year) cannot be precisely matched with the `filing` table that only contains the year of birth. By extracting the year of birth from the birth date field using `lubridate`, we can conform the birth date information and facilitate a match.

```
# Load package
pacman::p_load(lubridate)
```

```
# Extract year and assign a variable of the same name.
filers$birth.year <- as.numeric(year(mdy(filers$dob)))
```

Table 5.3: Duplicate records.

first.name	last.name	birth.year	filing.year	income	clean.first.name	clean.last.name
Maxine	Cho	1987	2012	109500	maxine	cho
MaxineJ	Cho	1987	2011	139500	maxinej	cho
Alex	Coyle	1965	2011	42123	alex	coyle
Alex	Coyle	1965	2011	42123	alex	coyle
.	.	.	.	.	.	.
.	.	.	.	.	.	.
.	.	.	.	.	.	.
Alex	Coyle	1965	2011	42123	alex	coyle

```
filing$birth.year <- as.numeric(filing$birth.year)
```

5.4   De-duplication

Duplicate records can be Achilles' heel of a data science project. They can easily overemphasize inconsequential patterns in the data. When online applications collect data, for example, users may click the submit button multiple times, creating a disproportionate number of duplicate records. To a rookie data analyst, there may appear to be an enormous amount of activity when the majority of records are duplicates. The problem can be further compounded when an attempted match results in a Cartesian product. De-duplicating records is a crucial step in keeping an accurate accounting of events.

If we take a closer look at the filings in Table 5.3, we find an inordinate number of duplicate records for 2011 tax filings for "Alex Coyle"—21 in all! Let's remove duplicates by specifying a *primary key*—an identifier or combination of identifiers that uniquely identify each record in a table. The primary key can be a single unique identifier such as a social security number, when available. In this case, one is more likely to specify a *composite key*, or an identifier constructed from multiple fields used in concert. For example, we could use combinations of

- First Initial + Last Name + Date of Birth.
- First Name + Last Name + Year of Birth.

Each composite key achieves a different degree of uniqueness. Social security numbers would have higher uniqueness than first and last name alone. The choice of a primary key will invariably require some trial and error.

The *key* should reflect the principal unit of analysis contained in each table. The `filers` data, for example, captures person-level information. A natural composite key could contain `clean.first.name`, `clean.last.name`, and `birth.year`. The `filings` data reflects tax filings over time for individual people, thus the composite key could be `clean.first.name`, `clean.last.name` and `birth.year`, and `filing.year`. Notice that three of four fields can be used as the basis of a $1:m$ match.

Having defined the keys, de-duplication is as simple as applying the `distinct` function in the `dplyr` package. The function expects a data frame along with a list of identifying fields. Each table has been reduced in size, with the `filing` table reduced by 40% in size and a lesser reduction observed in the `filer` table.

```
# Load dplyr package
pacman::p_load(dplyr)

# De-duplicate filing
f_table <- filing %>%
distinct(clean.first.name, clean.last.name, birth.year, filing.year)
```

Table 5.4: Types of joins in the dplyr package.

Join Type	Description
left_join	Left outer join returns all rows of X along with any matched rows of Y and its columns.
right_join	Right outer join returns all rows of Y along with any matched rows of X and its columns.
full_join	Full outer join returns all rows of X and Y with both matched and unmatched results.
inner_join	Inner join returns all rows of X and Y that were matched.
anti_join	Anti-joins return rows of X that did not match Y.
semi_join	Semi-joins return rows of X that matched Y, retaining only X columns.

```
# Deduplicate filers
p_table <- filers %>%
distinct(clean.first.name, clean.last.name, birth.year)
```

The ideal process would de-duplicate data once. In practice, we often do not truly know which records are duplicates. De-duplication thus requires vigilance and constant improvement throughout the life of a dataset.

5.5 Deterministic record linkage

With each table processed and de-duplicated, we can now try our hand at deterministic record linkage. However, before doing so, it is helpful to think about what precisely we hope to achieve from the linkage exercise. Do we want to find the intersection or overlap of two datasets? Do we need to find which records are not matched? Each of these goals can be addressed by a different *join operator*.

Join Operators. Each join is conducted pairwise between two sets of data. The first dataset X is referred to as the *left-hand side table*. This typically is the table with the largest size universe and sometimes referred to as the *backbone* to which all tables are joined. We cover backbones in greater detail later in this section. The dataset Y is to be joined to X and is referred to as the *right-hand side table*. Join operators can produce four types of outputs (see Figure 5.2):

- A *Left Outer Join* (or "Left Join") shows which records in Y overlap with X while retaining all records from X. In practice, left joins help us see how much of a sample (Y) overlaps with a known population (X). For example, the response rate of a survey can be calculated from a left join where X is the sampling frame and Y are records from respondents.
- A close cousin of the left join is the *Right Outer Join* (or "Right Join"), which identifies which records in X overlap with Y while retaining all records from X.
- *Inner Joins* only retain matched records. In set theory, this is the equivalent to an intersection $X \cap Y$. In practice, inner joins are a quick way to identify overlapping records. If the no-fly list and a passenger list were joined, for example, TSA and airlines staff would be most interested in the inner join.
- *Full Outer Joins* (or "Full Join") retain all records from both datasets X and Y, regardless of whether a record is matched. In set theory, this is the equivalent of a union $X \cup Y$. The output is the most comprehensive accounting of records, showing the overlap and coverage gap, which in turn can help understand sources of differences between datasets. For example, two organizations that want to compare their user bases may want to conduct a full join to produce an ecosystem-level view of their reach.

Each of these join operators can be applied using the `merge` function in base R, but more conveniently through the `dplyr` package (see Table 5.4). In addition to the four types of joins described above, `dplyr` implements *filtering joins* such as *anti-joins* that return unmatched records and *semi-joins* that return matched records in Y with the columns from X.

Match rules. Deterministic record linkage is dependent on constructing match rules by creatively combining join operators and intuition on the usefulness of match variables. The rule of thumb is to first match using

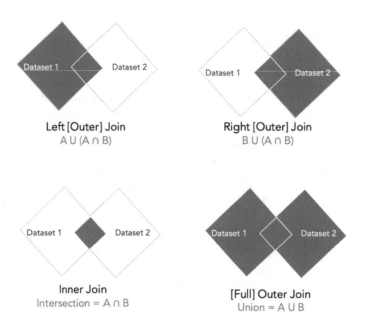

Figure 5.2: Types of joins. Diamonds represent datasets.

the most precise information—using more identifying fields can give us more confidence in the accuracy of the match. As matches are exhausted with the "best" information, follow-up waves of matching loosen the precision of match variables, allowing the use of approximate identifiers (e.g., Soundex, partial names).

Suppose our objective with the tax data is to report on the following:

- What proportion of the people in the filers table are in the tax filings table?
- What proportion of the people in the tax filings table are in the filers table?

These questions sound the same. While the numerators will be the same (number of matched records), the denominators will differ as they are drawn from different datasets. We walk through a simple three-step process below. First, we de-duplicate both tables ensuring the unit of measure is distinct people.

```
# De-duplicate the filings by person
people <- f_table %>%
distinct(clean.first.name, clean.last.name, birth.year,
                   .keep_all = FALSE)


# De-duplicate filers
tax <- p_table %>%
distinct(clean.first.name, clean.last.name, birth.year,
                 .keep_all = FALSE)
```

Next, we add a dummy variable to each table, which will come in handy for analyzing which records were matched after conducting a join.

```
# Create indicator
people$flag.p <- 1
tax$flag.t <- 1
```

Match Wave #1. Our first wave of matching relies on all available identifying variables, returning the results

Table 5.5: Unmatched records.

Birth Year	First Name	Last Name
1964	stephan	colbert
1951	richerd	hawking
1924	rube	amor
1994	pedro	oconnell
1987	maxine	cho
1987	maxinej	cho

of an inner join (`inner_join`)—only matching records are returned. If field names are not supplied, R will assume that variables with the same names in each dataset are match keys. The output table contains five columns: our three identifiers (first name, last name, and year of birth) along with two dummy variables.

```
# Inner join
xy <- inner_join(people, tax)
```

Now, we can estimate the proportion of people who have filed any of their previous taxes ($\frac{n_{\text{matches}}}{n_{\text{people}}}$) is slightly over 57%, whereas the proportion of people whose taxes were filed and are captured on the person list ($\frac{n_{\text{matches}}}{n_{\text{tax}}}$) is approximately 89%. In other words, there is a data coverage gap.

```
# How much of people table has been matched
sum(xy$flag.p == xy$flag.t, na.rm = T) / nrow(people)

# How much of the tax filings table has been matched
sum(xy$flag.p == xy$flag.t, na.rm = T) / nrow(tax)
```

An anti-join also can help assess the quality of the pre-processing and offer clues for how to improve the match rate. Our last statement retrieves a list of unmatched records, uncovering that at least two entities may have duplicates due to spelling errors (Table 5.5). Such a result is not entirely surprising when conducting deterministic record linkage as the method is relatively brute force, ignoring the spelling variations that are all too common in real-world data. We bridge some of these gaps when considering fuzzy matching techniques and probabilistic record linkage later in this chapter.

```
# Find unmatched filers
outstanding <- anti_join(people, tax)
```

Match Wave #2. As a follow-up attempt, we relax the match criteria by using partial information, such as the first three letters of the first name, last name, and birth year. So that we avoid matching on previously matched records in the `people` table, the second wave uses the `outstanding` persons table. Follow-up was successful in identifying an additional two names, bringing the total matches to $n = 10$.

```
# First three letters
outstanding$clean.3fn <- str_sub(outstanding$clean.first.name, 1, 3)
  tax$clean.3fn <- str_sub(tax$clean.first.name, 1, 3)

# Inner join
xy2 <- inner_join(outstanding, tax,
                  by = c("clean.3fn", "clean.last.name", "birth.year"))
```

At this point, the analyst would need to decide how to treat unmatched records. *Is this match result reasonable? Should unmatched values be imputed or left alone? Should another round of matching be attempted?* Perhaps *comparison functions* can help squeeze a few more matches from the data.

5.6   Comparison functions

An exact match should arguably merit our greatest confidence, but it will undoubtedly leave some portion of the data unmatched. A spelling error of even one character, such as "Stephen Colbert" versus "Stephen Culbert", can mean the difference between a match and a miss. Other than human intervention, a *fuzzy matching* strategy can help. Rather than searching for exact matches on an unique identifier, comparison functions enable fuzzy matching by quantifying the similarities between two strings while minimizing the influence of individual spelling variations. Here, we highlight two approaches: *edit distances* and *phonetic algorithms*.

5.6.1   Edit distances

The names "John Smith" and "jon Smith" are quite similar, yet software are unable to make the connection. Humans, on the other hand, can spot similarities. The pair of strings share an overwhelming majority of characters and only a couple of edits are required to convert one name to the other, such as turning the capital "J" to lower case and removing the "h". These qualities are suggestive of a match, or at least that these names are related. Alternatively, we can quickly see that the names "John Smith" and "Nikolai Rimsky-Korsakov" are nothing alike. These string comparisons encapsulate the spirit of *edit distances*—quantifying similarity as the number of transformations required to convert one string to the other. It does not take much to see how edit distances can apply in other contexts beyond name matching and spell-checking. In bioinformatics, edit distances can quantify how similar two gene sequences are. In sociology, edit distances quantify and relate the trajectories of people's lives. The potential uses and implications of edit distances are far reaching.

Levenshtein Distance is one of the most common edit distances that counts the number of *additions*, *substitutions*, or *deletions* that are required to transform one string to another string (Navarro 2001). The algorithm calculates distance by performing a systematic search for the *minimum* number of edits required to transform one string into another. In other words, there might be many ways to transform one string into another, but we are interested in only the most efficient path. Many of the most popular baby names as published by the US Social Security Administration (SSA) have the same pronunciation but different spellings (Social Security Administration 2017). Using Levenshtein distances, we can quantify their similarity:

- *Maya*: Mya (1), Myah (2), Miya (1)
- *Muhammad*: Mohamed (3), Mohammed (1), Mohammad (1)
- *Cameron*: Camron (1), Kameron (1)
- *Catherine*: Katherine (1)
- *Lily*: Lillie (3), Lilly (1)

Let's see Levenshtein distances in action when comparing two vectors of names `inputs` and `ref_table`. Using the `stringdist` package, the `stringdistmatrix` function calculates the edit distance between each combination of string values drawn from a pair of vectors, providing facility to specify the type of edit distance through the `method` argument. For Levenshtein Distance, we simply specify `method = "lv"`. As seen in Table 5.6, the name "Bill" is most similar to "Billy" with only one required edit. "Billie" and "Willy", both nicknames that are also derived from "William", also have relatively similar Levenshtein distances. However, the name "Lill" is not conceptually related yet it has the same Levenshtein distance as "Billy". *What happened?*

```
# Load library
pacman::p_load(stringdist)


# Two sets of names
inputs <- c( "Bill", "Warren")
candidates <- c("Billy", "Lill", "Wally", "Willy",
                "Billie", "William", "Georgie")


# Calculate Levenshtein distances
bill_dist =stringdistmatrix(inputs, candidates,method = "lv")
```

Table 5.6: Levenshtein distances for "Bill" and "Warren".

	Billy	Lill	Billie	Willy	Wally	William	Georgie
Bill	1	1	2	2	3	4	7
Warren	6	6	6	5	4	6	6

Table 5.7: Jaro-Winkler distances for "Bill" and "Warren".

	Billy	Lill	Billie	Willy	Wally	William	Georgie
Bill	0.033	0.167	0.056	0.217	0.367	0.274	1.00
Warren	1.000	1.000	0.556	0.476	0.317	0.493	0.46

While Levenshtein distance is widely used for applications like spell checking, they were designed with a particular concept of similarity in mind. The comparison function's assumptions have significant bearing on whether it is appropriate in different contexts.

The *Jaro-Winkler Distance* (Winkler 1990) is another widely used edit distance that focuses on the overlap between strings and the similarity of prefixes. The measure is an extension of Jaro Distance d_j as

$$d_j = \begin{cases} 0 & \text{if m} = 0 \\ \frac{1}{3}\left(\frac{m}{s_1} + \frac{m}{s_2} + \frac{m-t}{m}\right) & \text{in all other cases} \end{cases}$$

where $|s_1|$ and $|s_2|$ are the number of characters in a pair of strings, m is the number of common characters that are in common between the string pair, and transpositions t are defined as $\lfloor \frac{max(|s_1|,|s_2|)}{2} - 1 \rfloor$. In effect, the measure quantifies the overlap given a pair of string values, normalized on a range between 0 and 1 in which the latter indicates an exact match. A Jaro Distance can also be represented as a *Jaro Similarity* by flipping the scale $1 - d_j$.

The Jaro-Winkler (JW) distance extends the Jaro distance to account for similarities in the prefix characters of a string. The formula adds additional terms to give weight to the beginning of strings:

$$d_{jw} = d_j + lp(1 - d_j)$$

in which the Jaro distance d_j is adjusted by the length l of a common prefix at the beginning of a string (maximum is four characters). The scalar p adjusts the importance of similar prefixes to a range of $0 \leq p \leq 0.25$ in which $p = 0$ indicates that prefixes are unimportant for the comparison in question. Like Jaro similarities, JW distances often can be reported as *similarities*, defined simply as $1 - d_{jw}$ where 0 is an exact match.

Let's now re-examine the "Bill" vector by setting $p = 0.125$, which is equivalent to placing half the weight on prefix comparison. The result is notably improved (Table 5.7), ranking similar looking names toward the top of the list.

```
# Calculate Levenshtein distances
bill_dist =stringdistmatrix(inputs, candidates,method = "jw",p = 0.125)
```

5.6.2   Phonetic algorithms

The names "Billy" and "Billie" sound the same as do "John" and "Jon", yet their edit distances are similar but not exactly the same. *Phonetic algorithms* are designed to transcend spelling differences and emphasize the role of pronunciation—if two names sound the same, they should be coded the same. *Soundex*, for example, was developed to identify homophones, which are names that are pronounced the same but may

Table 5.8: Soundex letter encodings.

Letter	Encoded Value
B, F, P, V	1
C, G, J, K, Q, S, X, Z	2
D, T	3
L	4
M,N	5
R	6

have different spelling. The trick is to transform a string into a phonetic encoding that retains information on comparable sounds while removing minute deviations in spelling ("Soundex System" 2007). Despite its creation in the early 20th century, the steps that underlie Soundex are remarkably robust. Below are the basic encoding steps:

1. Keep the first letter of a string.

2. Drop the following letters: A, E, I, O, U, H, W, and Y.

3. For each of the following letters, encode with the associated number as seen in Table 5.8:

4. If a name has double letters, drop one (e.g., Keep one *r* in *Torres*).

5. If sequential number encodings are the same, keep only one. The name "Pfister", for example, is coded as P236 (P, F ignored, 2 for the S, 3 for the T, 6 for the R). The name "Jackson" is coded as J250 (J, 2 for the C, K ignored, S ignored, 5 for the N, 0 added).

6. If a name has a prefix (e.g., Van, Di) that is separated by a space, encode the name with and without the prefix and use both sets of encodings for search purposes. For example, "Van Doren", "Di Caprio", etc.

7. Consonant separators follow two rules: (1) If the letters "H" or "W" separate two consonants with the same Soundex code, the consonant to the right of the vowel is not coded. (2) If a vowel (A, E, I, O, U) separates two consonants that have the same Soundex code, the consonant to the right of the vowel is coded.

The output of Soundex is a four character code: the first character is a letter followed by three numbers—essentially compressing distinguishing sounds into a concise form. It is important to note that Soundex has a number of known weaknesses. The procedure was developed for Anglophone names assumptions, thus it will have difficulty with different transcription systems (e.g., Chinese and Russian names), silent consonants (e.g., Dei*gh*ton), and name equivalence (e.g., Jeff and Jeffrey) among others. A number of alternative phonetic algorithms, such as *metaphone*, *Caverphone*, and the *New York State Identification and Intelligence System* (NYSIIS), have been developed over the years to improve the accuracy and generalizability of encodings.

Let's see how the encodings perform on our "Bill" data using the `phonics` package, which implements a number of phonetic algorithms such as Soundex, Metaphone, NYSIIS, and Caverphone. We again compare the name "Bill" with a vector of names using both phonetic encodings and edit distances (see Table 5.9). Once again, Levenshtein distances are shown to be insensitive to the prefixes, whereas Jaro-Winkler distances offer a visible improvement in identifying likely matches. In contrast, phonetic algorithms provide a cluster label—all names that sound similar share the same encoding. The names "Billy" and "Billie" are perfectly matched with "Bill" when using Soundex (B400) and Metaphone (BL). "Wally" and "Willy" share the same encodings (W400 and WL), but these names originate from different proper names (Walter versus William). These results illustrate that the approximations made by phonetic algorithms can uncover missed matches, but also increase the risk of mismatches.

Table 5.9: Encodings and distances relative to the name "Bill".

Names	Levenshtein	Jaro-Winkler	Soundex	Metaphone	NYSIIS	Caverphone
Bill	0	0.000	B400	BL	BAL	P11111
Billy	1	0.033	B400	BL	BALY	PL1111
Lill	1	0.167	L400	LL	LAL	L11111
Billie	2	0.056	B400	BL	BALY	PL1111
Willy	2	0.217	W400	WL	WALY	WL1111
Wally	3	0.367	W400	WL	WALY	WL1111
William	4	0.274	W450	WLM	WALAN	WLM111
Georgie	7	1.000	G620	JRJ	GARGY	KK1111

```
# Load phonics package
pacman::p_load(phonics)

# Names
bill <- c("Bill", "Billy", "Lill", "Billie",
          "Willy", "Wally", "William", "Georgie")

# Apply algorithms
stringdist("Bill", bill, method = "jw", p = 0.125)
stringdist("Bill", bill, method = "lv")
soundex(bill)
metaphone(bill)
nysiis(bill)
caverphone(bill)
```

5.6.3   New tricks, same heuristics

With these new fuzzy tricks, we should ask *how does one match using edit distances and phonetic encodings given the fuzziness?*. The last mile of fuzzy matching requires judgment calls and heuristics—a kind of premium version of deterministic record linkage. For example:

- Should the shortest distance be considered to be a match or the shortest distance within a threshold?
- If any values are within the threshold, what if two names have the same distance value?
- If two or more phonetic encodings are the same, which is right?

These fuzzy matching methods are new tricks that can increase our match rate, but they also come with the added risk of increasing the false matches rate. One approach to place guide rails on the matching process is *Blocking*, which reduces the search space to plausible candidates, then only conducts matching on records whose fields exactly match on *blocking fields*. These fields contain accurate but less specific information, like geography, sex, first initial of name, phonetic encodings among others. Once the search space has been narrowed, then edit distances can rank potential matches.

We put this logic to the test in the example below, searching for matches for names in the `inputs` vector drawing from `candidates`. Soundex is applied to create a blocking variable, then the `amatch` function in the `stringdist` package retrieves the index position of the closest matching names as measured by Jaro-Winker similarly setting a fence of `maxDist = 0.2`. The match results are packaged into a data frame for review.

```
# Step 1: Soundex encodings
soundex_inputs <-
soundex(inputs)
```

```
soundex_candidates <-soundex(candidates)

# Step 2: Match within blocks
out <- data.frame()

# Loop through each input name
for(i in 1:length(inputs)){

# Subset to a block of soundex
short_list <- candidates[soundex_candidates == soundex_inputs[i]]

# Match using JW distance
res <- amatch(x = inputs[i], table = short_list,
              method = "jw", p = 0.125, maxDist = 0.2)

# Store results
out <- rbind(out,
data.frame(original = inputs[i],
matched = short_list[res]))
    }
```

It may seem reasonable to employ a set of human-tuned heuristics to match data, but consider this: datasets can have many variables on which matches can be made. In real-world data, data are almost certainly dirty. Should all variables still be treated the same? Also, how does one use multiple matching variables in a fuzzy framework? The code for the tax example will quickly become more complex, filled with judgment calls. If inaccurately matched, there's a social cost of producing data with false match and not producing enough matches—*we can introduce a bias into our data, which in turn biases insights, and biases policy.*

5.7   Probabilistic record linkage

Probabilistic matching cranks up the sophistication of matching, removing the need for complicated heuristics and human overhead while accounting for variable data quality. The most popular framework, as described in Fellegi and Sunter (1969), calculates the probability of a match for every possible pair of records drawn from two sets of data given the agreement among matching variables. The match probability can then serve as the basis for classifying which records are likely matches and which are not. In other words, probabilistic record linkage can automate a significant proportion of the data joining process, but automation usually requires more complex machinery to work in the background.

There are many moving parts to the framework. For simplicity, we lay out the approach through our filers X and filings Y data. Each sample has a sample size, n_X and n_Y, which means we compare up to $n_X \times n_Y$ pairs of records. For each record pair (i, j), where i is in X and j is in Y, we compare each of k match variables and store their agreement in a *comparison vector* $\gamma(i, j)$:

$$\gamma(i,j) = \begin{bmatrix} \gamma_1(i,j) \\ \vdots \\ \gamma_k(i,j) \end{bmatrix}$$

Each value in $\gamma(i, j)$ is a discrete label such as "Same", "Different", or "Similar", classified from setting thresholds on similarity measures. In our example, we compare birth year (Year), first name (FN), and last name (LN) using Jaro-Winkler similarity, then set thresholds of 0.08 and 0.12 that discretize each comparison as discrete labels. For each pair, the set of three agreement labels is stored in $\gamma(i, j)$. For example, in Table 5.10, a pair of names like "Stephen Colbert" and "Arthur Duran" would be stored as

Table 5.10: Comparison and classification of record pairs.

filings table			Filers Table			JW Distance			Agreement		
Year	FN	LN	Year	FN	LN	Year	FN	LN	Year	FN	LN
1964	stephen	colbert	1985	franz	li	0.250	1.000	0.452	Different	Different	Different
1964	stephen	colbert	1964	stephen	culbert	0.000	0.000	0.083	Same	Same	Similar
1964	stephen	colbert	1956	arthur	duran	0.125	0.460	1.000	Different	Different	Different
1964	stephen	colbert	1965	alex	coyle	0.104	0.536	0.157	Similar	Different	Different

$$\gamma(i,j) = \begin{bmatrix} \gamma_{\text{First Name}}(i,j) = \text{Different} \\ \vdots \\ \gamma_{\text{Last Name}}(i,j) = \text{Different} \end{bmatrix}$$

Some datasets will inevitably have missing values that cause errors in matching. In deterministic record linkage, for example, records with missing values may be dropped from consideration, potentially skewing insights from the data. We can keep track of missingness using an additional *missingness vector* $\delta(i,j)$, marking which variable k in record pair (i,j) is missing values ($\delta_k(i,j) = 1$) or have complete values ($\delta_k(i,j) = 0$).

Once again, we find ourselves with the problem of deciding which records are matches M non-matches U. Per Enamorado, Fifield, and Imai (2019), the comparison and missingness vector are inputs into an estimation procedure to tease out the chance that a pair of records is a match (ξ_{ij}):

$$\xi_{ij} = Pr(m_{ij} = 1 | \delta(i,j), \gamma(i,j))$$

The precise calculation involves estimating a model through an Expectation-Maximization (EM) algorithm, incorporating the probability π_{km} of the agreement level in variable k given a match m and the match probability of all comparison pairs λ. For a more in-depth treatment, see Enamorado, Fifield, and Imai (2019). A probability is flagged as a match by setting a threshold S. Values above S ($\xi_{ij} \geq S$) are considered to be matches, whereas values below are non-matches ($\xi_{ij} < S$).

In Practice. In only a few lines of code, the `fastLink` package rapidly deploys a probabilistic record linkage model. The example builds on the `p_table` and `f_table` from the previous example, using a mix of cleaned fields (first name, last name, birth year) as well as encoded fields (Soundex first name and last name).

```
# Load fastLink package
pacman::p_load(fastLink)

# Augment tables with comparison functions
# Filers table
p_table$soundex.first.name <- soundex(p_table$clean.first.name)
p_table$soundex.last.name <- soundex(p_table$clean.last.name)

# Filings table
f_table$soundex.first.name <- soundex(f_table$clean.first.name)
f_table$soundex.last.name <- soundex(f_table$clean.last.name)
```

To get started with matching, the `fastLink` function scales matching to large datasets. Not only does the function conduct matching, it makes it easy to examine match quality, tune match thresholds, and retrieve results. At a minimum, the function expects two data frames and a vector of variable names on which to match. The developers set a number of sensible defaults:

- A list of variables need to be specified under `stringdist.match` if fuzzy matching is desired.
- The string comparison function defaults to Jaro-Winkler distance with the p parameter set as $jw.weight = 0.1$. Other distances measures such as Levenshtein and Jaro are also supported.
- The match threshold S can be tuned using *threshold.match* (default = 0.85).

```
# Matching function
match_results <- fastLink(
                    dfA = p_table, dfB = f_table,
                    varnames = c("clean.first.name", "clean.last.name", "birth.year",
                                  "soundex.first.name", "soundex.last.name"),
                    stringdist.match = c("clean.first.name", "clean.last.name"),
                    dedupe.matches = FALSE)
```

With the linkage model trained, we tune the match threshold to maximize the match rate. In this case, the match probabilities are so high that the choice of threshold has no influence on the match results. Interestingly, the match rate exceed 100%, which suggests that some matches are $m : 1$—perhaps two similarly spelled names were linked to the same entity.

```
 summary(match_results, threshold = seq(0.7, 0.95, 0.05))
```

```
 knitr::kable(summary(match_results, threshold = seq(0.7, 0.95, 0.05)),
              caption = "Linkage rates by threshold.",
              booktab = TRUE)
```

Lastly, matches are returned as a single data frame using the `getMatches` function. By default, `fastLink` will automatically de-duplicate matches, which is not ideal for matching people to transactional records. In this case, we should retain all tax filings, thus we re-run `fastLink` and set `dedupe.matches = FALSE` before retrieving matches.

```
# Retrieve matches
out <- getMatches(dfA = p_table,
                  dfB = f_table,
                  fl.out = match_results)
```

5.8   Data privacy

Identifying matching records can pose an information exposure risk. In an age where information security and privacy are constantly at risk, stricter practices need to be developed to safeguard people's information during the data science process. After matching records, we advocate for the use of *cryptographic hash functions* so that identifiers cannot be plainly read. A hash function is a mathematical algorithm that converts an arbitrarily long string into a fixed size code known as a *hash*. Using one of the most commonly available cryptographic hash algorithms known as "MD5", we convert an identifier into a unique hash.[1] For example, "Duran Duran" becomes 77d167993f17e0f169e425ddc5b0c1a3.

Simply encrypting an identifier into a hash does not safeguard the information. Hashes can be cracked if the heuristic is exposed, thus vigilance in data stewardship is a must. We recommend the following steps be taken:

1. After linkage is complete, construct an identifier vector `temp_key`.
2. Use the `digest` function in the `digest` package to apply the MD5 algorithm to each identifier in each person and tax filings tables. The `digest` function will turn any R object into an MD5 hash. As the `digest` function can only produce one hash at a time, we use a special apply function `sapply` to loop through each value of `temp_key` to encrypt the unique hashes.

---

[1]Hash functions are suitable for within organizational use in a secure environment, but are not safe for de-identifying individuals in publicly disseminated data.

3. After the hash is created, delete the person identifiers in the dataset (e.g., first name, last name, etc.) as well as the `temp_key`.

By taking these steps, we leave little to no personally identifiable information in the datasets, making a more secure dataset for analysis. These steps are illustrated in the code below:

```
# Construct identifier using stringr for the person table
temp_key <- paste(p_table$soundex.first.name,
p_table$soundex.last.name,
p_table$birth.year)

# Load MD5 hash library
pacman::p_load(digest)
p_table$hash <- sapply(temp_key, digest, algo = "md5")

# Keep de-identified information
p_table <- p_table[, c("hash")]
rm(temp_key)
```

5.9 DIY: Matching people in the UK-UN sanction lists

*DIYs are designed to be hands-on. To follow along, download the DIYs repository from Github (*https://github.com/DataScienceForPublicPolicy/diys*). The R Markdown file for this example is labeled* **diy-ch05-record-linkage.Rmd***.*

Creating and maintaining lists of high-risk individuals and organizations is part of any national security apparatus. In fact, a number of nations and governing bodies publish such lists of enemies of the state and their aliases so that companies and people conducting global commerce can follow international sanctions. For example, a number of countries and international organizations maintain sanctions lists:

- *US Consolidated Screening List.*
- *UK Financial sanctions targets.*
- *UN Sanctions List.*
- *EU Sanctions List.*

These datasets pertain to the serious business of global security, but also present an opportunity for record linkage using real-world data. Every day, companies shift through customer lists and marketing lists to understand their customers. Nonprofit development managers compare the lists of donors to marketing lists to find potential donors who resemble their current support base. And organizations operating in the interest of global security compare names of intended recipients of shipments against sanction lists.

For simplicity, we will focus only on the UN and EU lists to learn the extent to which they have common identified risks. We begin by loading a `Rda` file containing all watch lists. Note that these lists are constantly updated, thus the files extracted for this exercise are a snapshot from one point in time.

```
load("data/watch_lists.Rda")
```

The watch lists capture similar attributes, but often are organized quite differently. Let's examine the case of Saddam Hussein, the former President of Iraq (Table 5.11). Both the EU and UN lists contain his first name and last name, birth date, and citizenship. There are slight deviations in how the names are recorded, but we can see how the lists could be matched after basic string manipulation.

Based on visual inspection, the following five fields contain similar information useful for matching: `firstname`, `middlename/secondname`, `lastname/thirdname`, `wholename`, and `birthdate`. Using the procedures laid out in this chapter, we can easily compare the performance of deterministic record linkage versus probabilistic record linkage.

Data Preparation. Different organizations treat names differently. Some may concatenate parts of first

Table 5.11: Comparison of EU and UN lists.

EU Variables		UN Variables	Example
eu.id	2	un.id	172
id	13	rec_type	individual
legal_basis	1210/2003 (OJ L169)	id	6908048
reg_date	2003-07-07	version_number	1
lastname	Hussein Al-Tikriti	firstname	SADDAM
firstname	Saddam	secondname	HUSSEIN
middlename		thirdname	AL-TIKRITI
wholename	Saddam Hussein Al-Tikri	un_list_type	Iraq
gender	M	ref_num	IQi.001
title		listed_on	2003-06-27
func_role		comments	
language		designation	
birthdate	1937-04-28	citizen_country	Iraq
birth_place	al-Awja, near Tikrit	list_type	UN List
birth_country	IRQ	date_updated	
passport_number		alias	Abu Ali
passport_country		alias_quality	Low
citizen_country	IRQ	country	
pdf_link	http://eur-lex.europa.e	birthdate	1937-04-28
programme	IRQ	birthdate_type	EXACT
		birth_place	al-Awja, near Tikrit
		birth_country	Iraq
		passport_number	
		passport_country	
		sort_key	
		sort_key_mod	

and middle names while others keep them separate. To minimize the influence of recording error, all parts of a name from the UN dataset are concatenated into a `wholename` field, similar to the EU data, and filling any `NA` value with empty quotes. This will prevent `NA` values from being treated as "NA" strings. To establish a baseline for matching performance, we check for possible matches without any additional text standardization.

```
# Fill empty quotes
eu[is.na(eu)] <- ""
un[is.na(un)] <- ""

# Create wholename
un$wholename <- paste(un$firstname, un$secondname, un$thirdname)

# Number of overlapping records
print(sum(un$wholename %in% eu$wholename))
```

Sadly, only $n = 4$ records from thousands match. With only a modest level of effort, the `textStandard` function from earlier in this chapter can correct for typographical differences. A second attempt at matching returns much improved results, reaching $n = 668$ matches.

```
# Clean
un$clean.wholename <- textStandard(un$wholename)
eu$clean.wholename <- textStandard(eu$wholename)

# Overlap
print(sum(un$clean.wholename %in% eu$clean.wholename))
```

Next, the `birthdate` values also need to be standardized. Some UN records contain YYYY-MM-DD while others oddly capture minute-level detail (likely a recording error). The EU records are a mix of year and YYYY-MM-DD along with commentary on the data quality. Using `stringr` functions, we remove text from the `birthdate` fields as well as construct a four-digit `birthyear` field.

```
# Clean UN to YYYY-MM-DD records
un$birthdate <- str_extract(un$birthdate, "\\d{4}-\\d{2}-\\d{2}")

# Remove text from YYYY
eu$birthdate <- str_replace(eu$birthdate, " \\(approximative\\)", "")

# Extract birth year using regex and stringr
un$birthyear <- str_extract(un$birthdate, "\\d{4}")
eu$birthyear <- str_extract(eu$birthdate, "\\d{4}")
```

Deterministic record linkage. Our deterministic linkage strategy starts with the most precise information, then relaxes matching requirements in each of two subsequent matching waves:

- first: `clean.wholename` and `birthdate`;
- second: `clean.wholename` and `birthyear`; then
- third: `clean.wholename` only.

In each wave, we make note of which wave a match was made, giving credit to more precise matches.

```
# Abridged tables
eu_short <- eu[, c("eu.id", "wholename", "clean.wholename", "birthdate", "birthyear")]
un_short <- un[, c("un.id", "wholename", "clean.wholename", "birthdate", "birthyear")]

# Match waves
wave_1 <- inner_join(eu_short, un_short, by = c("clean.wholename", "birthdate"))
wave_2 <- inner_join(eu_short, un_short, by = c("clean.wholename", "birthyear"))
wave_3 <- inner_join(eu_short, un_short, by = c("clean.wholename"))
```

```
# Combine into master table and de-duplicate
det_key <- rbind(cbind(wave_1[, c("eu.id", "un.id", "clean.wholename")], match = 1),
cbind(wave_2[, c("eu.id", "un.id", "clean.wholename")], match = 2),
cbind(wave_3[, c("eu.id", "un.id", "clean.wholename")], match = 3))
det_key <- det_key[!duplicated(det_key[,1:3]), ]
```

From three waves of matching, a total of $n = 676$ matches (`nrow(det_key)`) were identified. This is only a fraction of the UN list ($n = 1046$–`nrow(un_short)`) and the EU list ($n = 2016$–`nrow(eu_short)`). While the most precise matches accounted for only $n = 146$ (`nrow(wave_1)`), the bulk of matches were identified in the second wave. The third wave using `wholename` alone offers few gains over the first two waves.

Probabilistic record linkage. There will undoubtedly be missed matching opportunities due to variation in spelling. Probabilistic record linkage can close the gap using the same three fields, but cast a wider net by allowing a partial fuzzy match on the `clean.wholename` field.

```
prob_link <- fastLink(eu_short, un_short,
            varnames = c("clean.wholename", "birthdate", "birthyear"),
            stringdist.match = c("clean.wholename"),
            partial.match = c("clean.wholename"))
```

A comparison of classification thresholds indicates that there is little difference in the match rate from $.75 \leq \xi \leq 0.9$ – model's estimates are robust. The match rate is significantly greater than the deterministic approach.

```
# Optimal cutoff
summary(prob_link, threshold = seq(0.75, 0.95, 0.05))
```

We use `getMatches` to recover the dataframe of matches, containing attributes from the `eu` data along with a row index for the corresponding match from the `un` data. In total, the model identified $n = 915$ matches—most of which are $1 : 1$ matches and a smaller number ($n = 22$) are $1 : m$ matches. Virtually all deterministic matches are also present in the probabilistic match results, thus the latter can do the job of the former and much more. There are clearly more matches generated through probabilistic matching, but some may wonder if a model-based approach actually surpasses human intuition: *Are the matches accurate?*

By drawing a random subset of non-exact matches as seen in Table 5.12 there is clear evidence that the matching model is robust to spelling differences, excess and missing information, and order of names. The proof is in the pudding. Probabilistic record linkage is arguably a superior strategy to deterministic record linkage, but in practice, it does not hurt to try both.

```
# Recover matches
prob_match <- getMatches(dfA = eu_short, dfB = un_short,
                fl.out = prob_link, threshold.match = 0.85)

# Recover UN IDs, whole name and birth year for visual inspection
prob_match$un.id <- un_short$un.id[prob_match$`dfB.match[, names.dfB]`]
prob_match$un.wholename <- un_short$wholename[prob_match$`dfB.match[, names.dfB]`]
prob_match$un.birthyear <- un_short$birthyear[prob_match$`dfB.match[, names.dfB]`]
```

5.10   Beyond this chapter

5.10.1   Best practices

In this chapter, we have seen that matching can be an intensive process. Every decision in the matching process, whether in the data processing or linkage strategy, can mean the difference between modest and impressive match rates. The code that we have laid out in this chapter can serve as a springboard for

Table 5.12: Sample results from probabilistic record linkage.

Name (UN)	Year (UN)	Name (EU)	Year (EU)
Ali Ben Taher Ben Faleh	1986	Ali Ben Taher Ben Faleh Ouni Harzi	1986
Ibrahim Ali Abu Bakr	1966	Ibrahim Ali Abu Bakr Tantoush	1966
Jo Yong Chol	1973	Yong Chol Cho	1973
Choe Chun Yong	NA	Ch'oe Ch'un-Yo'ng	NA
Ali Barzan Ibrahim	1981	Ali Barzan Ibrahim Hasan Al-Tikriti	1981
Kim Mun Chol	1957	Kim Mun-Ch'o'l	1957
Haqqani Network (Hqn)	NA	Haqqani Network	NA
Son Jong Hyok	1980	Son Min	1980
Hamid Hamad Hamid	1960	Hamid Hamad Hamid Al-'Ali	1960

constructing your own data linkage model. We close this chapter with some considerations on how record linkage fits within a public policy context.

One of the most requested record linkage questions is *how many records overlap between two lists?* This is a simple question that leads to misleading answers. Policy and strategy audiences expect a singular number that can stand up to scrutiny, but as we have seen in this chapter, we cannot know this answer for sure. In fact, reporting the overlap between lists does not address the existential questions policy audiences are interested, but rather reflects the effectiveness of matching strategies—it is an important distinction to make. A reasonable compromise is to report the range of match rates that correspond to match thresholds as used in probabilistic record linkage. By presenting the range of possible matches, we can communicate the uncertainty that comes with matching and that it is not a precise science.

The process of record linkage can be technically challenging, yet it does not take much to deflate confidence— a data user simply needs to identify a mismatch, then each additional mismatch can snowball the loss of confidence. *Spot checking a randomly selected set of 100 matches and conducting a visual scan* can help quality control match results, or at least aid in articulating the quality of matches.

There is not a single best way of conducting record linkage as it is dependent on the use case that the matches support. The only thing we know for sure should be how the matches will be used—this is the north star of record linkage. Before starting the matching process, be clear about how the data will be used as this will determine how the data will be processed, the type of relational model that should be expected, the cost of false positives and false negatives, and the appropriate matching strategy. In national security, false negatives may be more costly than false positives in the short run in order to prevent attacks and incidents, but too many false positives in the long run can lead to an erosion of trust. Alternatively, a social services organization can be weary of too many false positives due to limited resources to review all matches, thus focusing on exact matches can ensure a higher hit rate and a more manageable caseload. Investing time to understand the concerns of the ultimate data user will go a long way toward garnering support and trust.

Record linkage can resemble more of an art than science and it is a necessary step for expanding our understanding of phenomena. To be successful, we first need clarity of purpose. Otherwise, matching can generate useless or even harmful noise.

5.10.2 Further study

How sophisticated record linkage practices are differs from one organization to the next. Rules and heuristic-based linkage are very much prominent for ad hoc analyses while probabilistic linkages are necessary for rapid, large-scale, recurring analyses. Governmental agencies rely on a wide range of techniques to accomplish the task. One thing that is consistently important across all domains is upholding strong privacy practices— critical for maintaining the Public's trust in governments' use of data. The health and medical research, for example, have invested significant resources to advance record linkage capabilities in a way that protects privacy in accordance to HIPAA. Demographers and social scientists who work on large-scale administrative records continuously research methods for preserving privacy when linking records, especially across two or more large databases. Indeed, the methodologies for privacy-preserving record linkage will play an important role as data becomes larger and more available. For in-depth treatment on privacy and ethics, review Chapter 14 (*The Ethics of Data Science*).

Chapter 6

Exploratory Data Analysis

6.1 Visually detecting patterns

Are your phone's sensor data a privacy risk?

Smart phones are a modern wonder that allow society to stay connected and to enhance interactive experiences with the world around. Each interaction between a user and a phone is dependent on a sophisticated array of sensors. The accelerometer, for example, measures movement on a phone in three dimensions allowing for the content on the screen to be oriented when the device changes positions. Each motion, action, and gesture of a user is indicative of different kinds of activity and may even uniquely identify a person—clear implications for privacy and security. In fact, Kwapisz, Weiss, and Moore (2010) found that a person's movement can serve as a biometric marker like a fingerprint while researchers at Stanford University found that imperfections in the accelerometer itself can uniquely identify phones (Owano 2013). The rich detail generated from accelerometers can also have higher order policy implications. The City of Boston, for example, launched an effort in 2011 to use smartphone accelerometers to detect potholes. The Street Bump initiative utilized a free Android app to detect anomalies when Bostonians drive through Boston (Moskowitz 2011). In the commercial space, companies like Fitbit and Apple have developed physical fitness activity monitoring that helps users healthy and fit.

Do sensor data really carry such identifiable and insightful signal? It turns out that even a small sample of accelerometer data is an enormous amount of information that can be exploited. Let's take a look at 6.5 minutes of exercise data and graphed at a frequency of 5 hertz (five readings per second) in Figure 6.1. *Can you visually identify distinct patterns? How would you quantify a pattern?*

Four distinct clusters of activity emerge in Figure 6.2:

- Activity close to zero acceleration indicate periods of rest;
- Values that fluctuate around 0.2 indicate walking;
- Sprinting activity hovers around 0.6 +/- 0.2—large amplitude; and
- Walking downstairs is the fourth and most volatile activity.

Visual inspection of data is a central tenet of *Exploratory Data Analysis (EDA)*. Heavily promoted by legendary mathematician John Tukey, EDA emphasizes the analysis of data through visual and graphical means in order to inform hypotheses and validate assumptions. Humans have mastered the ability of detecting patterns by simply looking and inspecting an object. We do ourselves a great service as data scientists to use our vision to quickly intuit patterns in graphs.

In a public policy, EDA is sometimes both the start and end of a data project. Policy audiences are interested in a high-level story and the visualizations that are produced in the course of EDA provide a tangible product

J. C. Chen et al., *Data Science for Public Policy*, Springer Series in the Data Sciences,
https://doi.org/10.1007/978-3-030-71352-2_6

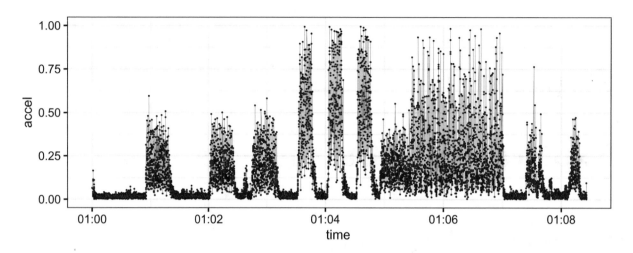

Figure 6.1: Net acceleration collected from a mobile phone, sampled at a rate of 5 hertz (readings per second).

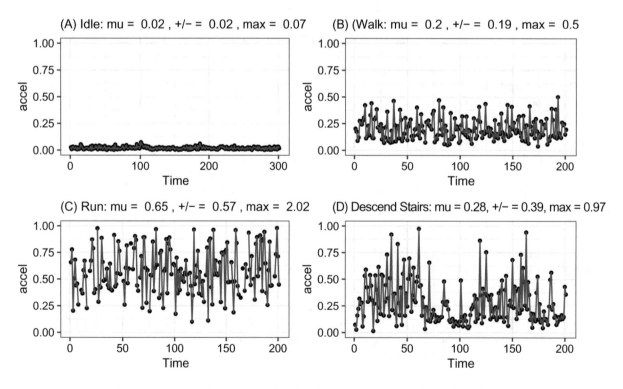

Figure 6.2: Acceleration profile for different types of physical activity.

to lend credence to a hypothesis. A well-conducted EDA can serve as a feasibility study, exposing critical flaws in the data or help determine if the data has any value. The results of the EDA can help green light a promising technical adventure or prevent a time-consuming disaster.

In this chapter, we present the basics of EDA and illustrate how graphical methods can expose the qualities of data. Throughout the chapter, we rely on `ggplot2`—a graphing package that is a favorite among data scientists. We begin with an overview of the objectives of EDA, then explore concepts of spread, similarity, time, and correlation.[1]

6.2 The gist of EDA

A few things to consider. In many ways, EDA is a "pre-flight checklist"—a chance to *kick the tires* on the data and a proposed project. Even before touching the data, consider formulating a set of questions that you would like to have answered. These answers should determine if you should proceed with a data project or if the EDA itself is the project. Table 6.1 enumerates a suggested list of five objectives and associated questions that structure an EDA.

With the increasing availability of data, it has become common practice to take data collected for one purpose and re-use it for another. Thus, a reasonable first step is to understand how the data were collected and if the data can support an objective other than what it was originally designed. If the data could be useful, then dive into the data at an operational level: check the qualities of each variable. Where needed, convert raw values to the right data type and clean values so the data is in an analysis-ready condition. With the data in usable shape, plot each variable's distribution to determine if it is well behaved. If the distribution of a continuous variable, for example, has two or more peaks, then see if any discrete variables can explain those different central tendencies. If there are long tails or noticeably large or small values, then perhaps there are outliers? If a variable has missing values, is it only a small proportion or the overwhelming majority? How should one deal with them? Among mostly complete variables, are any variables correlated? Is it possible that the most highly correlated contain redundant information?

There are, indeed, many twists and turns in each EDA. How you approach and address each quirk in the data is up to you—it can be relatively subjective, requiring each assumption to be documented for the sake of transparency. Nonetheless, the principal objective of the EDA process is to understand the data in a way that empowers you and your analytical efforts.

Table 6.1: A checklist for EDA.

Objective	Common Questions
Check the conceptual soundness of the data.	How were the data collected? What are potential sources of recording error? What does each variable mean? Does the unit of measurement make sense for your analytical goals?
Process the data so that it is useful.	Are the data types appropriate for analysis? (e.g., numeric as numbers, date values as date objects) What manipulations will you need to perform to get the data into usable shape? Should data be aggregated to see patterns at a higher level?

---

[1]There are many graphs presented in this chapter. To follow along, download the DIYs repository from Github (https://github.com/DataScienceForPublicPolicy/diys). The R Markdown file for this section is `diy-ch06-visuals.Rmd`.*

Objective	Common Questions
Assess the signal quality.	By plotting continuous values as histograms or kernel density plots, is there a central tendency (e.g., a peak)? Does the data fall into a commonly recognizable shape (e.g., normal curve)? Is the distribution tight or dispersed? Is the distibution unimodal (one hump), bimodal (two hump), multi-modal (many humps) or random? If there are categorical variables, how are the continuous variable distributions different when plotted by categorical? Are there categorical variables with small cells (e.g., a category with a small sample)? Can small cells be combined with larger cells to reduce small sample biases? If the data are recorded over time, does variables trend over time?
Identify problematic values.	When plotting the empirical probability distribution, are there potential outliers? (e.g., long tails) In time series, do some values plotted over the time spike well above the rest? Are values missing? Are missing values occassional or common? Is there a pattern to the missingness?
Identify important variables.	Which variables are correlated with one another? Can some variables be transformed to obtain a better correlation?

A short introduction to `ggplot2`. Virtually any graphing software can produce the visuals that facilitate EDA. In fact, base R is furnished with a rich set of elementary graphing tools. It is, however, hard to beat the `ggplot2` package as it is remarkably extensible with a large number of add-ons that make almost any visualization possible. Before we proceed further into EDA, we first provide a brief introduction to `ggplot2` and `gridExtra`. Let's install and load these two packages.

```
pacman::p_load(ggplot2, gridExtra)
```

The `ggplot` framework constructs visualizations in layers, which has the added benefit of making the code more readable and maintainable. Each discrete element of a plot is a layer (e.g., labels, lines, grids, etc.) that is be tacked onto a `ggplot` object. To illustrate how this works, we construct a line graph of the unemployment rate (`uempmed`) over time (`date`) using the US `economics` dataset, doing so one layer at a time.

p0. We begin with providing `ggplot` with the name of the dataset (`ggplot(data = economics)`) and specifying the input variables (`x = date, y = uempmed`) in the esthetic properties (`aes`). This information is assigned to a `ggplot` object p0. Although we have defined the input data, we have not yet specified which type of graph to render, thus p0 renders only an empty canvas.

```
p0 <- ggplot(data = economics, aes(x = date, y = uempmed))
```

p1. To add geometric elements, such as points and lines, we call on a `geom`. The specific `geom` for a line plot is `geom_line`. We can use a plus operator (`+`) to add new layers to p0 without re-defining the `aes` mappings. When rendered, p1 is an unstylized line chart.

```
p1 <- p0 + geom_line()
```

p2. To add points, we specify `geom_point` and set a `color` and point `size`.

```
p2 <- p1 + geom_point(colour = "red", size = 0.2)
```

p3. The plot can be stylized by creating a theme. Fortunately, the `ggplot2` package gives users a head start with preset themes such as `theme_bw`, `theme_minimal` and `theme_light`. The font, placement, and styling of every graphical element can also be customized to taste. We create a `custom_theme` that combines `theme_bw` and adjust plot and axis titles to `size = 10`. In p3, we add these themes to p2 along with axis labels. These stylings occupy can large amount of real estate in code. To economize on space, we will re-use the `custom_theme` for the remainder of this chapter.

```
# Define custom formats
custom_theme <- theme_bw()  +
theme(plot.title = element_text(size = 9),
                    axis.title.x = element_text(size = 9),
                    axis.title.y = element_text(size = 9))

# Apply styles and labels
p3 <- p2 + custom_theme + ylab("Unemployment") + xlab("Time")
```

Using `grid.arrange` (`gridExtra` package), we place each graph in a 2 × 2 grid (Figure 6.3). While this example is mundane, the key insight is that any set of graphs can be quickly developed with minimal code and effortlessly laid out for a professional-grade presentation.

```
# Load package for grid layouts
pacman::p_load(gridExtra)

# Arrange graphs in two columns---effectively a 2 x 2 grid
grid.arrange(p0, p1, p2, p3, ncol = 2)
```

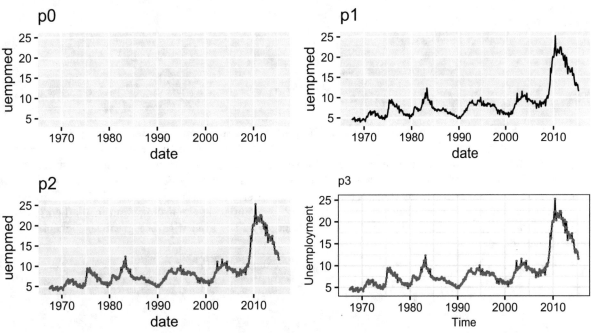

Figure 6.3: Effect of adding layers of graphical elements and styles in `ggplot2`. The 2 x 2 grid is laid out across two columns using `grid.arrange`.

6.3 Visualizing distributions

Originally published in *Science*, LaCour and Green (2014)—a groundbreaking-now-retracted article—showed how a simple 20-minute conversation with a canvasser could change voter attitudes on same-sex marriage. Within a year of publication, Broockman, Kalla, and Aronow (2015) used an assortment of distribution plots and hypothesis tests to flag irregularities in the LaCour and Green (2014) data and expose uncanny similarities with existing national surveys. The implication was that the study's data were not collected as stated, implying that the 20-minute intervention may also have been falsified. EDA played a pivotal role in exposing this intellectual dishonesty.

The incident illustrates how the qualities of a dataset can be scrutinize by simply graphing its distributions—

it is easy to fake a summary statistic, but it is far harder to fake a distribution. It is in the shape of the data that we can learn the most about a phenomenon. Both discrete and continuous variables have distributions that can be visualized. Using a cross-sectional dataset of the economic characteristics of $n = 3137$ US counties (`cty` stored in a .Rda file), we review a few graphing options for different types of data.

```
load("data/county_compare.Rda")
```

Discrete distributions. Discrete variables are collections of *classes* or *categories*. For example, each US county is located in a `region`. Looking across all counties in the US, each belongs to one of four regions.[2] Thus, when we look at the number of counties in each `region`, we can analyze its *multinomial* distribution as there are three or more distinct categories. A variable with only two discrete classes has a *binomial* distribution.

To work with discrete distributions, we need to tabulate each class in a variable. Using `dplyr`, let's calculate an assortment of summary statistics to add context regional economic patterns: the total number of counties, total employment, average county employment, and median county employment per region.

```
# Load dplyr
pacman::p_load(dplyr)

# Summarize by region
regional <- cty %>%
                group_by(region) %>%
                summarize(tabulation = n(),
                      total = sum(all.emp),
                      average = mean(all.emp),
                      p50 = median(all.emp))
```

Each of these four measures is plotted using `geom_bar`, specifying the `stat = "identity"` option to indicate that we will supply a set of pre-computed summary statistics.[3] As shown in Figure 6.4, the US' economic story varies quite a bit depending on how we cut the numbers. The region with the largest number of counties is the South, but this does not account for the amount of economic activity. When we compare regions based on total employment (Graph (B)), the South is still the largest, but the relative differences are far less. As we continue to normalize the data even more by comparing the average and median employment numbers (Graphs (C) and (D)), the US Northeast emerges as the area with the greatest concentration of the employment.

```
reg0 <- ggplot(regional, aes(x = region, y = tabulation)) +
            geom_bar(stat = "identity", fill = "navy", width = 0.8) +
            ggtitle("(A) Total counties") + ylab("Number of Counties") +
            custom_theme

reg1 <- ggplot(regional, aes(x = region, y = total / 1000000 )) +
            geom_bar(stat = "identity", fill = "navy", width = 0.8) +
            ggtitle("(B) Total employment") + ylab("Employment (Millions)") +
            custom_theme

reg2 <- ggplot(regional, aes(x = region, y = average)) +
            geom_bar(stat = "identity", fill = "navy", width = 0.8) +
            ggtitle("(C) Average county employment") + ylab("Employment") +
            custom_theme

reg3 <- ggplot(regional, aes(x = region, y = p50 )) +
```

---

[2]The designation of a region depends on classification system. The U.S. Census Bureau, for example, has regional divisions and sub-divisions.

[3]With `ggplot2`, it is also possible to supply the unaggregated data using `stat_count`.

```
        geom_bar(stat = "identity", fill = "navy", width = 0.8) +
        ggtitle("(D) Median employment")  + ylab("Employment") +
        custom_theme

grid.arrange(reg0, reg1, reg2, reg3, ncol = 2)
```

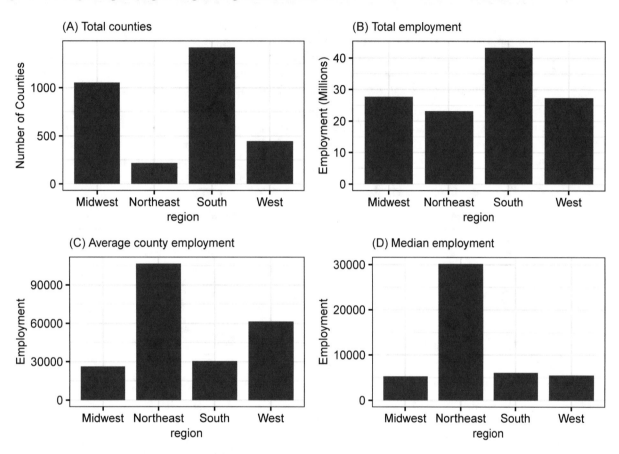

Figure 6.4: An assortment of bar chart by region.

Continuous distributions. *Histograms* are ideal for exposing the shape of a continuous distribution. These plots resemble bar charts with one minor difference: rather than plotting the frequency of categories in discrete variables, histograms convert a continuous variable into a series of equal-sized *bins*. A bin's height represents how many observations fall into the bin's interval.

The ideal case for any variable is the presence of signal. In a histogram, a clear indication of signal is if there is a peak—a central tendency at which the mean or median is more likely to reside. A common misconception is that continuous data will always follow a normal curve (Gaussian distribution). For the most part, very few variables in the wild truly follow this data stereotype and when data does not follow this expected pattern, it is hard for many to understand. For example, the percent of a county that is college education (ba) does not follow a symmetrical curve.

```
# Assign data to ggplot
us_cty <- ggplot(cty)

us_cty1 <- us_cty + geom_histogram(aes(x =ba), colour = "navy", fill = "blue") +
        ggtitle("College Educated") +  xlab("% College Educated") +
        ylab("Number of Observations") + custom_theme
```

We can re-apply this code snippet to three other continuous variables in our dataset to produce similar graphs for comparison. Graph (A) in Figure 6.5 shows the percent of a county that is college-educated (ba), which peaks at 18% and has a right tail that is slightly longer than the left tail. Median county income (inc) in Graph (B) also exhibits a similar distribution shape. Both (A) and (B) are fairly well-behaved variables. However, total employment (all.emp) in Graph (C) is highly skewed. When presented with a skewed distribution, take some time to investigate observations in the *long tail* to understand how they differ from the peaks. In some cases, there is a reasonable explanation for these outlying observations. In these graphs, for example, the tails are comprised of counties with the largest cities in the United States such as Cook County (Chicago), Los Angeles, the Five Boroughs of New York City, among other large metropolitan areas. In fact, the top 20% of counties account for 85.9% of the employment in the country. Alternatively, if there is not a discernable pattern, then perhaps these observations are *anomalies* or *outliers*.

The first three graphs illustrate when *signal is present*—there is a clear central tendency. How does a histogram look when a variable lacks signal? Perhaps the most common cause is a random uniform distribution (runif()) as seen in Graph (D). Every bin is approximately the same height. If we were to randomly draw an observation from Graph (D), there is an equal chance that it could originate from anywhere in the distribution. In contrast, most observations tend to cluster around the peak in the first three graphs thus there is a higher chance that a randomly drawn observation is closer to that region.

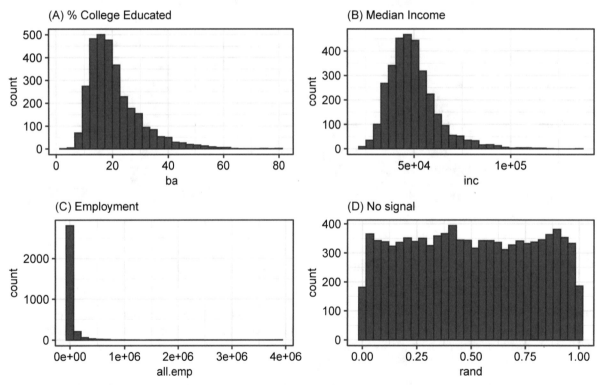

Figure 6.5: Histograms that show various forms of signal.

For small samples, histograms will appear jagged and may prove challenging to interpret. An elegant alternative is the *kernel density plot*. The plot relies on *kernel density estimation* (KDE) to calculate the chance of observing a given value based on proximity to all points within a bandwidth. The bandwidth is a window size that indicates how much of a variable should be included in the calculation, where larger bandwidths produce smoother graphs with less noise, while smaller bandwidths produce more volatile results. The implication is that the shape of the curve can change if we change the bandwidth—the pattern is an approximation that can be manipulated.

How do histograms and kernel densities compare? To see the differences, we generate two kernel density plots using different bandwidths using geom_density. Notice that the distributions in the kernel density plots are represented by continuous lines, which are far smoother than the more bumpy histogram in Graph (A), whereas Graph (B) uses a default value for the bandwidth that minimizes noise, Graph (C) reduces the bandwidth to 1% of the default and exhibits far more noise (Figure 6.6).

```
# Histogram
h1 <- us_cty +
        geom_histogram(aes(ba), colour = FALSE, fill = "blue")  +
        ggtitle("(A) Histogram") +
        custom_theme

# Kernel Density
k1 <- us_cty +
        geom_density(aes(ba), colour = FALSE, fill = "blue") +
        ggtitle("(B) Kernel (default bandwidth)") +
        custom_theme


# Kernel Density
k2 <- us_cty +
        geom_density(aes(ba), adjust = 0.01,  colour = FALSE, fill = "blue") +
        ggtitle("(C) Kernel (1%  bandwidth)") +
        custom_theme

# Juxtapose two graphs
grid.arrange(h1, k1, k2, ncol = 3)
```

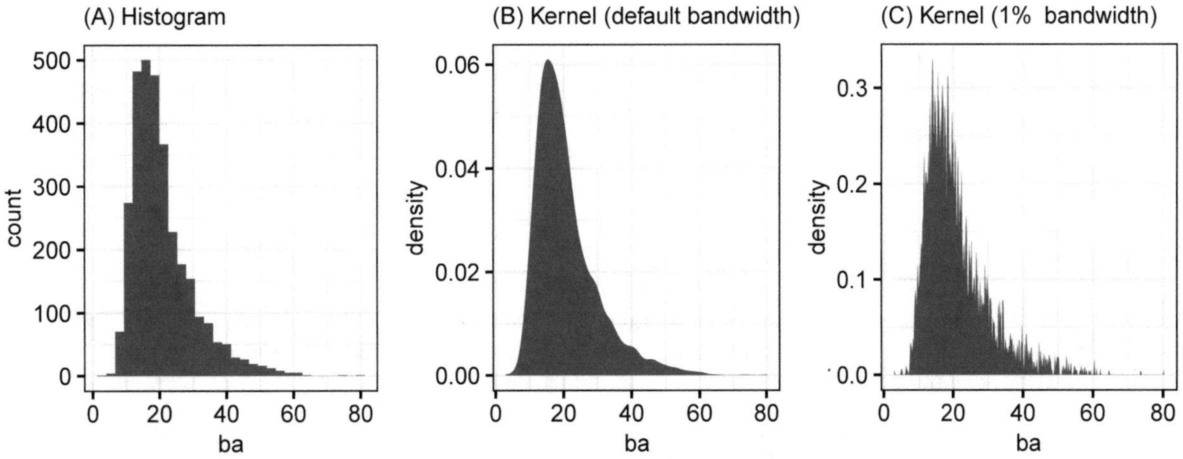

Figure 6.6: Histograms and kernel density plots showing county-level estimates of percent college educated.

Combining discrete and continuous distributions. By incorporating discrete variables, we can enrich the insights drawn from continuous variables. Imagine the differences that could emerge when plotting a kernel density plot for each level of a discrete variable. However, overlaying multiple kernel density charts can easily overwhelm the viewer—too much information on one canvas. The `geom_density_ridges` (ggridges package) lays out kernel density plots in an innovative format: each kernel density is plotted on a separate row, resembling a series of mountain regions. This format makes it easy to compare levels of the discrete variables while preserving the horizontal axis scale of the continuous variable. Figure 6.7 plots poverty rate by region, showing that the US South has a higher level of poverty relative to other regions.

```
# Load ggridges
pacman::p_load(ggridges)
```

```
# Plots by group
us_cty + geom_density_ridges(aes(x = pov, y = region),
                    fill = "blue", colour = "white", scale = 2, alpha = 0.9) +
   xlab("Poverty Rate") + ylab("Region") + custom_theme
```

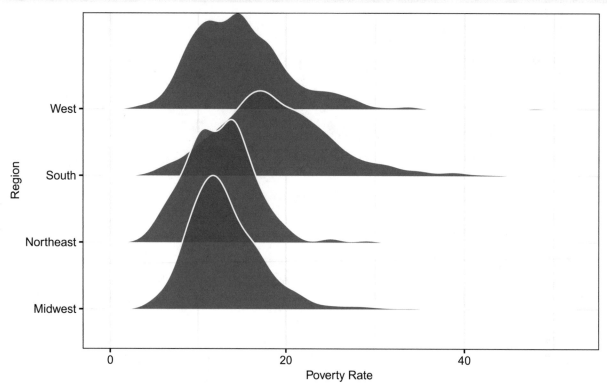

Figure 6.7: Kernel density plots by region.

6.3.1   Skewed variables

Many statistical models used in public policy assume that the underlying variables are normally distributed. If they are not, then there is a chance that insights are not generalizable. We have already observed, however, that it is fairly uncommon to have a normally distributed variable—many variables exhibit some skew.

Many variables will be left-skewed (i.e., long tail to the left with a peak to the right) or right-skewed (i.e., long tail to the right with a peak to the left). Any skewness means a variable is asymmetrical. Highly skewed variables, such as Graph (A) in Figure 6.8, can be challenging to analyze as the core mass of the distribution is compressed into a small area that is hard to visually inspect. For right-skewed distributions, consider applying a *logarithm transformation* (or *log transform* for short) which has the effect of compressing the long tail (see Graph (B)). A square root transformation is far less drastic and is able to spread the center mass of the distribution over a greater area. Left-skewed distributions (not shown) can be transformed by squaring a variable (e.g., x^2).

In practice, the principal use of variable transformations is to adjust data to meet modeling assumptions. One could apply a log transformation blindly through any adjustments to data may arbitrarily move a variable closer or farther from an ideal theoretical distribution. Quantile-Quantile plots (QQ-plots) help assess how closely a variable follows a known theoretical distribution (e.g., Normal distribution) at each quantile. In `ggplot2`, a pair of functions are used to make this comparison:

- `stat_qq` produces the QQ plot. By default, the function makes comparisons with a Normal distribution, but can be adjusted to compare with other distributions.

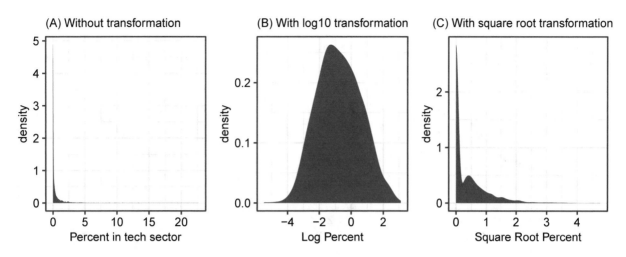

Figure 6.8: Kernel Density Plots for percent of workforce in tech sector, with and without a log_{10} transformation applied.

- `stat_qq_line` computes the line of equality—a benchmark that indicates when two distributions are identical at each quantile.

Figure 6.9 compares the percent college-educated (vertical axis) with a normal distribution (horizontal distribution). The straight blue line represents the line of equality generated from `stat_qq_line`—when the education rate exactly matches a theoretical normal distribution. The points represent specific values of the college-educated rate and how it maps to the same quantile in a theoretical normal distribution. The untransformed distribution only aligns with the theoretical distribution at center mass, while values at the tails diverge. By applying a square root transformation in Graph (B), the tails move closer to the line of equality but still fall short of target, while a log transformation is able to approximately normal distribution.

In all cases, variable transformations can improve a variable's properties by forcing it to appear more "normal". These adjustments come at the cost of a less intuitive set of variables.

```
# Raw data
qq1 <- ggplot(cty, aes(sample = ba)) +
          stat_qq() + stat_qq_line(col = "blue") +
          ggtitle("(A) College Educated versus Normal") +
          custom_theme

# Transformed
qq2 <- ggplot(cty, aes(sample = ba ^ 0.5)) +
          stat_qq() + stat_qq_line(col = "blue") +
          ggtitle("(B) Square Root Transformation") +
          custom_theme

# Transformed
qq3 <- ggplot(cty, aes(sample = log(ba))) +
          stat_qq() + stat_qq_line(col = "blue") +
          ggtitle("(C) Log Transformation") +
          custom_theme


# Juxtapose
grid.arrange(qq1, qq2, qq3, ncol = 3)
```

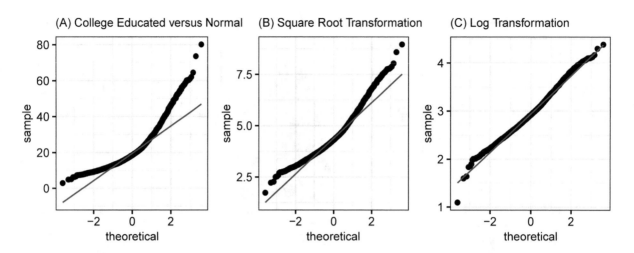

Figure 6.9: A QQ plot comparing transformed and untransformed education rates versus a theoretical normal distribution.

6.4   Exploring missing values

Missing values are a reality of working with real world and pose logistical challenges for data projects. Most analytical techniques expect complete data records. Linear regression—a favorite in public policy—can only be estimated with complete samples, thus most implementations will only use complete records. Many machine learning algorithms will not be able to train a model with any missing value. If all observations in a dataset are missing at least one value, then the value of the entire dataset may be at risk. For this reason, missing values are sometimes dropped, while other times they can be imputed (i.e., filled with an approximation). Exactly how one *treats* missing values depends on understanding the qualities of missingness. In this section, we explore topics in dealing with missing values—how does missingness appear in data, how does one analyze its qualities, and how to treat it.

6.4.1   Encodings

On the surface, missing values seem to have an obvious meaning—*there is no information available.* But as one works with a variety data sources, missingness signals different data quality issues in different contexts. The US Census Bureau's American Community Survey (ACS), for example, captures fine details about life in the United States. The dataset is carefully designed to ensure that virtually any demographic (e.g., gender, age, race) and economic context (e.g., income, region, employment) is accurately represented. Suppose we found a set of blank values in the variables "Year of naturalization" and "Means of transportation to work". How should one interpret those blanks? *It depends on the dataset and how data is encoded.* "Not Applicable"—or when the concept being measured does not apply to the survey respondent—are represented as blank cells. This is a different concept than "Not Specified"—or when the respondent chooses not to respond to the question or data is not provided for other reasons. In the ACS, we can interpret blank cells for both "Year of naturalization" and "Means of transportation to work" as someone who is born in the United States and does not travel to work as the respondent is not in the labor force.

These encodings are not universal, however. Table 6.2 compares a selection of encodings from different data sources.[4] In many data sources, "Not Specified" is encoded as a blank cell (R treats these empty cells as NA); however, blank cells can also represent other types of missingness. Whereas "Not Applicable" is encoded as

---

[4]The NNBS dataset is a *General Household Survey* produced by the Nigerian National Bureau of Statistics in collaboration with the World Bank to measure (Nigeria National Bureau of Statistics 2019). The example is drawn from survey questions relating to banking access in file "sect4a1_plantingw4". The EU LFS is the European Union's Labour Force Survey, which is conducted by national statistical agencies of EU member countries and maintained by Eurostat (Eurostat 2020). The ACS dataset is the American Community Survey, administered by the U.S. Census Bureau (U.S. Census Bureau 2018a). The NYC 311 SR dataset contains complaints and service requests made to the City of New York (NYC Department of Information Technology and Telecommunication 2020).

a blank in the ACS, the EU LFS uses the value 9. Meanwhile, high-frequency datasets that are updated as new information is made available over time, such as NYC's 311 Service Requests (NYC 311) capturing customer service interactions, do not distinguish between "Not Applicable", "Not Specified", and "Not Yet Known"—all recorded as blanks. Since requests require time to complete, "Not Yet Known" tends to be updated as more is known. One thing is clear: blank cells represent some form of missingness, but what they represent depends on how the data producer designs the dataset.

One encoding is missing from this picture: *zero*. Unlike missing values, *zero* is reserved for instances where no known activity occurred. Someone who did not work a salaried job does not have income. Two countries that have not engaged in trade do not have any recorded trade. Thus, zero can oftentimes be interpreted as zero. *When in doubt, consult the data dictionary.*

Table 6.2: Comparing encodings for a selection of datasets.

Dataset	Example Variable	Not Applicable	Not Specified (Non-Response)	Not Sure	Not Yet Known
ACS	Year of Naturalization	Blank	Imputed	-	-
ACS	Employment Status	Blank	Imputed	-	-
EU LFS	Permanency of the Job	9	Blank	-	-
EU LFS	Professional status - Employment status	9	Blank	-	-
NYC 311 SR	Closed Date - When a request has been completed	-	-	-	Blank
NYC 311 SR	Location Type - Where is	Blank	Blank	-	Blank
NNBS	Did [NAME] open their account him/herself?	-	Blank	-99	-
NNBS	Did [NAME] search for information before getting bank account?	-	Blank	-99	-

6.4.2 Missing value functions

R has functionality to detect missing values and other problematic values. In Table 6.3, we list a few functions that detect specific types of missingness, namely `NA`, `NaN`, and `Inf`.[5] Each of these functions accept data objects (e.g., single values, vectors, matrices, and data frames) and return equal-sized logical objects (e.g., `TRUE`, `FALSE`) that indicate the location of specific types of missing values.

Table 6.3: Functions for detecting different types of problematic values.

Function	Description	Example Target Value
is.na	Returns logical value for each element indicating if `NA` is present.	NA
is.nan	Returns logical value for each element indicating if impossible value is present.	FALSE / 0
is.infinite	Returns logical value for each element indicating if infinite value is present.	1 / 0
complete.cases	Returns logical value for each row indicating if complete.	NA

Let's see how these functions behave when applied to a vector of hypothetical hourly wages with inconsistent

---

[5]`Inf` values are not truly missing values, but can prove to be problematic. We include these values for awareness.

encodings. The results have been compiled in Table 6.3, comparing the returned logical values with the corresponding input data. Whereas `is.na` flags both `NA` and `NaN` values, `is.nan` is only sensitive to impossible values as is `is.infinite` with infinity values. Unlike the "is." functions, the `complete.cases` function evaluates whether a row is free of `NA` and `NaN` values. In other words, `complete.cases` checks if a row is complete. As missing values can be dispersed across multiple variables in a data frame, this function comes in handy when determining how many records are complete *across all variables*.

```
# Create sample values
hourly_wage <- c(14, 15, NA, 15, 20, NaN, -99, Inf)

# Test for missing values
miss_na <- is.na(hourly_wage)
miss_nan <- is.nan(hourly_wage)
miss_comp <- complete.cases(hourly_wage)
miss_inf <- is.infinite(hourly_wage)
```

While the conditional statement (`hourly_wage == -99`) can evaluate non-missing values, it returns `NA` wherever values are missing. This highlights the need to be vigilant when identifying missing values that are encoded as negative integers: *Clean and harmonize the data before treating missing values.* In this example, we edit -99 values as `NA` for consistency (Table 6.4).

```
# Conditional statement to detect -99
miss_enc <- hourly_wage == -99

# Recode -99 as NA
hourly_wage[hourly_wage == -99] <- NA
miss_broad <- is.na(hourly_wage)
```

Table 6.4: Output of missing value functions.

Missing	14	15	NA	15	20	NaN	NA	Inf
NA	F	F	T	F	F	T	F	F
NaN	F	F	F	F	F	T	F	F
Infinite	F	F	F	F	F	F	F	T
-99	F	F	NA	F	F	NA	T	F
Complete Cases	T	T	F	T	T	F	T	T
-99 adjusted	F	F	T	F	F	T	T	F

6.4.3 Exploring missingness

Quantifying missingness. We can apply arithmetic functions to the logical vectors to summarize the missing values. For example, we take the `sum` and `mean` below to calculate the number and proportion of `NA` values.

```
# Count the number of TRUE values when NA and NaN are present
sum(miss_na)

# Proportion missing
mean(miss_na)
```

When summarizing a vector with `NA` values, we need to provide the arithmetic functions explicit instructions on how to treat missingness.[6] For example, the `miss_enc` vector contains two `NA` values, which causes `sum` and `mean` to return `NA` values.

---

[6]Note this applies not only to the logical values from missing value functions but also to any vector with `NA` values.

```
# Sum returns an NA
sum(miss_enc)
```

```
## [1] NA
```

One option is to remove `NA` values by specifying `na.rm = T` with select functions.[7] This forces `R` to return an estimate based on non-missing values, but may also be inaccurate. In this example, suppressing `NA` values yields an accurate tally for `-99` values only.

```
# Suppress NA values
sum(miss_enc, na.rm = T)
```

```
## [1] 1
```

Alternatively, we could impute what values *would* have been if they were not missing. In this case, replacing `NA` values with `TRUE` gives a reasonable estimate of the number of `NA`, `NaN` and `-99` values.

```
# Replace NA value with guess
miss_enc[is.na(miss_enc)] <- TRUE

# Return result with imputation
sum(miss_enc)
```

```
## [1] 3
```

Should `R` skip over the `NA` values or find some way to include them? It depends on the objective. Treating the missing values is easy, but doing so in a way that is reliable and accurate requires analysis of the nature of missingness.

Analyzing missingness. Rubin (1976) lays out distinct types of missingness. Each type has a specific statistical definition and properties and, in turn, should be treated differently. In some cases, focusing on complete values can bias an analysis as the missingness is systematic. In other cases, missingness is completely random and does not adversely impact the analysis. Below, we describe these different grades of missingness, not in their full statistical splendor, but at a high conceptual level:

- *Missing Completely at Random* (MCAR): When missing values are uncorrelated with observed variables in the data. For example, the non-response pattern for a survey question about income is not correlated with other variables captured in the survey such as education level or poverty. When data are MCAR, the missing values are not likely to bias the analysis.
- *Missing at Random* (MAR): When missing values depend on observed variables in the dataset. For example, non-response to an income question is correlated with respondents' education level. When data are MAR, the missing values can lead to a biased analysis if left untreated or unadjusted.
- *Missing Not at Random* (MNAR): When missing values are correlated with the concept that is measured and also do not meet any of the criteria for MCAR and MAR. For example, survey respondents do not respond to the income question if they have higher incomes. In these cases, a researcher cannot directly identify the reasons for MNAR and instead requires deep subject matter expertise to identify the precise mechanisms causing missingness.

Whereas MNAR relies more on subject matter expertise to detect, MCAR and MAR can be tested through a relatively simple hypothesis testing approach: *Create a dummy variable to mark missing values, then use the dummy to test differences using T-tests for continuous variables and Chi-squared tests for discrete variables.* Wherever significant differences appear, we can conclude that missingness is correlated with observed variables.

To illustrate this testing strategy in action, we modify the `cty` dataset to simulate MCAR and MAR.

Detecting MCAR. In the example below, we randomly select and remove approximately 50% of the `ba` variable, create a `missing` dummy variable, then conduct hypothesis tests between the dummy variable and

---

[7]Many functions have the ability to ignore `NA` values. When in doubt, check the Help section for documentation.

other variables in the dataset (e.g., income inc, poverty pov, and the region in which the county is located). As none of the tests are significant at any standard levels (e.g., 1%, 5%, or 10%), we can infer that the values are likely missing completely at random.

```
# Re-load dataset
load("data/county_compare.Rda")

# Create copy of cty
cty_mcar <- cty

# Simulate MCAR with 20% randomly missing
set.seed(123)
cty_mcar$ba[runif(nrow(cty_mcar)) <= 0.5] <- NA

# Create missing value dummy variable
cty_mcar$missing <-is.na(cty_mcar$ba)

# Test covariates
t.test(inc ~ missing, data = cty_mcar)
t.test(pov ~ missing, data = cty_mcar)
chisq.test(factor(cty_mcar$region), cty_mcar$missing)
```

Detecting MAR. In this second example, we apply the same procedure except we simulate dependence between variables by randomly remove approximately 50% of the ba variable *below the first quartile of the income variable* inc. In other words, we make the missingness appear as if lower income counties are less likely to have education data available. The hypothesis tests confirm the presence of dependence as all tests return statistically significant results at the 1% level. This finding should not be a surprise as income, socioeconomics, and geography are all intrinsically intertwined. Because the missing value pattern is correlated with other variables, we can infer that the values are only missing at random.

```
# Create copy of cty
cty_mar <- cty

# Simulate MAR
set.seed(123)
cty_mar$ba[cty_mar$inc < 41073 & runif(nrow(cty_mar)) <= 0.5] <- NA

# Create missing value dummy variable
cty_mar$missing <-is.na(cty_mar$ba)

# Test covariates
t.test(inc ~ missing, data = cty_mar)
t.test(pov ~ missing, data = cty_mar)
chisq.test(factor(cty$region), cty_mar$missing)
```

Identifying the nature of the missing values is just the first step. There are different strategies for treating MCAR versus MAR.

6.4.4   Treating missingness

Treating missing values is necessary in order for an analysis to move forward. Otherwise, analyses would rarely be completed. Fortunately, there is a broad range of treatment options as shown in Table 6.5. Some options only apply to MCAR rather than MAR as bias can have a significant impact on the quality of adjustment. Other options operationalize a dataset by deleting incomplete records, while others rely on imputation. Ultimately, it is worth the investment to test and compare the results from multiple options if possible.

Table 6.5: Missing value treatment options.

Treatment Option	Description	MCAR	MAR
Listwise deletion	Drop any record with a missing value. Also known as complete case analysis.	x	
Drop variable	Drop variables with missing values. Most appropriate when the number of missing values is quite large and concentrated in one variable.	x	x
Mean or median imputation	Substitute missing values with the variable's mean or median. Mean is suitable for symmetrical continuous distributions, while the media is suitable for skewed continuous variables.	x	
Mode imputation	Substitute missing values with the variable's mode—suitable for discrete variables.	x	
Model-based imputation	Impute missing values using a prediction model such as a regression trees or k-Nearest Neighbors (kNN). The models take advantage of correlations observed among complete cases to approximate missing values.	x	x
Multiple imputation	Impute using model-based imputations, doing so multiple times then averaging results to remove possible biases.	x	x

Deletion is the quickest way to force data into a usable form. Generally, we approach deletion in two ways: listwise deletion (row-wise) and variable deletion (column-wise).

Listwise deletion only retains complete records—sometimes referred to as complete case analysis—and is only appropriate when data are MCAR. Because there is not a discernable pattern in missingness, whittling the dataset only to complete records will not likely yield biased results—the complete records are, on average, no different than records with missing values. While listwise deletion may seem to be an obvious choice, there are situations where *every record* has at least one missing value. A complete case analysis lead to *all* records being dropped.

In R, we can filter the dataset to include only complete cases using one of the following functions: `na.omit` or filtering using `complete.cases`. The `na.omit` function scans all columns and drops any row that has an `NA`, returning a filtered object. A slightly more verbose option relies on `complete.cases` to identify the position of complete records, then subset the data frame using row indexes of complete cases. Both options yield the same results. In the example below, both approaches filter `cty_mcar` to $n = 1555$ complete records.

```
# Option 1: na.omit
option1 <- na.omit(cty_mcar)

# Option 2: filter using complete.cases
option2 <- cty_mcar[complete.cases(cty_mcar), ]
```

Variable deletion is a reasonable option when a variable is missing too many values. Precisely how much missingness is acceptable is a personal judgment call, however. For example, if the `ba` variable is complete in 5% of a sample, this variable does not merit one's attention and can be discarded. Alternatively, if 80% of the variable is complete, then perhaps the variable can be salvaged. In actuality, removing just one variable might not remedy the problem, especially in datasets with many missingness dispersed across multiple variables. When we set a minimum completeness threshold (i.e., a criterion that indicates the lowest acceptable percent complete for a variable to be retained), we see trade-offs in the number of variables dropped and the number of rows retained. As we increase the completeness threshold, we will likely have fewer retained variables, but more complete records. These trade-offs are dataset-specific, thus we recommend conducting a missingness analysis to quantify these relationships.

In the example below, we simulate a dataset with $n = 10000$ and 100 variables. Each variable can have up to 25% missing values (see `pct_miss`) that are missing completely at random.[8]

```r
# Simulate 100 variable dataset with n = 10000
n <- 10000
k <- 100
pct_miss <- 0.25

# Each variable has between 0% and 25% missingness
df <- data.frame(matrix(nrow = n, ncol = 0))
  for(i in 1:50){

# Create random Poisson distribution
set.seed(i)
df[[paste0("x",i)]] <- rpois(n, runif(1, 0, 500))

# Substitute missing values with -9999
df[runif(n) <= runif(1, 0, pct_miss), i] <- NA
}
```

With the data ready, we will calculate the number of variables retained and number of complete records for different levels of variable completeness. This requires a vector `completeness`—constructed using `apply` to calculate the proportion of missing values for each column—in order to efficiently identify which columns are complete above a `threshold`. To map out the trade-offs, we loop over percent thresholds from 1% to 99%, recording complete records (`n.complete`) and retained variables (`num.vars`).

```r
# Summarize completeness by column using apply
completeness <- apply(df, 2, function(x){mean(!is.na(x))})

# Calculate complete case sample size when
# Only variables > X% completeness are retained
counts <- data.frame()

  for(threshold in seq(0.01, 0.99, 0.01)){

# Whittle sample to variables complete more than X%
temp <- df[,which(completeness >= threshold)]
temp <- temp[complete.cases(temp),]

# Tabulate summary
counts <- rbind(counts,
data.frame(min.pct.complete = 100 * threshold,
                   num.vars = sum(completeness >= threshold),
                   n.complete = nrow(temp)))
  }
```

When graphed, the trade-offs are quite striking in this MCAR scenario. As shown in Figure 6.10, Graphs (A) and (B) illustrate the number of complete records and variables retained as the completeness threshold is increased. As the threshold increases, we find fewer variables have enough complete records to meet the cutoff. The inverse is true for complete records: As we drop incomplete variables, the number of complete records increases. Graph (C) illustrates the trade-off. If we truly did not have a preferred set of variables, then the inflection point in the curve (around 25 variables, 3750 observations, threshold $>$ 87) is the point of perfect balance. If certain variables are valued more than others, however, we can adjust the sample along this curve. While $n = 3750$ may be an unsatisfying outcome for a dataset of $n = 10000$, it is the *lower bound*. As we will see in the following subsection, imputation can help fill data gaps so that we minimize the number

---

[8]Consider experimenting with the `pct_miss` parameter to understand the trade-offs.

of incomplete records.

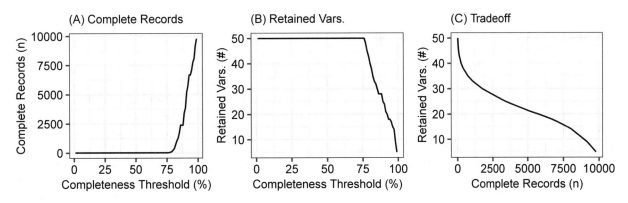

Figure 6.10: Mapping of variable and record completeness as a function of variable completeness threshold.

Imputation. Rather than discarding records, imputation fills each variable's missing values with a guess. How we construct arrive at the guess is very much a rabbit hole of complexity.

The simplest approaches involve *mean or median imputation,* which are appropriate for when data are MCAR. By substituting missing values with a central value, we make the conscious decision to preserve one quality of the original dataset—either the mean or median—but at the cost of disturbing other qualities. Symmetrical continuous values can be imputed with a mean, while the median is best for skewed distributions. For discrete values, the mode is the logical choice. In the example below, we impute continuous values in the `cty_mcar` dataset.

```
# Fill ba variable with mean
cty_mcar$ba_mean <- cty_mcar$ba
cty_mcar$ba_mean[which(cty_mcar$missing)] <- mean(cty_mcar$ba_mean, na.rm = T)


# Fill ba variable with median
cty_mcar$ba_med <- cty_mcar$ba
cty_mcar$ba_med[which(cty_mcar$missing)] <- quantile(cty_mcar$ba_med, 0.5, na.rm = T)
```

When data are MAR, however, we can increase the sophistication of our imputations by leveraging *model-based imputation* techniques that expose correlations *between* missing values and other variables. In many ways, model-based imputation works the same way as a predictive model: a model learns the relationship between each variable and all other variables using only complete records. Those learned patterns are then applied to where covariates are not missing in order to approximate the missing values.

While we will cover prediction in Chapter 10, we briefly illustrate how to access model-based imputation techniques using the `caret` package—a favorite for machine learning applications. In particular, we focus on the `preProcess` function that imputes missing values in data frames. For this example, we use a modeling technique known as *bagged regression trees* to find multivariate patterns that can inform imputation (`method = "bagImpute"`) and apply it to the `cty_mar` data frame. Model-based imputation requires some careful thought about which variables are relevant to missing values. In this case, we include all variables that are believed to have a correlation, excluding identifier variables such as `fips`, `state`, and county `name`.

Once the `preProcess` function finds the underlying correlations and stores the patterns as a model object `cty_mar_model`, we predict missing values in the `cty_mar` data frame and output a cleaned dataset `filled`.

```
# Load caret
pacman::p_load(caret)


# Train regression tree imputation model
cty_mar_model <- preProcess(cty_mar[,-c(2:4)],
```

```
                                                method = "bagImpute")

# Fill in dataset
filled <- predict(cty_mar_model, cty_mar)
```

Thus far, we have only considered imputation as a means to fill a missing value with the best guess. However, there is uncertainty in the guesses. We may never know the true value. Furthermore, we may have happened upon an imputation by chance. This view of imputation from the lens of random variables implies that a parameter estimates (e.g., a mean, a slope between two variables) should have upper and lower bounds that reflect uncertainty. *Multiple imputation* (MI) is a popular approach in the social and medical sciences. The technique recognizes that imputation generates m sets of imputations. Each dataset is analyzed following an identical approach, then the results are combined giving a mean and uncertainty estimate for each parameter estimate. While MI does increase the robustness of an analysis, it is technically demanding and beyond the scope of this text. For an in-depth treatment on the subject, we recommend the seminal text on the subject *Multiple Imputation for Nonresponse in Surveys* by Donald B. Rubin.

How do all of these imputation techniques compare? In Figure 6.11, we evaluate how well each imputation approach is able to approximate the original `ba` distribution. For the MCAR dataset, the mean and median imputations (Graphs (D) and (E)) preserve specific qualities (e.g., central tendency) while distorting the shape of the distribution. In contrast, the model-based imputation applied to the MAR dataset more successfully approximates the shape of the original distribution (Graph (F)). While model-based approaches can be applied to both MCAR and MAR datasets, MAR datasets are far more suitable given the presence of correlations that can be exploited.

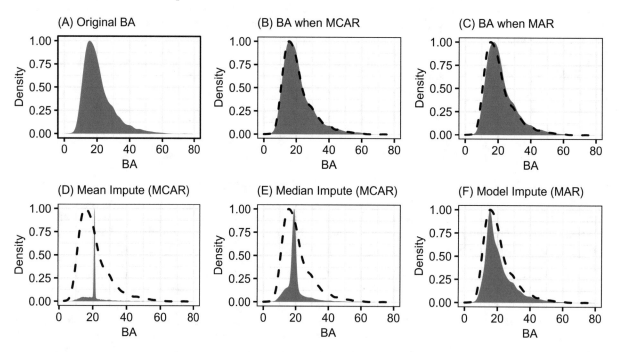

Figure 6.11: Distributions of original BA variable and its MCAR and MAR versions compared with imputation results. Dashed lines in panel (B) through (F) are a reference to the original distribution. Densities have been scaled.

When to use imputation versus deletion? There is no perfect solution, but ultimately the choice depends on the analytical objectives. If data are missing completely at random, listwise deletion offers a speedy conclusion to treating missing values, assuming that the number of complete records is large. Otherwise, it is likely one will need to combine imputation and deletion.

For use cases such as analyzing high-level trends or anticipating outcomes using prediction models, imputation can operationalize the data. Of course, this comes at a cost. As more imputation is applied to the data, we should be cautious about interpreting fine-grained relationships as imputed data may introduce patterns into the data. Imputation, however, is not suitable when analyses require *exact* measurement. For example, financial accounting requires precise numbers as imputation distort figures and detecting missing values is important for identifying errors and gaps in accounting systems.

6.5   Analyzing time series

Many datasets are time series that record information over regular intervals of time. Basically, all stock ticker prices, economic data, weather data, and physical sensors are time series.

Time series are special as their sequential nature dictate how one should work with the data. There are certain types of graphs that are better suited for time series. While the distribution could be plotted using a histogram, time series driven by population growth generally increase, thus the graph would show a crude approximation of growth. Instead, line graphs (`geom_line`), point graphs (`geom_point`), and area graphs (`geom_area`) capture the sequential nature of the data. In Figure 6.12, we plot the total number of shipping containers that traveled through the Port of Long Beach in California (Port of Long Beach 2019).

```
# Load data
polb <- readr::read_csv("data/polb.csv")

# Set data and create plots
teu <- ggplot(polb, aes(x = date, y = total)) + custom_theme
teu_line <- teu + geom_line(colour = "blue") +
              ggtitle("Line") + xlab('Date') + ylab('Total')
teu_area <- teu + geom_area(fill = "blue") +
              ggtitle("Area")  + xlab('Date') + ylab('Total')
  teu_point <- teu + geom_point(colour = "blue")  +
              ggtitle("Point") + xlab('Date') + ylab('Total')

# Render
grid.arrange(teu_line, teu_area, teu_point, ncol = 3)
```

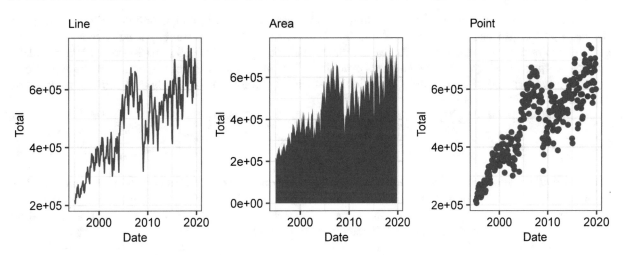

Figure 6.12: Comparison of line, area, and point graphs for Port of Long Beach's monthly TEU volume.

Because time series have a special structure, we can analyze layers of information embedded within it. A time series (Y) can be decomposed into three components:

$$Y = T + S + R$$

Where the T is a *trend*—the prevailing direction of the data, S is *seasonality* or patterns that recur on a regular interval (e.g., monthly seasonality), and R is the *noise* component (sometimes referred to as "irregular" or "residual")—the leftover information that is not explained by the trend or seasonality.[9] Raw time series, such as the POLB container volumes, are *not-seasonally adjusted* (NSA), retaining all components of a time series as evidenced by the peaks and troughs that occur on a regular annual cycle.

Successful policy narratives are typically crisp, yet seasonality and noise can muddle the story. Data are often *seasonally adjusted* to remove the seasonal component so there is less ambiguity when communicating the direction of trends.[10] Furthermore, decomposing a time series may prove useful for identifying outliers by inspecting unpredictable shifts in the residuals.

Seasonal decomposition. One widely used technique is *Seasonal Decomposition of Time Series by Loess* (STL) as developed in Cleveland, Cleveland, and Terpenning (1990). The technique uses two loops to iteratively extract the trend, seasonal component, and residual. Because seasonality can change over time, STL has the capability of finding one stable seasonal pattern for the entirety of a series versus one that evolves over time.

The technique has been implemented as the `stl` function in Base R, which expects a time series to be converted into a special time series object using `ts`. These time series objects carry metadata that are pertinent to working with time series, namely the frequency and starting period of the series. Excluding the date in Column 1 of `polb`, we convert five variables into a `ts` object (`polb_stl`) with a monthly frequency (`freq = 12`) and a starting date of January 1995 (`start = c(1995, 1)`).[11]

To put `stl` to the test, the only argument required other than the variable of interest is `s.window` that controls how quickly seasonality evolves. When `s.window = "periodic"`, the seasonal pattern is assumed to be roughly constant over the entire series while a smaller window size allows the pattern to change quickly. Below, we examine the total container volume shipped through Long Beach (`total`) and make explicit our assumption that seasonality occurs on a periodic basis—or the same as the 12-month cycle.

```
# Convert the polb object to time series excluding Column 1
polb_ts <- ts(polb[,-1], freq = 12, start= c(1995, 1))

# Apply STL - Total
polb_stl <- stl(polb_ts[,"total"], s.window = "periodic")
```

Each `stl` object contains a `time.series` table with the seasonal, trend, and remainder. Figure 6.13 plots the `polb_stl` object using the `autoplot` function (`forecast` package). Assuming constant seasonality, the monthly container volume can swing between -59,000 (February) and +41,000 (August) relative to trend, or a range of approximately 100,000 containers over the course of a year. A closer inspection of the *noise* component shows large peaks and troughs, some look suspiciously large: *Are there potential outliers, particularly in 2002 and 2015?*

```
# Plot STL
pacman::p_load(forecast)
autoplot(polb_stl) + custom_theme
```

The *remainder* is not particularly useful when measured in containers as it does not consider what is typical variability for a series. Scaling the remainder as a z-score ($z = \frac{(x - \mu_x)}{\sigma_x}$) maps it into an easily interpretable form. Figure 6.14 plots the z-score as a time series, placing *fences* at $|z| = 3$ to mark records that lie beyond 99.7% of the mass of a standard normal distribution. Indeed, the hypothesized outliers sit just beyond the fences, making it easy to isolate outliers: *The drop in container volumes in 2002 and 2015 occur during port labor disputes* (Greenhouse 2002; Isidore 2015).

---

[9]Some series follow a multiplicative formulation. For simplicity, we focus on the additive case.

[10]There are challenges with seasonal adjustment, however. The process of decomposing a time series can be subjective and requires analyst judgment. There does not exist a universal definition of what truly constitutes trend or seasonality. Ultimately, whether a series is "well-adjusted" is dependent on trust in the process.

[11]All variables should be numeric values.

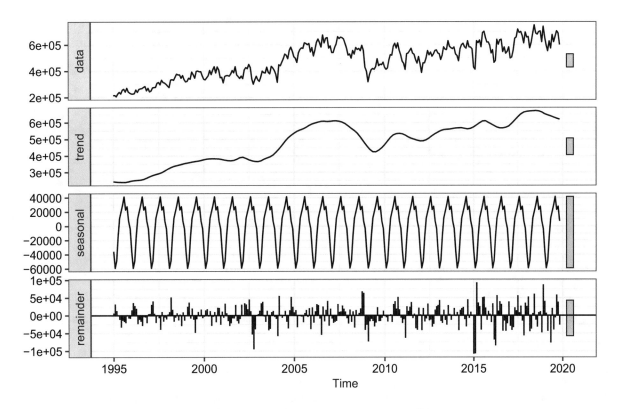

Figure 6.13: Seasonal decoposition using STL.

Pro Tips: Not all time series will be seasonal, but STL will always find a seasonal pattern by design. When conducting seasonal decomposition, consider comparing the size of the seasonal component to the trend. If the absolute maximum of the seasonal component is infintissimally small relative to the trend, then even if a seasonal pattern existed, it might not merit analytical attention.

```
# Extract series
polb$scaled <- scale(polb_stl$time.series[,3])

# Set data and z-score thresholds
ggplot() +
    geom_point(data = polb, aes(x = date, y = scaled), colour = "grey") +
    geom_point(data = subset(polb, abs(scaled) >= 3),
               aes(x = date, y = scaled), colour = "red", size = 2) +
    geom_hline(yintercept = c(-3, 3), linetype = "dashed") +
    ylab("Z-Score") + custom_theme
```

6.6   Finding visual correlations

In addition to visually detecting patterns, EDA is useful for finding relationships across variables. Policy and decision makers have limited time and can only focus on a distilled set of important factors. To help structure the problem space, simple bivariate plots are an efficient option to identify correlations.

Scatter plots are the typical starting point. An observation is mapped into a two-dimensional area with one variable on the vertical axis and the other on the horizontal axis. The intersection of the values is a point. In R, scatter plots are easily rendered using `geom_point` (`ggplot2` package). Despite their widespread use, there is still room for creativity when designing a scatter plot, especially when faced with a large dataset. The insights derived from a plain vanilla scatter plot can easily be crowded out—*a plethora of points conceal*

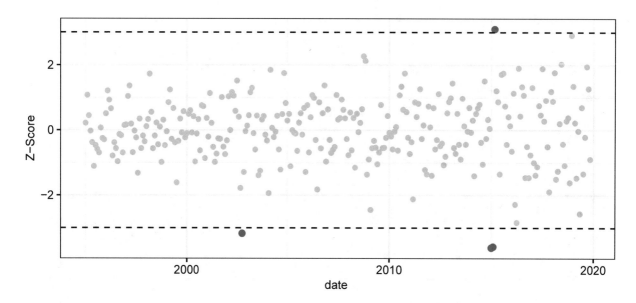

Figure 6.14: Z-score of the noise component for the total container volume at the Port of Long Beach.

meaning.

The simplest of tweaks to the `size` and opacity (`alpha`) of points can work wonders. In Figure 6.15, we compare a basic and stylized plot using the `cty` dataset. Graph (A) illustrates the crowding problem—it is difficult to see the center mass of the point cloud. By lowering the `alpha` value, we make each point more transparent. When two or points overlap, the opacity values add up and allow the data to naturally highlight dense regions.

```
# Set base data for plot
p <- ggplot(cty, aes(x = ba, y = inc / 1000)) +
        ylab("Median Income") + xlab("Percent college educated") +
        custom_theme

# Standard scatter plot
p1 <- p + geom_point() +
        ggtitle("(a) Traditional scatter")

# Scatter plot with transparency
p2 <- p + geom_point(size = 0.5, alpha = 0.2, colour="navy") +
        ggtitle("(b) Scatter (alpha = 0.2)")
```

Interpreting shapes. The shape of point clouds are good approximations of correlations. The Pearson's Correlation Coefficient is the standard metric for measuring the strength and direction of the relationship between two variables, X and Y. It is given as the covariance ($cov(X, Y)$) divided by the product of the standard deviations σ_X and σ_Y:

$$\rho_{X,Y} = \frac{cov(X,Y)}{\sigma_X \sigma_Y} = \frac{\sum_{i=1}^{n}(X_i - \mu_X)(Y_i - \mu_Y)}{\sum_{i=1}^{n}(X_i - \mu_X)^2 \sum_{i=1}^{n}(Y_i - \mu_Y)^2}$$

In Figure 6.16, we simulate the data for five scatter plots, each of which illustrates different values of $\rho_{X,Y}$. A value of $\rho = 0.7$ indicates a strong positive relationship as the point cloud moves up and to the right—when one variable increases, the other does as well. When $\rho = -0.9$, two variables are said to have a negative or inverse correlation in which one will increase, while the other will decrease. An absence of a correlation ($\rho = 0$) appears as an amorphous point cloud without direction.

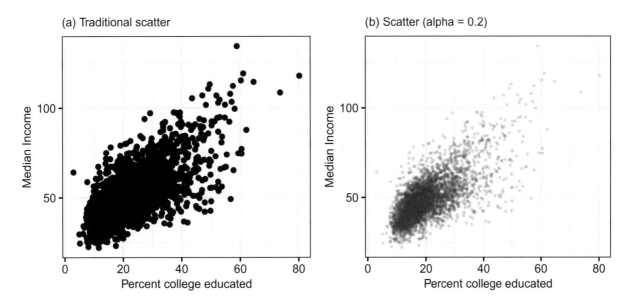

Figure 6.15: A comparison between a basic scatter plot and a stylized plot for county-level variables.

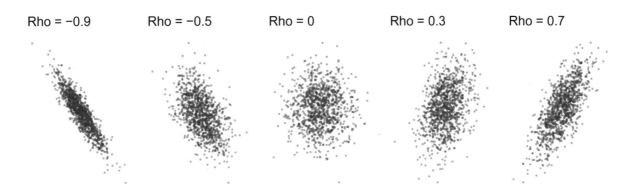

Figure 6.16: Scatter plots illustrating correlations of varying strength.

Note that the previous plots exemplify well-behaved *linear* relationships. When the shape of a plot is anything other than a "fuzzy" line, keep digging as there may be other forces at play. Figure 6.17 illustrates five common types of patterns in scatter plots:

- *Non-linearities* indicate two variables are related but in a non-constant fashion. In other words, the point cloud bends.
- *Discontinuities* occur when there is a break in the trend, oftentimes manifested as step change (a sharp vertical shift in the point cloud) or slope change. A discontinuity is a great opportunity to learn about a possible event or policy change. In social science, discontinuities are highly prized as they can facilitate causal inference, but could also indicate a change or error in how data is collected.
- *Slope changes* are a type of discontinuity in which the slope of the point cloud tilts up or down at a clear *kink*.
- *Stepwise* patterns is a special case of a discontinuity in which the point cloud is vertically shifted at two or more thresholds. While stepwise patterns can also be highly prized discontinuities, they may also simply be an indication that different tranches of data collection came online at different times or even analytical error—one of the variables is discrete values rather than continuous.
- *Distinct clusters* indicate that two variables can be treated as coordinates that can identify the concentration of observations rather than as a straight linear relationship.

Whatever the explanations may be for these empirical quirks, scatter plots expose intricate patterns in the data that would otherwise be overlooked by a summary statistic. If not adequately handled, these quirks can introduce unexpected biases and inaccuracies to a data science project.

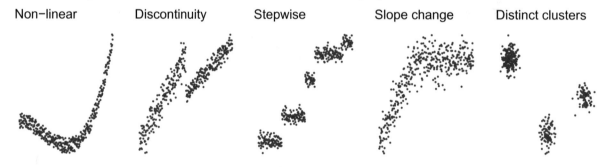

Figure 6.17: Common patterns that arise from scatter plots.

6.6.1   Visual analysis on high-dimensional datasets

Modern datasets are increasingly high-dimensional (containing many variables), which poses a challenge to examining every bivariate relationship. But what if EDA can be scaled to large datasets, data scientists could focus on promising leads so efforts are more effective. Graph matrices use small multiples or many small graphs, to summarize entire datasets on one canvas. We consider two types: *correlation matrix* clusters similarly behaving variables and a *Scatter Plot of Matrices* (SPLOM) to provide details for a select group of variables.

We illustrate graph matrices using the Federal Reserve Bank of Philadelphia's Real-Time Dataset for Macroeconomists (Federal Reserve Bank of Philadelphia 2019).[12] The dataset has 59 variables highly important metrics such as the number of housing starts in the US (`hstarts`) and the industrial production indices (`ipt` and `ipm`)

```
load("data/econ_vintage.Rda")
```

A **Correlation Matrix** produced using the `cor` function can be stylized to highlight and cluster similar continuous variables. Using the `corrplot` package, each Pearson correlation is color-coded to indicate

---

[12]Economic numbers are published as vintages, meaning that a given quarter's data will be revised each time a new release is made available. The data is "real-time" as the Philadelphia Fed archives the data based on each vintage so that the history of an estimate can be traced.

direction—blue is positive and red is negative—and the size of its circle indicates magnitude. As the correlation matrix is symmetrical around the diagonal, we only need the upper half of the correlation matrix (`type = "upper"`). Variables should be sorted into co-located related variables. One approach clusters variables together using a hierarchical clustering algorithm[13] by specifying `order = "hclust"`, which produces distinct pockets correlations that are easy to isolate for review. Alternatively, the matrix can be sorted by common direction of correlation using the angular order of the eigenvectors (not shown)—a more esthetically pleasing choice.

As shown in Figure 6.18, the correlation matrix exhibits five clusters distributed along the diagonal, each of which contains a cluster of positively correlated variables. The upper left cluster, for example, contains a set of economic consumption variables, such as producer price index for finished goods (`pppi`) and personal consumption of goods (`pcong`).

```
# Calculate correlations
pacman::p_load(corrplot)

# Produce correlation matrix that calculates pairwise relationships
cor_mat <- cor(econ[,-1], use = "pairwise.complete.obs")

# Render plot
corrplot(cor_mat, type = "upper",
         tl.col = "darkgrey", tl.cex = 0.5,
         order = "hclust")
```

A **Scatter Plot of Matrices** can evaluate the consumption variable cluster in more depth. Rather than reviewing each bivariate relationship one at a time, SPLOMs concisely summarize univariate and bivariate patterns in a graph matrix.

The `pair.panels` function (`psych` package) constructs a SPLOM (Figure 6.19) in which each plot matrix cell contains a Pearson correlation scaled by the size of the value, histogram or scatter plot. The histograms along the diagonals indicate that all but one variable have a clear peak. Most of the scatter plots below the diagonal indicate that all variables in the cluster are indeed positively correlated with a few exhibiting slight non-linearities (e.g., `nconhh` versus `ncongm`).

It is evident that SPLOMs generate a large amount of information to be packed onto one canvas, therefore, it is recommended using this tool on large screens to give enough space to render the patterns. If this is not an option, use SPLOMs in smaller subsets of up to 10 continuous variables. Ultimately, the use of any of these graph matrices requires context and inspiration—the goal of the data science project is to motivate how graphs should be interpreted.

```
# Load psych package
pacman::p_load(psych)

# Extract set cluster of variables
pos <- c("pcong", "pconhh", "nconhh", "ncongm", "pconhhm", "pcpi", "pppi")
abridged <- econ[, pos]

# Render plot
pairs.panels(abridged, method = "pearson")
```

6.7   Beyond this chapter

Most data projects start with an Exploratory Data Analysis. They present an opportunity to inform policies and analytical strategies using data, requiring careful attention to be paid to how insights are communicated. Flashy graphs can be easily created but can just as easily fail in communicating relevant information to an

---

[13]We will revisit hierarchical clustering in Chapter 11.

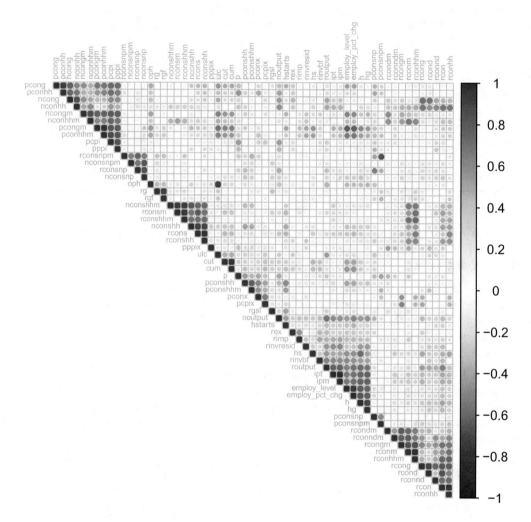

Figure 6.18: Color-coded correlation matrix sorted by clusters of correlated variables. Only the upper triangle of the correlation matrix is shown.

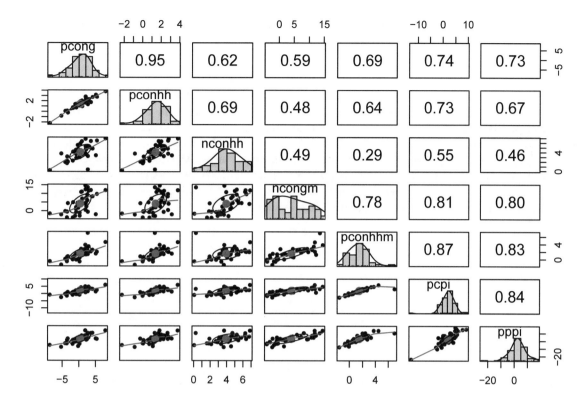

Figure 6.19: Scatter plot of matrices (SPLOM) showing univariate histograms, bivariate scatter plots, and Pearson correlation coefficients.

intended audience. Undertake EDA so you gain an intuitive understanding of the patterns and quirks of the data, but tailor your communication strategy to your intended audience.

Communication is key. When communicating to technical audiences, more information is better. Having detailed graphs in hand makes it easier to answer challenging questions and inspire confidence. Policy audiences, in contrast, like high-level stories—narratives that paint a picture. Too much information can overload policy audiences and even cause them to lose confidence in the analysis. *What is the appropriate balance to strike?* Keep the following best practices in mind:

Every plot should have a clear one-line takeaway. Data scientists should make it easy for their audiences to re-tell their stories with high fidelity. Hold yourself to the following standard: *only present graphs that lend themselves to a clear one-sentence takeaway.* If you need to think through the interpretation of your own graph, imagine the difficulty faced by a non-technical audience.

As for design, *simpler is better.* Graphs should be neat, minimalistic, and well-labeled. Consider using fewer colors, but avoid the default styles. Constructing your own custom styling will subtly signal effort and care put into each graph. Avoid busy visuals, however, that does not mean less volume. For a masterful treatment of information display, review Tufte (1993).

Not all information is relevant. What matters to a statistician will invariably differ from a policy advisor. There are only a few types of graphs that are intuitive to all audiences. Time series line graphs illustrate increasing or decreasing trends, while bar graphs show categorical distributions – both show *context*. But graphs about data quality such as time series outliers plots, quantile plots, kernel density plots, and correlation matrix heatmap will likely be pertinent to more data-savvy audiences like statisticians, economists, and other researchers.

Visualization. We recommend investing effort to master the `ggplot2` library to hone your craft and stylistic choices. Online reference materials contain a rich variety of examples (see https://ggplot2.tidyverse.org/ reference) that can be copied and executed in console. Experimenting with the varieties of `geom` functions will go a long way in building strong visualization muscle memory. While online references are useful, consider a more thoroughly worked collection of hands-on examples in *R Graphics Cookbook: Practical Recipes for Visualizing Data* by Winston Chang—a standard starting point for `ggplot2`.

As EDA can be the beginning and end of a data science project, a static graph can at times seem anti-climatic to audiences. In recent years, `R`'s core functionality has been extended to interactive web visualizations making it possible to rapidly produce sleek, tangible deliverables. Rather than writing code in HTML, CSS, and JavaScript, data scientists can write `R` code that outputs visualizations that can be rendered in web browsers. For example, the `flexdashboard` package makes it possible to layout a dashboard in RMarkdown (see https://rmarkdown.rstudio.com/flexdashboard), then render an HTML file. The various visual elements in the dashboard can be sourced from any number of *HTML Widgets* (see https://www.htmlwidgets.org/) such as the `leaflet` package for interactive web maps or the `plotly` package for a broad variety of graphs. For more on visualization products, see Chapter 14—Data Products.

Chapter 7

Regression Analysis

7.1 Measuring and predicting the preferences of society

Linear regression is the statistical workhorse of the social and physical sciences. At its core, regression is a fairly simple method that fits a line (or a curve) through a set of data points. But the technique can be much more than just fitting a line. It can be thought of as a way to quantify how a variable y (a *outcome*, *dependent*, or *target* variable) tends to change as *input variables* X (the *explanatory* or *independent* variables) change. In fact, many models we build follow the basic form:

$$y = f(X)$$

Or *y is a function of a set of input variables* X. Using this basic framework, policy researchers have been able to dissect some of the most challenging social issues that we face today.

One of the most useful properties of regression is its ability to "control" for different factors in order to distill the story about a single variable.[1] Housing prices, for example, are extraordinarily complex as they reflect a market's willingness to pay for an amalgam of housing and neighborhood characteristics. A house has attributes that add to and detract from its value, such as its size, materials, qualities, etc. Its proximity to amenities like transit, parks, and commercial districts also influence its value. Together, these layered effects make housing prices a perfect medium through which policy researchers can study how people value virtually anything.

Harrison and Rubinfeld (1978) famously analyzed the willingness to pay for clean air. By using aggregate data for the US Census tracts for the Boston metropolitan area, the study associated the median value (W) of owner-occupied housing units with the Nitrous Oxide concentration (NOX), household income (INC), and persons per dwelling unit (PDU). The contribution of each variable—known as a *coefficient*—reflects the effect of one additional unit of the input variable on the target variable. In the context of economics, a regression that associates the price of a good with its observable qualities are known as a hedonic price model. By applying a hedonic price model, the researchers found that households were willing to pay more for better air quality ($189 for one part per million), but higher income households were more willing to pay twice as much as lower income households.

$$W = -581 + 189\ \mathrm{NOX} + 12.4\ \mathrm{INC} - 119.8\ \mathrm{PDU}$$

The air quality study was observational—one in which the researchers did not have control over which subjects (units of analysis) are assigned to treatment and control groups. While these studies have less

---

[1]Some quantitative researchers may opt for the word "adjust" rather than "control". As we will later see, including a variety of factors in a regression *adjusts* the model to account for sources of information.

power to claim causality than experimental studies, the air quality study was influential in its time. It illustrated that society's values are engrained in housing prices, but require some clever math and study design to unlock it.[2] Since Harrison and Rubinfeld (1978), research techniques for policy have advanced as policy researchers have recognized that identifying a policy effect is quite challenging—an observed effect could be easily confounded by other unobserved factors. Thus, regression alone is not sufficient to claim causality.

In the late 1990s, Black (1999) took on one of the more challenging problems in public policy: *estimating the value of school quality*. In the United States, housing prices differ from one district to another, but quantifying how much is attributable to school quality was a speculative affair. To distill the value of school quality, Black turned to a quasi-experimental approach known as regression discontinuity design, which assumes that observations immediately on either side of a sharp threshold are no different than one another. Thus, for houses located immediately at a school district boundary, the difference in the housing prices can be attributed to the effect of school quality. Focusing on Massachusetts' housing market, Black estimated a hedonic price model to control for differences in housing characteristics, finding that within a 0.15-mile area around the district boundaries, parents are willing to pay 2.5house was given a 5not only serves as a milestone in policy research but showed the flexibility of regression techniques.

Regression can even go beyond quantitative storytelling. By shifting the focus from *why* to *who* and *what*, a regression model can *predict* what may come to pass with a reasonable degree of precision. Interestingly, housing price prediction is one of the best known and most visible examples of regression. Tech companies such as Zillow and Redfin predict sales prices to help inform prospective buyers, set expectations, and add utility to their real estate service offerings. While the underlying techniques are proprietary, early versions of Zillow's models have incorporated regression models (Thind (2016)), while recent developments rely on sophisticated machine learning techniques to improve housing price prediction (Humphries (2019)).

This chapter is the first of seven chapters on models that can surface insights from data. Regression is the go-to crowd pleaser—it often delivers reasonable results and is the gateway to more advanced techniques. We begin with simple linear regression, examining the case of associating a target variable with a single input variable. Then, we consider how regression can be used to test hypotheses in the data—a common application in policy research. As the world is a complex place, we expand into multiple regression and investigate how to address quirky patterns that tend to appear in data. We close the chapter with a discussion on selecting from a set of regression models and the pros and cons of more complex regression models.

7.2   Simple linear regression

Because linear regression focuses on fitting a line through a set of points, it is helpful to recall the formula of a line:

$$y = mx + b$$

where m gives the slope of the line and b gives the y-intercept.[3] In the land of regression, we tend to write the equation slightly differently:

$$y_i = \beta_0 + \beta_1 x_{i1}$$

where β_0 denotes the y intercept, and β_1 refers to the slope. The subscript i is an index that indicates the ith observation in the dataset. Figure 7.1 plots a line with an intercept of 5 and a slope of 0.5, i.e., $y = 5 + 0.5x$.[4]

For the moment, let's focus on the idea of *simple linear regression*, which relates one target variable (y) to one input variable (x). The pair of variables are plotted in Graph (A) of Figure 7.2. Our eyes can fit a line

---

[2]In fact, the underlying dataset is a standard "toy" dataset used for teaching statistics and regression.

[3]Recall the classic definition of slope: rise over run (or the change in y divided by the change in x).

[4]Regression jargon: You will generally hear/see "intercept" rather than "y-intercept".

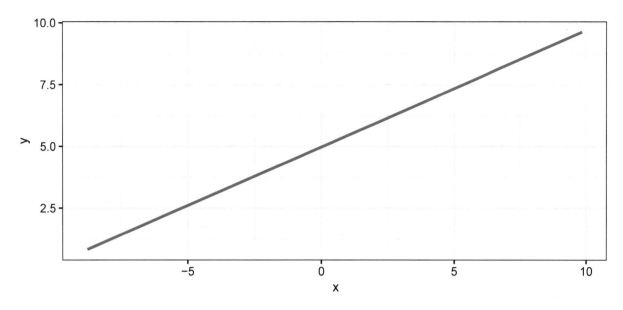

Figure 7.1: A line.

through the points. If we assume that y and x are related, we can describe that relationship by a simple line—i.e., an intercept and a slope. But in theory, there are infinite potential lines that could fit through a set of points. So how does linear regression pick one line out of all of these possibilities? Let's draw three for good measure (see Graph (B)).

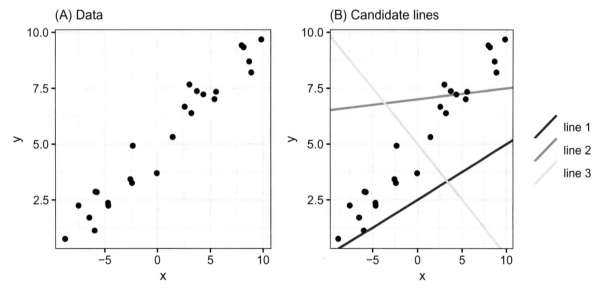

Figure 7.2: Simple linear regression summarizes the relationship of two variables y and x. Graph (A) shows a scatter plot of some data, whereas Graph (B) illustrates three candidate lines that could fit the data, albeit imperfectly.

How do we choose the *best* line for a given dataset? Line 3 looks to be clearly worse than the other two lines, but *why?* And how would we choose between Line 1 and Line 2? Is there a better line out there? All of these questions push us toward a bigger question: How can we formalize the concept of *fit* so that we can choose a line in a reasonable, repeatable, objective, and transparent manner?

The simple answer to this *big* question: *Use the data.* In fact, all methods we describe in the remainder of this text use the data to inform the best model. For any line that we draw, we can see how well it fits each point in the data but checking how *close* the line is to each point. If you think of the line as a prediction $\hat{y}_i$, then the distance between our proposed line and the actual data point gives us the *error* ε_i based upon the prediction. Let's draw each of these *errors* for Line 1 in Figure 7.3.

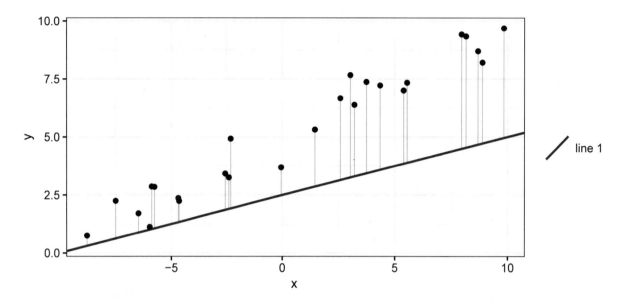

Figure 7.3: Diagnosing fit: Errors from line 1.

Each of the vertical gray lines connecting *line 1* to a black point from the dataset illustrates the error from predicting that data point with the line. In regression land, this error gets a special name: the *residual*. Formally, we define the residual as the difference between the observed value (y_i) and the predicted value $(\hat{y}_i)$, i.e.,

$$\varepsilon_i = y_i - \hat{y}_i$$

where ε_i refers to the residual. The predicted value for y (i.e., $\hat{y}_i$) comes from plugging in the associate value of x into the line's equation. For example, the equation underlying line 1 is $\hat{y} = 2.5 + 0.25x$. To produce a prediction for observation $(x = 0, y = 3.75)$, we plug in $x = 0$ (and get $\hat{y} = 2.5$).

Looking at the residuals from line 1, it seems clear that there's room for improvement. For instance, if the whole line were shifted up, we would reduce the size of all residuals. In addition, if we increased the slope—making it steeper—we would also reduce nearly all of the residuals. This type of reasoning is essentially what the machinery of linear regression works to choose the line that best fits a dataset: *these methods attempt to minimize some measure of error.*

7.2.1   Mean squared error

In the classical linear regression model, the goal is to minimize the *mean squared error*. Mean squared error (or *MSE*) is essentially what the name implies: it is the mean of the squared errors (the squared residuals), i.e.,

$$\text{Mean squared error (MSE)} = \frac{1}{n}\sum_{i=1}^{n}\varepsilon_i^2$$

Why do we square the error term? One might think that the objective should instead be to minimize the total (summed) error or take the mean of the residuals. However, there is some clever logic behind the MSE:

1. When we sum a positive and negative residual, they effectively cancel each other leaving little for a model to use to fit the line. You may have many positive errors and one huge negative error, yet your sum (or mean) could be very close to zero. To get around this issue, data scientists will occasionally use mean (or total) absolute error—replacing the square of the residual with the absolute value of the residual (i.e., $|\varepsilon|$).
2. Square errors turn out to be very mathematically tractable, meaning they are easy to work with in matrices. Summed absolute value does not have this advantage.
3. *MSE* is more sensitive to values farther from the center of the data (for the input variables). For some people, this sensitivity to outliers is a problem. For others, it is a selling point. We will provide more discussion later this chapter and let you decide.

For now, minimizing the MSE will be our *objective function.*

7.2.2   Ordinary least squares

The secret sauce of a linear regression model is it chooses the line of *best fit* by finding the line that minimizes the mean squared error. The method that gives this best-fitting line—from the perspective of minimizing the mean squared error– is called *ordinary least squares* (or *OLS* for short). OLS provides the line that minimizes the mean squared error for your dataset. Even if there is any other candidate line that looks promising, OLS will find the smaller MSE.[5] In Figure 7.4, OLS demonstrates its prowess over the data.

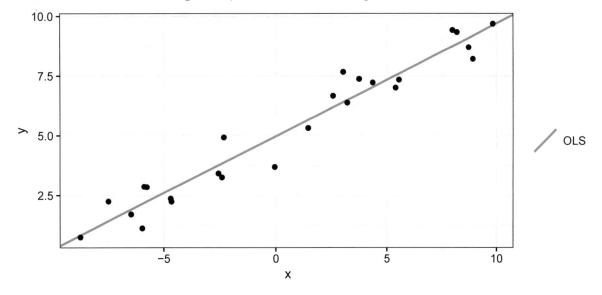

Figure 7.4: Diagnosing fit: OLS is best (for minimizing MSE).

While OLS does an impressive job at fitting a line through a dataset, how does OLS actually come up with this line? Where do the numbers come from?

While we will rarely need to calculate the OLS intercept and slope coefficient by hand, in the case of one target variable and one input variable, writing the math actually offers some insights into the magic behind OLS. But first, a bit of notation. It is standard notation to use Greek letters as the *unknown parameters* in our model. Given a simple linear regression, the model takes the form:

$$y_i = \beta_0 + \beta_1 x_{i,1} + \varepsilon_i$$

where y denotes our target variable (or dependent variable), β_0 references the (*unknown*) intercept, x refers to our single input variable (or independent variable), β_1 gives the slope of the line, and ε_i represents unobserved

---

[5]If the goal is to minimize the mean absolute error, then OLS is not best. Change the objective function and the algorithm will find the new winner.

errors in the model. We want to estimate β_0 and β_1, but we only observe y and x. Our task is to estimate β_0 and β_1 using only what we "know" (i.e., our data: y and x). Finally, we will refer to estimators by adding a hat to the parameter that the estimator estimates. For instance, $\hat{\beta}_1$ estimates β_0.

OLS estimates the slope coefficient β_1 using the correlation between y and x (recall that *correlation* measures the strength and direction of the linear relationship between two variables), the standard deviation of y, and the standard deviation of x.[6] Specifically,

$$\hat{\beta}_1 = \text{Cor}(x, y)\frac{s_y}{s_x}$$

which says that the OLS estimate for the slope coefficient comes from multiplying the correlation between x and y by the ratio of their standard deviations.

Let's think about the intuition of this formula for a minute. First, the coefficient is based upon the correlation between the two variables, which makes a lot of sense. When the correlation (the strength of the linear relationship) is strong (close to -1 or 1), the coefficient will be larger. When the correlation is small (a weak linear relationship with correlation near zero), the estimated coefficient moves closer to zero. Furthermore, the sign of the coefficient will match the sign of the correlation between y and x. This part also makes a lot of sense, as the correlation indicates whether y tends to increase or decrease as x increases—we want the estimated slope coefficient to match the correlation. Finally, the ratio of the standard deviations tells us how much y relative to x. If y covers a lot of distance, relative to a small distance in x, then we want a steeper line. If y barely changes (a small standard deviation) relative to the variation in x, then the slope should be approximately zero.

The formula for the OLS estimator of the intercept is simpler:

$$\hat{\beta}_0 = \overline{y} - \hat{\beta}_1\overline{x}$$

where $\overline{y}$ and $\overline{x}$ denote the mean of y and x, respectively.

One way to interpret this equation is that OLS will run its line (with slope β_1) through the point $(\overline{x}, \overline{y})$—a point that sits at the means of the two variables. We can calculate the y-intercept of the OLS regression line by starting at this point of the two means and following the line for $\overline{x}$ units until we hit the y-axis (Figure 7.5).

7.2.3   DIY: A simple hedonic model

DIYs are designed to be hands-on. To follow along, download the DIYs repository from Github (https://github.com/DataScienceForPublicPolicy/diys). The R Markdown file for this example is labeled **diy-ch07-hedonic.Rmd***.*

To put regression to the test, let's estimate a regression model. The basic linear regression function in R is `lm`, which stands for *linear model*. To run a regression with the `lm` function, we need to provide `lm` with a formula where the target variable on the left-hand side (LHS) is separated from the input variables on the right-hand side (RHS) by a *tilde* (i.e., "~""). If the variables are part of an `data.frame` or similar object, then we will also need to indicate for R the name of the object using the `data` argument in `lm`. For example,

```
lm(y ~ x, data = my_data)
```

will regress the variable `y` from the data frame `my_data` on the variable `x` (also from `my_data`). This phrase "regress y on x" is the standard way to refer to your *target* and *input* variables: *you regress your target variable*

---

[6]Also recall that the standard deviation is simply the square root of the variance. The (sample) variance of a variable $x = \{x_1, \ldots, x_n\}$ is $s_x^2 = \text{Var}(x) = \frac{1}{n-1}\sum_{i=1}^{n}(x_i - \overline{x})$ (where $\overline{x} = \frac{1}{n}\sum_{i=1}^{n}x_i$, i.e., the mean of x). Thus, the standard deviation of $x = s_x = \sqrt{\text{Var}(x)}$. The correlation between two variables x and y is given by $\text{Cor}(x, y) = \frac{\text{Cov}(x, y)}{s_x s_y}$, where $\text{Cov}(x, y) = \frac{\sum_{i=1}^{n}(x_i - \overline{x})(y_i - \overline{y})}{n-1}$ is the covariance of the two variables. It is also worth remembering that correlation is bounded between -1 (a strongly linear negative relationship) and 1 (strongly linear positive relationship), whereas covariance is unbounded.

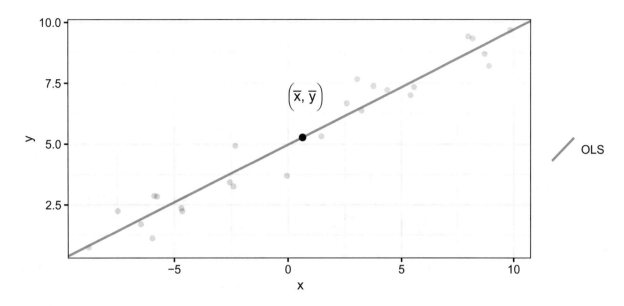

Figure 7.5: The OLS regression line passes through a point at the means of the two variables.

on your input variable(s). Note the terminology is different depending on the context. Social scientists use outcome and explanatory variable. Computer scientists use target and input feature. General data analysts use dependent variable and independent variable. But data scientists use target and input variables.

For this example, we use a sample of $n = 12687$ residential property sales from 2017 and 2018 in NYC.

```
# Load readr package
pacman::p_load(readr)

# Read in .csv of property sales
sale_df <- readr::read_csv("data/home_sales_nyc.csv")
```

While the dataset (`home_sales_nyc.csv`) contains a number of interesting variables, we focus on the relationship between a property's sales price (the aptly named `sale.price` variable in the dataset) and the property's size (measured in square feet, i.e., the variable `gross.square.feet`). Before running any analysis, let's plot the data to check if it is in good shape.[7]

```
ggplot(data = sale_df, aes(x = gross.square.feet, y = sale.price)) +
  geom_point(alpha = 0.15, size = 1.2, colour = "blue") +
  scale_x_continuous("Property size (gross square feet)", labels = scales::comma) +
  scale_y_continuous("Sale price (USD)", labels = scales::comma) +
  custom_theme
```

In NYC's expensive housing market, larger properties unsurprisingly cost more, but the relationship is not as strong as we would have expected. Let's run the regression. Our target variable is `sale.price` and our only input variable is `gross.square.feet`—both of which are contained in the data frame `sale_df`. We feed all of this information to the `lm` function, which we assign to an object named `reg_est`.

```
reg_est <- lm(sale.price ~ gross.square.feet, data = sale_df)
```

The `lm` function returns an `lm` class object, which is assign to an object named `reg_est`. If we enter the name of the object into the console, R will return the estimates for the intercept β_0 and the slope coefficient for `gross.square.feet`. We, however, are interested in the rich detail about virtually aspect of the regression, which can be accessed by applying the `summary` function to the `lm` object.

---

[7]For the `custom_theme` used to generate Figure 7.6, see Chapter 6.

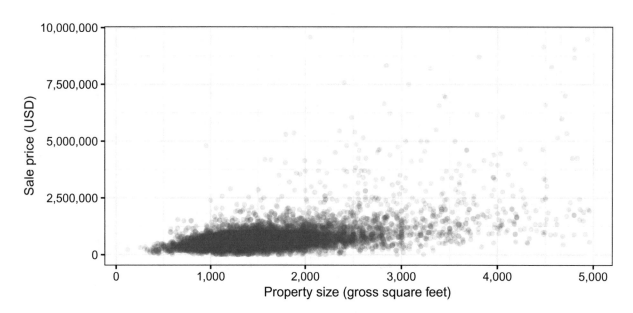

Figure 7.6: Residential property sales and property size, 2017–2018.

```
summary(reg_est)
```

```
Call:
lm(formula = sale.price ~ gross.square.feet, data = sale_df)

Residuals:
     Min       1Q    Median       3Q       Max
 -1700116  -212264    -44958   138638   8661923

Coefficients:
                    Estimate Std. Error t value Pr(>|t|)
(Intercept)       -42584.389  11534.260  -3.692 0.000223 ***
gross.square.feet    466.176      7.097  65.684  < 2e-16 ***
---
Signif. codes:  0 '***' 0.001 '**' 0.01 '*' 0.05 '.' 0.1 ' ' 1

Residual standard error: 463900 on 12666 degrees of freedom
Multiple R-squared:  0.2541,    Adjusted R-squared:  0.254
F-statistic:  4314 on 1 and 12666 DF,  p-value: < 2.2e-16
```

Figure 7.7: Regressionv output for simple hedonic model.

R prints a summary of the residuals (ε)—the error between the actual sales price and the linear model's fit. The size of the residuals is quite large indicating the model is lossy.

After the residuals, are gives the coefficient estimates and several other important statistics. Within the Coefficients section, R first prints the intercept (i.e., (Intercept)). Wait! We never asked for an intercept... or did we? Because intercepts are standard in statistics, econometrics, data science, and essentially every other empirical field, R defaults to using an intercept—it assumes you want an intercept. If you do not want an intercept, then you need to put a -1 in your lm equation.[8] You will often hear an intercept

---

[8]For example, you would write lm(sale.price ~ -1 + gross.square.feet, data = sale_df) if you wanted to estimate the

interpreted as the predicted value for the target variable when the input variable takes on a value of zero. However, this interpretation can often be misleading (Figure 7.7).

For instance, this interpretation would say that a property in NYC with approximately zero square feet would be worth negative 46,000 dollars. This interpretation is obviously absurd but what about the statistics makes this interpretation so absurd? The problem is that this interpretation of the intercept generally ignores the fact that we only observe data for a certain set of values for the target and input variables—we do not observe properties smaller than 120 square feet[9], and we do not observe any sales below $10,000.[10] Making predictions/interpretations outside of the range of your data is called *extrapolation*. Extrapolation is not necessarily bad—sometimes you have to make an educated guess about a data point that is unlike any data you have previously seen. However, when you extrapolate, you should be aware that you are moving outside of the range of data on which you fit your model—and the relationships between your target variable and input variable(s) may look very different in different ranges of the data.

Next: The estimate slope coefficient on `gross.square.feet`. The estimated coefficient on square feet is approximately 466.2. First, consider the *sign* of the coefficient. The fact that this coefficient is positive indicates that properties with more gross square feet (bigger properties) tend to sell for more money. No surprises here. The actual value of the coefficient tells us that *in this dataset* each additional gross square foot correlates with an increase in sale price of approximately 466 dollars.

But how confident should we be in this estimate? The next three columns provide us with tools for statistical inference—telling how precise our estimate is. The column labeled `Std. Error` gives the *standard error*, the column labeled `t value` gives the (Student's) t-statistic for testing whether the coefficient is different from zero, and the column `Pr(>|t|)` gives the p-value for a two-sided hypothesis test that tests whether there is statistically significant evidence that our estimated coefficient differs from zero (i.e., testing the null hypothesis $\beta_1 = 0$). Common practices within statistics and social sciences suggests that p-values below 0.05 indicate sufficient statistical evidence to reject the null hypothesis. Here, because our p-value is much less than 0.05, we would say that we find statistically significant evidence of a non-zero relationship between a property's sales price and its gross square feet at the 5% level.

7.3 Checking for linearity

Why might we care about the relationship between price and property size? We will often encounter normalized estimates like *price per square foot* that communicate the value of real estate within a market. It makes the assumption that the price per square foot is a constant dollar amount, but if the relationship between sale price and square footage is not exactly linear—meaning that larger properties cost more (or less) per square foot than smaller properties—then this measure is flawed. It is important to know whether this measure is indeed flawed as construction, real estate and many costly decisions are dependent on it.

How do we test this hypothesis?

If the relationship between a target and input bends even slightly, the relationship is non-linear—the price per square foot measure is not accurate. One simple trick to test for linearity is to include a quadratic term in the regression:

$$y_i = \beta_0 + \beta_1 x_{i,1} + \beta_2 x_{i,2}^2 + \varepsilon_i$$

In which the target (y_i) depends on the input x_i and its square x_i^2. Notice that we still use the term *linear regression* even when the relationship between the target and inputs are not strictly a straight line. The *linear* in *linear regression* refers to the fact that the right-hand side of the regression equation can be written as a *linear combination* of the unknown parameters (the βs) and the input variables. In other words, *linear* means that the βs are multiplied by the input variables.

---

relationship between the sale price and the property size without an intercept.

[9]Which is still amazingly small.

[10]When assembling these data, we restricted the sample to sales between $10,000 and $10,000,000 so as to avoid gifted properties and extreme values.

In the case of housing prices, we are mainly interested in whether there is evidence of a quadratic relationship between sale price and gross square feet by estimating the following regression:

$$\text{Price}_i = \beta_0 + \beta_1 \text{Footage}_{i,1} + \beta_2 \text{Footage}^2_{i,2} + \varepsilon_i$$

The key evidence is in the β_2 coefficient: *it needs to differ significantly from zero*. To operationalize this in R, our specification should include + I(gross.square.feet) to indicate that the regression should square the variable gross.square.feet. First, we estimate the regression model (saving the estimated model as reg_est2), then the review the results in Table 7.1).

A hypothesis test of the coefficient on the quadratic term (β_2 above) rejects the null hypothesis at the 5% level.[11] In practical terms, we have found statistically significant evidence that shows that sales prices are not constant across building area. Yet, the coefficient β_1 is a large negative value, while the size of β_2 is quite small. This may at first seem confounding, but it makes perfect sense once we consider that the *marginal* effect of building area is dependent on β_1, β_2, and the size of the building. Let's calculate the price per square foot at 1000 SF and 2000 SF:

$$\text{price per sf}_{sf=1000} = \frac{(-68.91 \times 1000) + (0.1284 \times 1000^2)}{1000} = 59.49$$

$$\text{price per sf}_{sf=2000} = \frac{(-68.91 \times 2000) + (0.1284 \times 2000^2)}{2000} = 187.89$$

In effect, the small but mighty β_2 coefficient overpowers the downward force of β_1. The estimate suggests that the price per square foot conditional on square feet is convex—it grows non-linearly with the size of the housing unit.

```
reg_est2 <- lm(sale.price ~ gross.square.feet + I(gross.square.feet^2),
               data = sale_df)


summary(reg_est2)
```

Table 7.1: Fitted linear model that regresses price per square foot on gross square feet and its square.

	Estimate	Std. Error	t value	Pr(>|t|)
(Intercept)	430405	24573	17.52	6.849e-68
gross.square.feet	-68.91	25.64	-2.688	0.007194
I(gross.square.feet^2)	0.1284	0.005921	21.69	1.835e-102

It can be hard to mentally visualize the regression. Let's plot housing prices by gross square feet, then overlay a simple regression (price regressed on square feet) and the quadratic model. The shape of the fitted regressions confirms the convex shape—which is satisfying from an analytical perspective, but more complicated for policy. The non-linearities in the price per square foot can prove challenging to communicate and operationalize policy.

In applied settings, the additional model complexity might not be worth the trouble. We must always ask if analytical splendor is meaningful in the grander scheme of public service. In this case, the quadratic model decreased the error by $8,400 (from $463,900 to $455,500)—only a slight increase in accuracy. While the quadratic model is *more* correct, choosing a *simpler* model can still communicate the gist. On the other hand, non-linearities may indicate that other influences are at play (Figure 7.8).

---

[11]The null hypothesis in this case is that $\beta_2 = 0$, which means there is not a quadratic relationship.

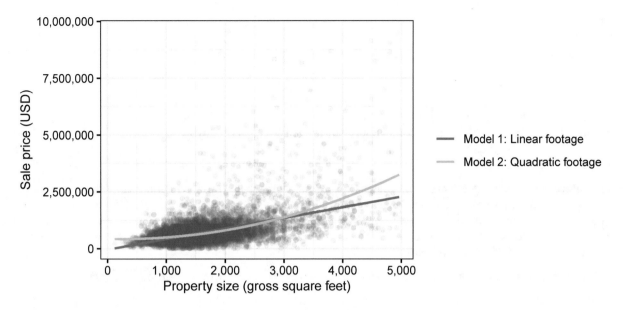

Figure 7.8: Comparing linear and quadratic models for sales price regressed on square feet.

7.4 Multiple regression

What if the regression were missing *something*? Maybe prices are not just about size, but maybe there are certain parts of NYC that are categorically more expensive than other parts of NYC. Maybe Manhattan is just more expensive than Queens. Or maybe apartments are different than non-apartments.

In many cases, several—likely *many*—factors predict the target variable. Whether the goal is to accurately predict the target variable (e.g., predicting property values) or explain the variation in the target variable (e.g., why some houses are worth much more than other houses), it is often helpful to bring in multiple input variables—a concept you will hear people refer to as **multiple regression**.

Aside from having more input (right-hand side) variables, multiple regression is pretty much the same thing as the simple-linear regression framework we just covered: you have a single variable as your target variable, and you estimate a set of parameters (β) by minimizing the sum of the squared residuals. The multiple linear regression model—now with k input variables—looks like

$$y_i = \beta_0 + \beta_1 x_{i,1} + \beta_2 x_{i,2} + \beta_3 x_{i,3} \cdots + \beta_{k-1} x_{i,k-1} + \beta_k x_{i,k} + \varepsilon_i$$

where y still represents our target variable, β_0 still gives the intercept, and ε still denotes the unexplained error in the model. We now have k input variables (x_1 through x_k) and $k + 1$ unknown parameters (β_0 through β_k).

Now that we've laid out our expanded framework for multiple linear regression, let's estimate expand our NYC property-value regression model.[12] Specifically, let's estimate the model

$$\text{Price}_i = \beta_0 + \beta_1 \text{Footage}_{i,1} + \beta_2 \text{Age}_{i,2} + \beta_3 \text{Year}_{i,3} + \varepsilon$$

which says that the sale price depends upon an intercept β_0,

```
lm(sale.price ~ gross.square.feet + age + sale.year,
    data = sale_df)
```

---

[12]We actually already ran a multiple regression model when we regressed the price on footage and footage squared.

```
##
## Call:
## lm(formula = sale.price ~ gross.square.feet + age + sale.year,
##     data = sale_df)
##
## Coefficients:
##         (Intercept)    gross.square.feet                      age
## -3.244745725486e+07    4.728285046508e+02    2.553170941379e+03
##            sale.year
##   1.596294907242e+04
```

The interpretation of the estimated coefficients changes a bit when you are in a multiple linear regression setting. We now interpret the coefficients as the relationship between y and x_i, *conditional* on the other variables. What do we mean by "*conditional*"? The estimate of the relationship comes from a model that includes a specific set of other variables. Thus, *conditional* means we've estimated the relationship between y and x_i *controlling for* the variables.

The multiple regression model that we have constructed makes several important assumptions:

$$\text{Price}_i = \beta_0 + \beta_1 \text{Footage}_{i,1} + \beta_2 \text{Age}_{i,2} + \beta_3 \text{Year}_{i,3} + \varepsilon_i$$

We assume that each square footage, age, and year of sale have a linear relationship with price. In other words, for every additional unit of the input variables, the expected change in price is dictated by the value of the coefficient estimates β_k. While this means we can tell the story behind each input variable, we also make the large assumption that the coefficient is constant across all values of the input. This linear assumption, as shown in Graph (A) in Figure 7.9, holds when data are distributed in a specific way. But real-world data is filled with curve balls as can be seen in Graphs (B) to (D). When linear regressions are not instructed to account for unfamiliar patterns, the models tend to do poorly. Fortunately, there are simple tricks to capture these other patterns.

7.4.1   Non-linearities

Regressions can systematically overshoot and undershoot the target in the presence of *non-linear relationships*. As shown in Graph (B) in Figure 7.9, the regression's prediction is a rigid straight line, while the observations weave in and out following a third-order polynomial relationship. To mold the model to the curve, we can approximate the shape of the relationship using a *Taylor expansion*, a concept in calculus that the shape of a curve can be approximated by some set of polynomials. To visual trick for identifying the order of the polynomial is to count the number of inflection points, then add one. If this were square footage, we can express the regression specification, which is a third-order polynomial:

$$\text{Price} = \beta_0 + \beta_1 \text{Footage} + \beta_2 \text{Footage}^2 + \beta_2 \text{Footage}^3 + \varepsilon$$

To apply this specification in R, we write this using the `poly` function that produces a matrix of k-degree polynomials:

```
# Third order polynomial term
reg_poly3 <- lm(sale.price ~ gross.square.feet + poly(gross.square.feet, 3),
                data = sale_df)
```

Adding polynomials can mitigate non-linear estimation biases, but also makes model evaluation more complicated. For one, just adding *any* polynomial might not improve model fit and generally, a more parsimonious specification is preferred. If we were to estimate the above specification, we would find that the additional polynomial is not statistically significant and only improved the R-squared by 0.01 relative to the second order. Interpretation of the coefficients is also complicated by the additional polynomial terms. To interpret

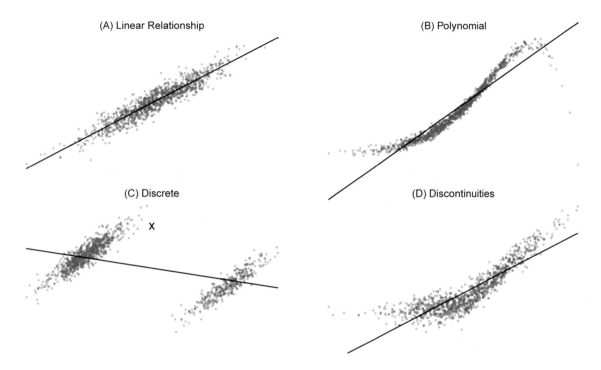

Figure 7.9: Examples of different types of relationships between the target (vertical axis) and input variables (horizontal axis) along with the linear regression prediction (solid line). Graph (A) illustrates a linear relationship, Graph (B) shows a polynomial relationship—a curve, Graph (C) shows the effect of a categorical variable, and Graph (D) depicts a sharp discontinuity or slope change at a threshold value.

the effect of *Footage*, we would need to account for the effects of all terms together. In short, try to keep your specification as simple as possible.

7.4.2 Discrete variables

The *presence of two or more subpopulations* in the data can create clusters within the target variable—it is multi-modal. As seen in Graph (C) in Figure 7.9, two groups appear as parallel groups of observations. The shape of the plot suggests that there is a positive relationship between x and y—approximately the same for each subpopulation (i.e., the same β for all), yet the regression erroneously suggests that the relationship is negative. In short, subpopulations are similar but different.

To rectify this situation, we need to find some way to distinguish between the parallel groups, and discrete variables hold the key. In a simplified housing market, let's suppose there are only two neighborhoods—one area is pricier than the other;

$$\text{Price} = \beta_0 + \beta_1 \text{ Footage} + \delta_1 \text{ Neighborhood} + \varepsilon$$

The new δ_1 parameter is the coefficient for a *dummy variable*, more commonly referred to as an *indicator variable*. Dummy variables are a binary vector that indicate whether a given observation belongs to a specific group, making it possible for a regression to distinguish between the two. In this case, if a house is in neighborhood P_1 (Neighborhood = 1) or neighborhood P_2 (Neighborhood = 0):

$$\text{Neighborhood} = \begin{bmatrix} P_1 \\ P_2 \\ P_1 \\ \vdots \\ P_2 \end{bmatrix} = \begin{bmatrix} 1 \\ 0 \\ 1 \\ \vdots \\ 0 \end{bmatrix}$$

We only need one vector to represent a two-class case as one district is measured relative to the other—the "zero" class is known as a *reference level*. For example if $\delta_1 = 1253$, then we can say that, holding all else constant, *houses in neighborhood P_1 are on average $1,253 higher than neighborhood P_2*. In effect, the value of δ_1 modifies the intercept β_0 in cases when Neighborhood = 1. In R, the `neighborhood` variable should either be converted into a `factor` *or* into a binary variable of zeros and ones:

```
reg_poly3 <- lm(sale.price ~ gross.square.feet + neighborhood,
                data = sale_df)
```

Categorical variables. Of course, there are multiple neighborhoods in a city, and a city as large as New York has hundreds. The `neighborhood` variable becomes a categorical with many classes. To flag each city for a regression, each of the – essentially a matrix of binary values:

$$\text{Neighorhood} = \begin{bmatrix} P_1 \\ P_{k-1} \\ P_1 \\ \vdots \\ P_k \end{bmatrix} = \begin{bmatrix} 1 & 0 & 0 & \dots & 0 \\ 0 & 0 & 0 & \dots & 1 \\ 1 & 0 & 0 & \dots & 0 \\ \vdots & \vdots & \vdots & \ddots & \vdots \\ 0 & 0 & 0 & \dots & 0 \end{bmatrix}$$

If we follow the logic in translating the vector, the $k - 1$ unique values in `neighborhood` are represented as columns in the matrix, but we leave out a category as a reference level against which we compare. This matrix expansion is referred to as a *dummy variable matrix* or *indicator matrix* in more formal mathematical terms. In computer science circles, representing categorical variables in this format is known as *one hot encoding*. Thus, when we add a categorical variable to the regression:

$$\text{Price} = \beta_0 + \beta_1 \, \text{Footage} + \delta \text{Neighborhood} + \varepsilon$$

It is equivalent to expanding the `neighborhood` variable as a set of dummy variables:

$$\text{Price} = \beta_0 + \beta_1 \, \text{Footage} + \delta_1 \, \text{Neighborhood}_1 + \dots + \delta_{k-1} \, \text{Neighborhood}_{k-1} + \varepsilon$$

Similar to a simple binary variable, the `neighborhood` variable should either be converted into a `factor` *or* expanded into a dummy variable matrix using `model.matrix`:

```
# Option 1: Categorical assuming factors
reg_cat <- lm(sale.price ~ gross.square.feet + neighborhood,
              data = sale_df)

# Option 2: Create dummy variable matrix, thenregress
# New data frame with price, footage and neighborhood
new_df <- data.frame(sale_df[, c("sale.price", "gross.square.feet")],
                     model.matrix(~neighborhood + 0, data = sale_df))

# Regress on all variables in new data frame
reg_cat <- lm(sale.price ~ ., data = new_df)
```

This is a (dire) warning: not all variables that look like numbers are actually numbers. In fact, many are not cardinal at all. A number of categorical variables are *coded* as numbers, such as zipcodes, area codes, interstates, and serial numbers. A particularly egregious case is that of the North American Industry Classification System (NAICS) codes—a way to classify companies by what they do at varying levels of detail. For example, the aircraft manufacturer *Boeing Co* has a six-digit NAICS code of *336411* ("Aircraft Manufacturing"), which is within sector *33* ("Manufacturing"). *McKinsey and Company* has a NAICS code of *541611* ("Administrative Management and General Management Consulting Services") within sector *54* ("Professional, Scientific, and Technical Services"). The authors have encountered one too many instances in which experienced analysts have treated NAICS as numeric values (assumes a linear relationship with the target), when in fact, they are categorical. Fight the temptation to treat all numbers as numeric. It helps to think about your data to avoid rather embarrassing moments.

7.4.3 Discontinuities

A sudden event or paradigm shift can cause a *discontinuity* in the data. Discontinuities usually appear in one of two forms: *step change* and *slope change*. As seen in Figure 7.10, a step change is a sharp vertical shift in the level of the data at a threshold whereas a slope change is a rotation of the trend at a threshold. Failure to detect a discontinuity will bias regression results. In some fields such as economics and epidemiology, discontinuities are highly prized for causal inference. As we will later see in Chapter 8, if the causal of the discontinuity is known, then we can derive an estimate of the effect size by examining observations immediately around the threshold.

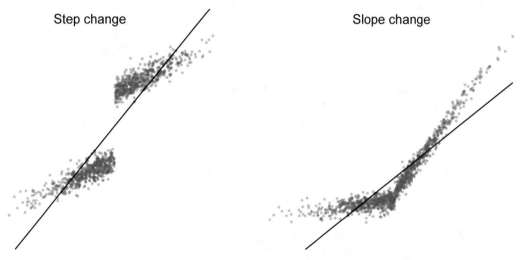

Figure 7.10: Two types of discontinuities: step change and slope change.

Accounting for a step change is essentially involves a dummy variable in the regression specification that indicates which records occur before (0) and after (1) the identified threshold. For example, a new Central Business District (CBD) in a quiet town could lead to a change in housing prices. If prices plotted over time showed a sharp step change, we could estimate a simple regression where we measure the relationship between sales price and time (a time trend variable is the date in numeric form), capturing the natural growth in a housing market. Then, we add a dummy variable (`Event`) to indicate which sales occurred before and after the creation of the CBD. The coefficient β_2 is the estimate on the step change—the effect of a new CBD, if any.

$$\text{Price} = \beta_0 + \beta_1 \text{Time} + \beta_2 \text{Event} + \varepsilon$$

Accounting for a slope change is slightly more complex. In addition to each `Time` and `Event`, an *interaction* term combines the effects of the dummy variable with the time trend. When we include the term Time×Event, the effect on the regression is an additional *modifier to the coefficient on time when Event* $= 1$, making time a piecewise function known as a *spline*:

$$\text{Price} = \beta_0 + \beta_1 \text{Time} + \beta_2 \text{Event} + \beta_3 (\text{Time} \times \text{Event}) + \varepsilon$$

In R, we can implement interactions fairly easily. In the example below, we first simulate an `event` with a time threshold (March 1, 2018) and add a linear slope change to the sales price (+\$5,000 per day). With the pieces set up, we interact `event` with `time.index` with an asterisk in the regression specification. Notice that we did not need to write out each of the terms as the `lm` function assumes that all three terms are needed.

```
# Create an event
sale_df$event <- 1 * (sale_df$sale.date > as.Date("2018-03-01"))
event_flag <- which(sale_df$event == 1)

# Add a simulated slope change
# Slope change of 5000 per day
new_price <- sale_df$sale.price[event_flag] + (sale_df$time.index[event_flag])* 5000
sale_df$sale.price[event_flag] <- new_price

# Specification with interaction
reg_inter <- lm(sale.price ~ event * time.index, data = sale_df)
```

7.4.4 Measures of model fitness

Measures of model fitness help identify the best model from alternatives. There are, however, a variety of different ways of measuring accuracy, each of which is motivated quite differently.

R-squared. Practitioners of public policy and government are most accustomed to metrics that are more precise and bounded, such as *R-squared* (R^2), sometimes referred to as the coefficient of determination. The values of R^2 are bound by between 0 and 1 (or 0% and 100%), making it easy to conceptualize how much of the total variance of the target variable is explained by the predictors in the model. More formally, the measure is given as:

$$R^2 = 1 - \frac{RSS}{TSS}$$

where

- *Residual Sum of Squares* (RSS) is $\sum_i^n \varepsilon_i^2$ where $\varepsilon_i = y_i - \hat{y}_i$
- *Total Sum of Squares* (TSS) is $\sum_i^n (y_i - \bar{y})^2$—or the variance of the target y

A value of $R^2 = 0$ indicates that a regression model does not explain the variation in the target y, while $R^2 = 1$ indicates a perfect fit. While this is convenient, more input variables can increase the R^2 – *just being present is not enough*. Thus, the *adjusted* R^2 penalizes model performance by requiring that each additional variable contribute more to a model's performance than just being included.[13] This concept is operationalized by accounting for the sample size n and the number of input variables k—more variables will dilute the R^2:

$$R^2_{\text{adjusted}} = 1 - \frac{(1 - R^2)(n - 1)}{n - k - 1}$$

When we compare the unadjusted and adjusted R^2, the former will be larger than the latter.

Error. The residuals from a regression can be summarized as various kinds of error that communicate different properties. The most common is the *Root Mean Square Error* (RMSE), or the square root of *MSE*. The measure is defined as:

$$\text{RMSE} = \sqrt{\frac{\sum_{i=1}^n (\hat{y}_i - y_i)^2}{n}}$$

RMSE emphasizes larger errors by squaring the errors, allowing us to avoid models that have larger misses. Furthermore, RMSE is more interpretable than MSE as it places error in the same scale as the target y. For example, a RMSE = 0.05 in one model is small relative to a $RMSE = 164$.

---

[13]While we have included Adjusted R^2 in this text and R reports it in regression outputs, friends do not let friends use Adjusted R^2. Mills and Prasad (1992) shows that the Adjusted R^2 is a poor measure for model selection.

Information Criteria measure the quality of a model, assuming there is a theoretical *true model* and any estimated model will lose some of the information content. Two information criteria see frequent use in the field: *Akaike Information Criterion (AIC)* and the *Bayesian Information Criterion (BIC, aka Schwarz Criterion)*. For both AIC and BIC, *the objective is to identify the model that produces that smallest value.*

AIC (Akaike (1974)), and its small sample counterpart AICc (Hurvich and Tsai (1993)), much like many other criteria, evaluates the log-likelihood of the data given the parameter estimates, $\hat{\theta}^{MLE}$, and uses a correction term $(2(k + 1))$ to dissuade models with unnecessary variables. For AIC, the penalty is simply an accounting of the parameters estimated in the model.

$$\text{AIC} = -2log(p(y|\hat{\theta}^{MLE})) + 2(k + 1)$$

AICc adds a secondary penalty which is a function of both k and N, to correct for small samples. This second penalty disappears asymptotically as the sample size grows.

$$\text{AICc} = -2log(p(y|\hat{\theta}^{MLE})) + 2(k + 1) + \left(\frac{2(k + 1)^2 + 2(k + 1)}{(N - (k + 1) - 1)} \right)$$

It is important to note the primary criticism of AIC is its *permissiveness of model expansion*—when the number of variables increases relative to the sample size, AIC becomes negatively biased. This leads to consistent over parameterization which is corrected for small sample sizes in AICc.

BIC is a measure of information content which as a parameter penalty which grows as the sample size increases (Schwarz and others (1978b)). This often leads to more parsimonious model selection than the alternatives (Raftery (1995)). BIC for any model can be calculated as,

$$\text{BIC} = -2log(p(y|\hat{\theta}^{MLE})) + klog(n)$$

This criterion is focused primarily upon selecting the model that is most probable and, to that end, the calculation for BIC is a rough approximation of the marginal probability of the data given the model.

Which should be used for model selection? The choice of which to use boils down to how much you value communicating the result to stakeholders (percentages are better) and if you want to manage large errors (squared errors are better). It turns out the two squared error measures (R^2 and RMSE) are two sides of the same coin: *maximizing R^2 is equivalent to minimizing RMSE has given a greater penalty on large errors.* Likewise, MAE and MAPE are yield similar results as one another giving equal weight to all errors, except MAPE is best for larger positive values.

7.4.5 DIY: Choosing between models

DIYs are designed to be hands-on. To follow along, download the DIYs repository from Github (https://github.com/DataScienceForPublicPolicy/diys). The R Markdown file for this example is labeled `diy-ch07-model-selection.Rmd`.

Let's move beyond the rules and diagnostics of regression and put it into action using the house price data. Our goal in this DIY is to estimate a set of regressions, evaluate the pros and cons of each, and select the "best" specification. The challenge, however, is that the best model might be different for each use case. The model that provides the clearest story may not necessarily be the one that produces the best model fit. We investigate these trade-offs in this section.

Let's estimate four regressions where each successive specification adds more variables and complex elements:

- Model 1 (`mod1`) regresses sales prices and building area;
- Model 2 (`mod2`) adds borough as a categorical variable;
- Model 3 (`mod3`) incorporates an interaction to estimate borough-specific slopes for building area;
- Model 4 (`mod4`) adds land area.

By gradually build up the model, we can show the relative effects and contributions of each variable giving a sense of what is important for interpretation and what is important in capturing the variation in the data.

```
# Simple regression
mod1 <- lm(sale.price ~ gross.square.feet,
           data = sale_df)

# With borough
mod2 <- lm(sale.price ~ gross.square.feet + factor(borough),
           data = sale_df)

# Interaction
mod3 <- lm(sale.price ~ gross.square.feet*factor(borough),
           data = sale_df)

# With Additional variables
mod4 <- lm(sale.price ~ gross.square.feet*factor(borough) + land.square.feet + age,
           data = sale_df)
```

How well does each model perform in practice? We can easily see performance when overlaying the fitted model on the actual series. In Figure 7.11, we plot the actual (black points) and predicted sale prices (red points) by gross square footage. Each graph's label contains the BIC score, which can be extracted from a model object by specifying:

```
# Obtain BIC
BIC(model_object)

# Obtain AIC
AIC(model_object)
```

In an ideal model, the red and black points would perfectly overlap. Model 1, however, is overly simplistic—it misses most of the point cloud. Model 2 is an improvement—by accounting for the naturally different price points by borough. Model 3's interactions between gross square footage and borough show that the price curve is different in each part of the city; However, the predictions still resemble five straight lines. By adding land area and building age in Model 4, the predictions cover more area. From visual inspection, it is clear that Model 4 offers the best performance. Furthermore, BIC is also at a minimum with Model 4 indicating that it is the best relative choice.

If we are interested in *predicting* the values is Model 4 good enough? There are specific guidelines for how to construct accurate predictions, which we will cover in the next chapter. However, using the models that we have before us, we can obtain the best-case scenario for model performance by applying **summary** to the model object **mod4**. The residual standard error ($RMSE$) is $RMSE = 384,415$. On the surface, this might not mean much without context, but when considering that the average housing price in the sample is $\bar{y} = 665,012$, the value is quite large. In practice, if Model 4 were applied in the wild and produced a prediction of $\hat{y} = 300,000$, the confidence interval on $\hat{y}$ would be include zero! In other words, the price could be anyone's guess. Good predictions have small intervals—reducing the uncertainty around the prediction.

Ultimately, our example above is illustrative of a *fitted* line—quite different than a *predicted* line. The former shows how the model behaves when given the data it was estimated on, whereas the latter provides a sense of out-of-sample performance. In the next chapter, we distinguish between estimation, causal inference, and prediction.

Despite the challenges with predicting housing price, we can still use the regression results to interpret patterns in the housing market. As we can see in Table 7.2[14], the size of the regression coefficients change quite dramatically as new variables and interactions are added. The game is to find some combination of input variables that produce decent model performance (R-squared, BIC, AIC, RMSE) and with stable

---

[14]We used the **Stargazer** package to co-locate the four model into a regression table much like in academic papers.

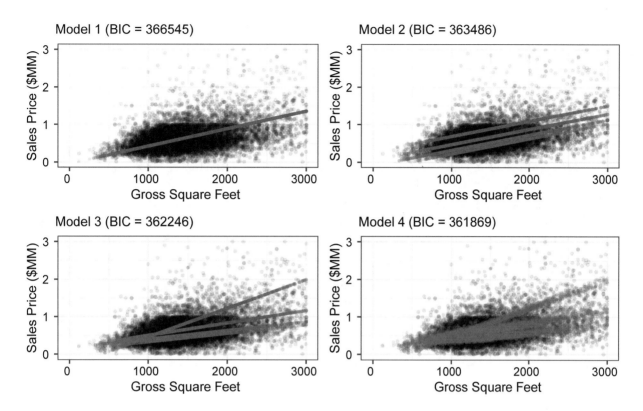

Figure 7.11: Comparison of the four specifications plotted against gross square footage. Black points represent the actual sales prices, while red points represented the fitted model.

coefficients (i.e., the values do not fluctuate wildly with additional variables). Here, we point out a few key insights:

Model 1 versus Model 2. The initial estimate for Gross SF is +466.18 in Model 1 but falls to 406.99 in Model 2 with the inclusion of Boroughs. When we combine this insight with the sharp increase in the R^2 (0.25 to 0.42), we can conclude that the first model was underfit—it did capture enough of the variability in housing prices.

Model 2. The Borough variables are estimated relative to Borough 1 (Manhattan)—one of the most expensive areas in the United States. The negative coefficients thus indicate that housing in other boroughs are, on average, worth $2 million less than units in Manhattan. Interestingly, the Constant holds the natural underlying value of property in Manhattan and suggests that even when the property has zero area, the base price is +$2.8 million. Boroughs are relatively coarse geographic units. Categorical variables with many levels (e.g., neighborhoods, zipcodes) can provide a more precise local estimate on price. But if *cells* in the data are too small (i.e., number of records per level in the categorical variable), predictions can become less stable and more noisy—there simply is not enough information for the regression to learn from.

Model 3. The introduction of interactions causes a large shift in the coefficients. The interpretation of Gross SF is different—it is the price in Manhattan, which explains the more than twofold increase in the value. To obtain the Gross SF for Boroughs 2 to 5, we need simply add the interacted coefficients. For example, to obtain Gross SF for Borough 2 from Model 3:

$$\theta_2 = \theta_{\text{GrossSF}} + \theta_{\text{GrossSF} \times \text{Borough 2}}$$

In other words, by adding $1,043.44$, and -817.22, the price per square foot is $226.22, which is markedly less than the price in Manhattan. The additional complexity increased the R^2 by 0.05—not as large as the change from Model 1 to 2, but is nonetheless useful.

Finally, in Model 4, we only see a modest improvement in model fit with the addition of the land area and building age variables. Unlike the other models, the size of the coefficients only shifts slightly, suggesting that the additional variables have limited effect. Each regression helped to inform a statistical narrative about housing prices in New York: *There is an allure of some boroughs (higher relative price of Manhattan). Space is relatively scarce and comes at a premium (price per square foot).*

7.4.6   DIY: Housing prices over time

DIYs are designed to be hands-on. To follow along, download the DIYs repository from Github (https://github.com/DataScienceForPublicPolicy/diys). The R Markdown file for this example is labeled `diy-ch07-time-series.Rmd`.

Time series data are the de facto choice for measuring and tracking trends over time. Most macroeconomic data, for example, are time series that summarize an economy's performance and provide a way to gage how global events influence the markets. In the United Kingdom, housing prices are closely monitored to understand affordability and the health of the economy. Estimated by the UK's Office of National Statistics using data from HM Land Registry, the UK's Housing Price Index (HPI) measures the price level of real estate by applying a hedonic model to distill the value of housing holding all else constant (HM Land Registry Public Data (2019)). But when the price levels are strung together as a time series, it is easy to see how the market evolves over time. As seen in Figure 7.12, the housing price index was relatively flat until 2013, then prices began to sharply increase – there is a clear upward *trend*. The price gently fluctuates over the course of each year following a regular, fairly predictable pattern—this is a sign of *seasonality*.

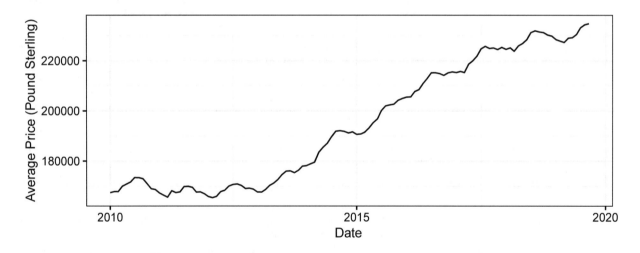

Figure 7.12: UK Housing Price Index (Jan 2010 to Oct 2019).

In this DIY, we illustrate how to tame a time series and extract insights seasonal patterns and construct a simple forecast. When working with time series, there are special rules that we need to follow. We thus begin by describing the *components* of a time series and the assumption of *stationarity*, then we dive into approaches modeling strategies. To start, let's import the HPI from `ukhpi.csv`.

```
# Load package
pacman::p_load(forecast,ggplot2)

# Load file
uk <- readr::read_csv("data/ukhpi.csv")

# Set data as a time series object
train <- ts(uk$avg.price,start = c(2010,1),freq = 12)
```

Table 7.2: Regression coefficient table for four specifications.

	Dependent variable:			
	Sales Price			
	Area	+ Borough	+ Interaction	+ Other Variables
	(1)	(2)	(3)	(4)
Gross SF	466.2***	406.0***	1,043.4***	1,032.6***
	(7.1)	(6.4)	(57.4)	(56.5)
Borough 2		−2,998,287.9***	−637,746.3***	−649,531.7***
		(61,499.6)	(198,071.6)	(195,070.0)
Borough 3		−2,530,787.4***	−1,050,643.6***	−1,096,470.7***
		(60,818.4)	(195,726.7)	(192,775.2)
Borough 4		−2,748,294.6***	−571,516.0***	−608,897.1***
		(60,608.0)	(195,120.9)	(192,157.4)
Borough 5		−2,902,616.7***	−567,646.9***	−596,639.5***
		(60,721.5)	(195,327.2)	(192,495.5)
Land SF				37.2***
				(1.9)
Building Age				−706.6***
				(153.6)
Gross SF * Borough 2			−817.2***	−866.3***
			(61.4)	(60.5)
Gross SF * Borough 3			−277.0***	−290.6***
			(58.7)	(57.8)
Gross SF * Borough 4			−712.3***	−761.1***
			(58.3)	(57.4)
Gross SF * Borough 5			−814.2***	−889.4***
			(58.4)	(57.7)
Constant	−42,584.4***	2,811,606.0***	743,371.4***	801,452.2***
	(11,534.3)	(62,840.6)	(194,493.8)	(192,296.1)
Observations	12,668	12,668	12,668	12,668
R^2	0.3	0.4	0.5	0.5
Adjusted R^2	0.3	0.4	0.5	0.5
Residual Std. Error	463,876.1	410,579.7	390,445.8	384,415.2
F Statistic	4,314.3***	1,802.6***	1,256.6***	1,097.3***

Note: $^{*}p<0.1$; $^{**}p<0.05$; $^{***}p<0.01$

Some ground rules. Recall the brief aside about Chapter 6. Any time series can be described in terms of three components:

$$Y = T + S + R$$

Where the T is a *trend*, S is *seasonality*, and R is the *noise* component—the leftover information that is not explained by the trend or seasonality. Each of these components holds a piece of the story of HPI.

Statisticians had discovered in the early days of time series that the data needs to be *stationary* in order for analyses to be generalizable. When a series drifts up or down unpredictably, it becomes a moving target—its mean and variance will change with the trend over time and its autocorrelation structure (how a series is correlated with its past values) could also change—it is *non-stationary*. A *stationary* time series exudes stability—it has a constant mean and variance and a stable autocorrelation structure over time.

How do we identify a stationary series in practice? First, we should introduce the mechanism for stabilizing a time series: *differencing*, or subtracting the previous period from the current ($y'_t = y_t - y_{t-1}$). Any time series object can be differenced in R applying the `diff` function to a vector or time series object.[15] Differencing, in effect, calculates a period-to-period change. Such a simple idea can collapse time series with a trend into an oscillating pattern with a mean of zero—it is quite powerful.

The question then is whether a series requires differencing. Fortunately, statisticians and economists have devised a battery of tests that build evidence that a series is non-stationary (or the contrary):

- *Graph the series.* Stationary time series do not exhibit a trend. HPI, in contrast, clearly grows over time, indicating possible non-stationarity.
- *Calculate σ.* If we take the difference of a series and find its standard deviation σ falls by half, the time series may be non-stationary. If we were to difference an already stationary series, σ could actually *increase*. The σ_{HPI} falls from $24{,}411$ to $1{,}242$ when differenced—more evidence of non-stationarity.
- *Correlograms* plot a series' autocorrelation function (ACF)—the correlation between a series and its past values known as *lags*. In Figure 7.13, the first bar measures the relationship between the current period (t) and last period ($t-1$)—the correlation between y_t and y_{t-1}. The second bar is the correlation between y_t and y_{t-2}, and so on. A stationary series' ACF quickly falls to zero with each subsequent lag. However, the ACF in Graph (A) falls slowly—-ever-so-slightly over 24 lags—adding to the evidence of non-stationarity. In contrast, the correlogram of the differenced series (Graph (B)) shows a seasonal ACF pattern—a much better behaved series than its raw version.

In short, the HPI is non-stationary and should be differenced. There are many other tests that lend more support, such as the Augmented Dickey Fuller Test (ADF test) and other Unit Root tests.

```
# (A) Calculate ACF for unemployment
acf_level <- autoplot(acf(train))


# (B) Calculate ACF for first difference
acf_diff <- autoplot(acf(diff(train)))


# Plot
grid.arrange(acf_level, acf_diff, ncol = 2)
```

Quantifying the seasonal patterns. Time series models can be quite complex with complex names like Autoregressive Integrated Moving Average (ARIMA) and Long Short-Term Memory (LSTM) algorithms. There are, however, simpler and deterministic approaches that make use of an ordinary regression framework. With a differenced time series, we can model seasonality through a *seasonal dummy* model: for a monthly series, we include 11 dummies, allowing the 12th month to be absorbed by the constant β_0:

$$y_t = \beta_0 + \delta_1 m_{t,1} + ... + \delta_{11} m_{t,11} + \varepsilon_t$$

---

[15]When differencing a series, we automatically lose one observation. For example, making a difference ($d = 1$) leaves $n - 1$ observations.

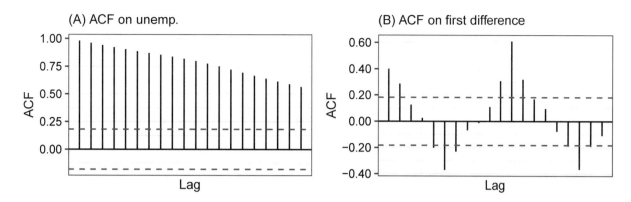

Figure 7.13: Correlograms of the unemployment series, unadjusted and differenced.

Notice that we have replaced the record index i with t for time index. Seasonal coefficients are the *average* of each month's price change and should be interpreted *relative* to the reference month. To put this into action, we need to first construct a differenced time series (`diff.series`), which then can be regressed on each month dummy.[16]

```
# Difference the series
uk$diff.series <- c(NA, diff(uk$avg.price))

# Run linear model
mod_diff <- lm(diff.series ~ factor(month), data = uk)
```

When we apply `summary` to the regression object, we can infer when housing prices are more expensive. Figure 7.14 visualizes the regression coefficients and their confidence intervals as a dot-whisker plot. Coefficients whiskers that cross the dashed line are not likely statistically significant at the 5% level. The plots indicate that peak prices occur in July (month 8), fetching on average £3,011 (+/- 737.2) more than the reference period, January. Interestingly, the patterns in September and October are less clear—the coefficients are not statistically significant at the 5% level. Thus, we can infer that houses sold during the summer fetch higher prices.

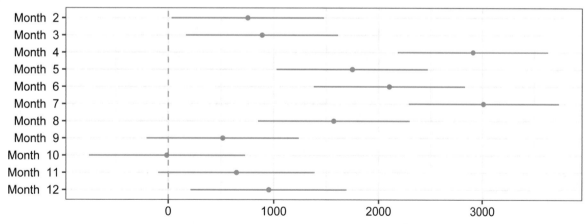

Figure 7.14: Dot-whisker plots visualize the coefficients and their confidence interval.

Mechanics of a forecast. Time series models can also be used to forecast the future price of housing. While seasonal dummies are easy to interpret, forecasting requires some data gymnastics.[17] The ingredients

---

[16]Because differencing yields $n - 1$ records, we add an `NA` value at the beginning of the differenced series so the transformed series can be assigned to a column in `uk`.

[17]There are simple functions for estimating and producing forecasts, but we have chosen to show the underlying process to

for constructing a seasonal dummy forecast are a *training* dataset that is used to estimate a model as well as an out-of-sample test dataset to apply the model.

First, we construct the out-of-sample set for the next 12 months (time periods 118–129) that contains the same variable names and data types as were used to estimate the regression. We add these 12 new periods to an abridged version of the uk data frame.

```
# Construct a forecast set
temp <- data.frame(time.index = 118:129,
                    month = 1:12,
                    avg.price = NA,
                    diff.series = NA)


# Add the new periods to the future data
fcst <- rbind(uk[, c("time.index", "month", "avg.price", "diff.series")],
              temp)
```

Often times in policy, practitioners will focus on the expected value $\hat{y}_t$ (the forecast line) while ignoring the forecast uncertainty. In ignoring the uncertainty, decisions are made without considering the risks. In fact, for every step in the forecast horizon (periods into the future), the uncertainty grows as no new information is being integrated into the forecast. But, we can make a reasonable guess that the forecast can fall within a *prediction interval*. For this reason, data scientists include both the lower and upper bounds of the prediction interval for context and to help temper expectations. Using the predict function, we *score* the fcst data frame and extract the forecast and its 95% prediction interval, then add these columns to the fcst data frame.

```
# Forecast the first difference
fcst <- cbind(fcst,
predict(mod_diff, fcst, interval = "prediction"))
```

The forecast on growth $\hat{y}'_t$ needs to be converted into an estimate of the HPI *level*. We cannot simply add the monthly growths, but instead need to "grow" the forecast ($\hat{y}'_t$) from a known price level (y_{t-1}). To do this, the price level for the first out-of-sample period ($t = 118$) can be obtained by adding the $y_{t-1} + \hat{y}'_t$, or taking the sum of the known price level in $t = 117$ and the forecasted growth for $t = 118$. For all subsequent periods, we build the forecast off of the forecasted level in the prior period:

$$\hat{y}_t = \begin{cases} y_{t-1} + \hat{y}'_t & \text{if } t = 118 \\ \hat{y}_{t-1} + \hat{y}'_t & \text{if } t > 118 \end{cases}$$

Implementing this calculation involves simple algebra and a for loop.

```
# Create yhats
fcst$yhat.level <- fcst$yhat.lower <- fcst$yhat.upper <-  NA

# Set levels for t = -1
fcst$yhat.level[117] <- fcst$yhat.lower[117] <- fcst$yhat.upper[117] <- fcst$avg.price[117]

# Recursively add the first difference forecast to the prior period's level
for(i in 118:nrow(fcst)){
fcst$yhat.level[i] <- fcst$yhat.level[i-1] + fcst$fit[i]
fcst$yhat.lower[i] <- fcst$yhat.lower[i-1] + fcst$lwr[i]
fcst$yhat.upper[i] <- fcst$yhat.upper[i-1] + fcst$upr[i]
  }
```

---

give a look inside the seemingly black box.

In Figure 7.15, the price forecast (red) is plotted along with the prediction interval (light red). By month 129 (September 2020), the price level is expected to exceed £240,000 with a prediction interval between £221,310 and £261,417. Notice that the forecast line extends the prevailing trend and retains the seasonal pattern. While it looks promising and plausible, keep in mind that the seasonal dummy forecast is deterministic—it is an extrapolation that assumes that trend and seasonality will follow a rigid, predictable pattern. The prediction intervals expand with the forecast horizon, serving as a reminder the uncertainty grows in the absence of data.

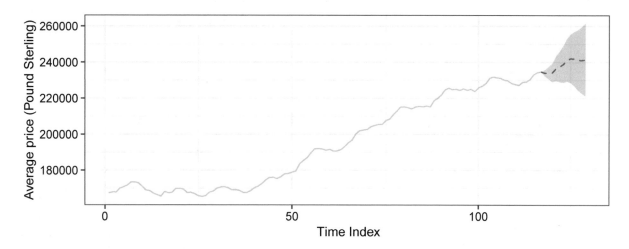

Figure 7.15: 12-month forecast for HPI (red) and historical series (gray).

7.5 Beyond this chapter

Regression still remains the go-to choice for working with continuous variables. The coefficients are interpretable and help build a story for policy settings. The trained regression model can produce predictions. However, to ensure the result captures the full story – including nuances in categorical variables, interactions, splines, transformations—requires careful investigation of the data's underlying detail. In order to get the most out of regression, consider these best practices:

Plan your objective. Regression is a flexible tool that can parse out the effect of input variables have on a target but also produce predictions. The former focuses on *how*, thus a parsimonious (i.e., fewer variable) specification tells a clearer story at the potential cost of overlooking finer details. The latter focuses on anticipating *who and what*, which can potentially lead a data scientist to include more variables into the regression to mold to the nooks and crannies of the problem at hand.

Keep your model simple, but not simplistic. It turns out that more complex regression—i.e., more variables— can lead to complications. In fact, regression can be quite finicky.

- There should always be fewer variables than observations, otherwise, the regression cannot find a solution for each coefficient. This is known as the *ill-posed problem*, which has historically required much manual variable selection to craft a well-specified model. In Chapter 9, we present an easy strategy to overcome the ill-posed problem in another form of regression.
- When there are *many variables but less than the number of observations*, a regression model can have reasonable fitness, but predictions can also be quite noisy. n some ways, too many variables are like having too many cooks in the kitchen: coefficients will fight for control. In contrast, having *too few variables* can fail to capture the underlying pattern in the data, thereby biasing predictions. Striking a balance sits at the heart of the *Bias-Variance Trade-off*—finding the right number of variables to adequately characterize the pattern in the data. In the following Chapter 8, we take a closer look at this trade-off and its implications on developing a well-behaved model.

Avoid multicollinearity. One of the potential complications of adding a large number of variables into a regression model is multicollinearity, or when two or more variables are highly correlated. When variables are too closely correlated, the regression is unable to decisively attribute weight to one variable versus another, which in turn leads to coefficients that are large and sometimes with an unexpected direction of effect (i.e., a negative coefficient when a positive one is expected) making it challenging to interpret the underlying patterns. Despite these *inflated coefficients*, multicollinearity delivers valid predictions as long as the relationship between the correlated the collinear variables remains stable. Otherwise, prediction will become erratic as the over-sized coefficients will exaggerate the effect of each variable.

Chapter 8

Framing Classification

8.1 Playing with fire

8.1.1 FireCast

How do you predict which buildings in a city are most likely to catch fire?

On an August afternoon in 2007, a fire broke out on the 17th floor of the then-vacant Deutsche Bank Building, a skyscraper situated across from the former World Trade Center in New York City. Seriously damaged after the 9/11 attacks, the building had been undergoing hazardous material abatement and controlled demolition, leading to fundamental changes to the building's functions and plans. Containment walls and negative pressure systems caused air to move in an unusual way. Disabled standpipes prevented water from circulating through the building. The result was a labyrinth of untold dangers. It should not have been a surprise that when the New York City Fire Department (FDNY) responded, the fire quickly escalated to a seven-alarm fire incident with 87 units and 475 firefighters on scene (Rivera (2007)). Struggling to navigate the maze to put water on the fire, fire crews resorted to unconventional methods like pulling hose up the side of the building. After 7 hours of battling the blaze, the fire was contained, not before two firefighters succumbed to cardiac arrest from heavy smoke inhalation (Dobnik (2007)). An after-action investigation found that the deaths could have been prevented had city agencies established information-sharing protocols and leveraged a risk-based strategy to mitigate and avoid hazards (Hearn, Scoppetta, and LiMadri (2009)).

Such a data-driven strategy would require major updates to how information was used. Since the 1950s, FDNY building inspections relied on index cards to keep track of inspection schedules. Risk was reliant on experience—perceived risk. The riskiest buildings were scheduled for inspection at most once a year, whereas the least risky buildings were inspected at least every 7 years. Of the one-million buildings in New York City, only one-third are inspection-qualified. Of those 300,000 buildings, FDNY had historically been only able to inspect at most 10% of the buildings in a given year. This meant that the 7-year schedule was aspirational at best. Indeed, NYC's risk mitigation strategy was due for improvements and an algorithmic approach could do the trick. By associating building characteristics with when fires did and did not occur, inspections could be more efficiently scheduled using the power of conditional probabilities.

In 2013, the FDNY set out to address the risk management problem by melding data and technology with their field operations. The idea of smart tech in the fire service was quite alluring. But in reality, it was also a herculean effort. Fortunately, the Commissioner and First Deputy Commissioner saw that technology had a role to play at FDNY. Aligned with Mayor Michael Bloomberg's vision of smart, data-driven government, they sought to set an example for the nation's fire services. To lead the charge, the Assistant Commissioner for Management Initiatives was tasked with leading the change management process, bringing together fire chiefs, fire officers, and information technology (IT) managers.

Government agencies have a track record of failed technology initiatives. Perhaps the government failure rate is not higher than any other sector, but it is certainly more visible. Part of the challenge is simply

© Springer Nature Switzerland AG 2021
J. C. Chen et al., *Data Science for Public Policy*, Springer Series in the Data Sciences,
https://doi.org/10.1007/978-3-030-71352-2_8

not knowing the territory. FDNY understood the odds and sought to stem the tide by building in-house technical capacity. The Assistant Commissioner tasked in-house software engineers to meticulously gather specifications that were then used to develop a digital scheduling platform. Decades-worth of index cards were digitized and integrated with dozens of databases capturing the state of reported activity within NYC buildings. But data on its own is not enough for risk targeting—it needs to be structured and converted into risk. FDNY hired a Director of Analytics to lead the research and development of prediction algorithms to recommend buildings for inspection and craft the scientific narrative to help socialize and garner support for its adoption.

On the operations side, buy-in was required. Trust is the substance that converts algorithmic intelligence into action on the ground. Anyone who has observed firefighters on the fireground (scene of the incident) will notice that it is a well-choreographed operation—every person knows their part. The same rank and file structure also meant that change needed to start with leadership. Fire departments have a culture of following a chain of command. Adopting a new system requires buy-in from the entirety of the firehouse, battalion, and department. The Assistant Commissioner worked with senior fire personnel including the Deputy Chief of Fire Operations and a number of Battalion Chiefs to amend Standard Operating Procedures (SOPs) to use a digital inspection system.

The end result was the Risk-Based Inspection System (RBIS), a firefighter-facing data platform that targeted buildings with the greatest risk of fire. Three times a week for 3-hour time blocks, fire officers logged onto RBIS to obtain a list of buildings scheduled for inspection. Buildings were selected using FireCast, a model developed in-house to predict fires at the building level. In early versions, the prediction model was a simple logistic regression—a method that is well known for easy interpretation while being able to associate 60 building characteristics to fires. But as the system advanced in its development, more complex algorithms were incorporated and integrated as much as 7,500 variables. Through FireCast, buildings no longer were assigned to a risk tier based on human perception, but rather a dynamic risk score that into account the latest fire ignition patterns.

The algorithm was able to identify buildings with fires over 80% of the time—a degree of accuracy that surpassed prior attempts. Upon implementing the new system, impacts were tracked through the number of violations issued. In the first month, the number of safety violations issued grew by +19% relative to what was possible using the index card system, indicating that FireCast was indeed identifying more buildings with observable risks. By the second month, violations were only +10% over trend, indicating that fire risk does in fact exist on a continuum. To measure efficacy, FDNY developed an indicator known as the Pre-Arrival Coverage Rate (PACR) that measures the proportion of buildings that experienced a fire that was inspected within some period (90 days) before the fire occurred. Likened to the film Minority Report, the PACR measures if FireCast could anticipate fires and if fire companies had the opportunity to evaluate priority buildings. Under FireCast, FDNY had achieved a PACR of 16.5%, which was an eight-fold improvement over an early attempt made by a Fortune 500 company (Roman (2014), Hamins et al. (2015)).

RBIS/FireCast was launched in 2014 and marked the beginning of a modern renaissance in fire prediction. Since then, new projects have emerged around the United States such as the Firebird open source system for Atlanta in 2016 (Madaio et al. (2016)) and a spatiotemporal fire prediction approach for Pittsburgh in 2018 (Singh Walia et al. (2018)).

8.1.2 What's a classifier?

The RBIS and FireCast is an example of a *classification* problem—a task in which observations are associated with a *class* or *label* based on its attributes. Labeled examples contain both factual (e.g., what happened, fire, group A) and counter-factual (e.g., what did not happen, no fire, group B). We cannot learn what makes buildings with fires different than buildings without fires—there's no ying without yang. By training a model to learn from *labeled examples*, we can predict the label of fresh observations.

Classification is nothing new in everyday life as we use our own mental classification models to contextualize the world around us. For example, marketers and advertisers are always looking to get product offerings in front of prospective customers. Marketers will often purchase lists of people and apply models based on past customer behavior in order to identify those who are most likely yield a sale. The criminal justice system has incorporated risk classification models like COMPAS to screen people being booked into jail for their chances of committing a crime in the future (Angwin et al. (2016)). On a more futuristic front, the technology behind self-driving cars uses a complex array of sensors and cameras that are processed by classifiers among other algorithms to distinguish between cars, people, motorbikes, and cyclists.

The same is true with fires.

By examining buildings that did and did not catch fire in the past, we are able to learn which characteristics are associated with a greater risk of fire. We can apply a learned pattern to new records to obtain the probability of fire, thereby giving fire fighters a sneak peek of what may happen. There can be thousands of variables that play a role in predicting fires. There are likewise classification algorithms that can learn and apply patterns.

In this chapter, we gently introduce classifiers, starting with a tour of the basic elements of any classifier, focusing on methods common in the social sciences. In particular, we introduce logistic regression—a cousin of linear regression—and describe how the technique can support data-driven narratives and prediction problems. While the technique is a mainstay of fields invested in causal inference, it has some notable handicaps, such as an inability to handle certain big data scenarios. We close the chapter by diving into regularized regressions, such as LASSO and Ridge regression, that are a machine learning spin on classical approaches.

8.2 The basics of classifiers

While classifiers are similar to regression problems, their anatomy and outputs are quite different. We begin this section by dissecting classifiers, laying out the expected inputs and typical outputs. Classifiers are designed to find variables that can *separate* one class from another. The way in which signal—or meaningful information that helps predict the target or outcome—manifests itself can easily be seen through visualization. Thus, we start this section by developing an intuitive, visual understanding of what constitutes signal in classification contexts and how accuracy is measured.

8.2.1 The anatomy of a classifier

If fires are truly predictable, we can employ supervised learning to map how input variables can distinguish buildings that had fires from those that did not. Given a binary outcome *Fire* we can determine class membership as a function of the building's characteristics and other inputs[1]:

$$Fire = f(\text{Building characteristics, Location, Complaints, ...})$$

Similar to regression, the extent to which each input variable contributes to predicting fire is learned from the data. How importance and contribution are measured differs from one technique to the other. Some methods produce an interpretable $\hat{\beta}$ estimate (e.g., logistic regression) while others are solely focused on extracting the most signal to produce a generalizable prediction of $\hat{y}_i$ (e.g., Random Forest). In all cases, a loss function helps a classifier gage if it has learned the patterns the best it can. Mean squared error is common for regression on continuous values, but a classifier's loss can emphasize different aspects of accuracy, represented through functions such as *log-loss*, *classification accuracy*, *F1-Scores*, among others. The output of a classifier is a probability that an observation belongs to a class (*g*). These probabilities are particularly useful for use cases focused on prioritization (e.g., inspection, document review) and risk monitoring. Converting probabilities into a predicted class can be an efficient solution to tag or label data whose disposition is yet to be known (e.g., add tags images, classifying documents).

---

[1]Note that in this example, we only use binary or dichotomous outcomes (e.g., Yes fire, No fire). In classification, outcomes can be multinomial outcomes in which multiple classes can exist (e.g., full fire, partial fire, smoke alarm, no fire).

8.2.2   Finding signal in classification contexts

In regression, the slope of a trend indicates the presence of signal. But what is signal in a classification context? A fire prediction algorithm needs to predict where fires and non-fires will be. A jail bail algorithm needs to be able to distinguish between those who are flight risks and those who are not. A political affiliation algorithm needs to be able to place prospective voters by political party. Much of the utility of a classifier is dependent on the qualities of its input data, in particular, its *separability*—or the degree to which one class can be distinguished using values of an input variable. Through basic exploratory data analysis, separability should be easy to spot if one knows what to look for.

When considering continuous variables, we must examine the overlap of one class's distribution with that of another. For example, if we are interested in determining if fires happen in larger buildings, we can plot the kernel density of buildings that have experienced fires versus buildings without (see Figure 8.1. A case of low separability (Graph (A)) would exhibit substantial overlap, indicating a variable has little ability to distinguish one class from the other. Highly separable variables (Panel (B)), in contrast, would only slightly overlap. But perhaps the neatest condition is perfect separability (Panel (C)). Because there is no overlap, we can precisely define the phenomenon at some threshold rather than using a model to approximate it—no model is required. Whether a family in the United States is in poverty, for example, need not be modeled if data contain both income and number of people in household as the government has set clear criteria. However, a model would be useful when either input variable (income or number of people) are missing.

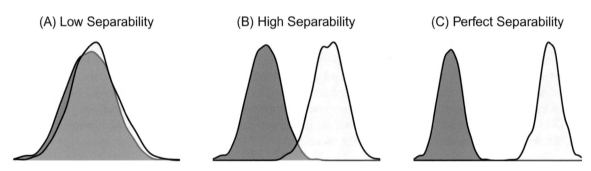

Figure 8.1: Separability can be easily seen in the distributions of a continuous variable when separated into two classes. These kernel density diagrams illustrate three scenarios: (A) Low Separability, (B) High Separability, (C) Perfect Separability.

When inputs are *discrete*, separability is most visible along the diagonals of a two-way contingency table of the target and input variable of interest. In the hypothetical in Table 8.1, we simulate a dataset containing building violations and fires. A separable variable will distribute most observations along the diagonal of the table: buildings that have a violation also have fires, and buildings without violations tend not to have fires. If past building violations alone were used to predict fires, we could calculate how often predictions lined up with actual fires. Buildings that were accurately identified as ones with fires are *true positives* (TP) and buildings correctly identified without fires are *true negatives* (TN). Together, 82.3% of the time.

In contrast, a discrete variable exhibiting low separability will distribute observations anywhere other than the diagonal. We can infer from Table 8.2 that whether a building has an elevator is a poor predictor of fires, misidentifying a disproportionate number of buildings. If this factor were the only input in a prediction model, accuracy falls to only 38%. Separability will make or break a classifier, and thorough and thoughtful exploratory data analysis can help evaluate if the data is fit for modeling.

8.2.3   Measuring accuracy

Much like test scores in school, we generally expect accuracy to be expressed on a scale from 0% to 100%. But in practice, a single summary statistic may hide the finer details in a model—there is not one catch-all measure. In fact, the types of classification accuracy measures vary from one use case to another.

Measures of classification accuracy

Table 8.1: A separable discrete variable.

	No Violation (F)	Violation (T)
Fire (T)	10.9	22.6
No Fire (F)	59.7	6.8

Table 8.2: A non-separable discrete variable.

	Elevator (F)	Elevator (T)
Fire (T)	2.6	30.9
No Fire (F)	6.0	60.5

Table 8.3: Structure of confusion matrix.

	Predicted (F)	Predicted (T)
Actual (F)	True Negative (TN)	False Positive (FP)
Actual (T)	False Negative (FN)	True Positive (TP)

Is the classifier accurate? How many records can we predict? What are the trade-offs? The objective that underlies these questions is more prediction-oriented, in pursuit of raw accuracy. Virtually all measures of classification accuracy measures are derived by first converting a model's outputs into predicted classes by setting a *classification threshold c*—a cutoff along the predicted probabilities that separates what the model believes to be positive from negative. The predicted and actual class labels can then be compared in a *confusion matrix*—a $n \times n$ table that compares actual classes as rows and predicted classes as columns. A two-class problem, for example, yields a 2×2 confusion matrix as seen in Table 8.3.

Each cell embodies different classification accuracy concepts. Ideally, the majority of predictions will be *true positives* or TP (a positive case was accurately predicted) and *true negatives* or TN (a negative case was correctly predicted). Their inverses—error—are also just as important. A *false positive* (FP), or Type I error, is a negative case misclassified as a positive. In applications, false positives are also referred to as *false alarms*, and fire departments are all too-well acquainted with fire alarms that did not involve a fire. There is some cost associated with actioning on FPs, but arguably the more damaging case is that of the *false negative* (FN), or Type II error. FNs occur when a negative prediction is actually a positive case. More colloquially, this can be viewed as being *blind-sided* as the model (and analyst) failed to foresee a problem.

When combined in creative ways, these four measures describe the quality of a classifier, which in turn *should* inform policies how algorithms are deployed. Let's take the example of the issuance of winter weather warnings given an incoming storm system. School districts may prefer to be over-warned as they are concerned for the safety of their pupils. Over-warning is an explicit preference for more TPs and a higher tolerance for FNs—people would rather be safe than sorry. The idea of over-warning is a matter of balancing between *True Positive Rate (TPR)* and *True Negative Rate* (TNR). TPR, also known as *sensitivity* or *recall*, is the proportion of actual storms that were predicted correctly:

$$TPR = \frac{TP}{TP + FN} = \frac{TP}{P}$$

While the TNR, also known as *specificity*, is the proportion of storm-free days that were correctly predicted.

$$TNR = \frac{TN}{TN + FP} = \frac{TN}{N}$$

The weatherman could in theory predict all storms perfectly ($TPR = 1.0$) but not predict any sunny days ($TNR = 0$)—clearly not a practical solution. But as we adjust the classification threshold to increase TPR (sensitivity), we will likely see more false positives which in turn reduces the TNR (specificity). Together, sensitivity and specificity are trade-offs.

An alternative pair of measures uses precision and recall (sensitivity). Precision, also *Positive Predictive Value* (PPV), is given as the proportion of predicted storms that are in fact storms:

$$PPV = \frac{TP}{TP + FP}$$

When combined with recall, we can evaluate if a model returns relevant results. The weatherman may be more likely to issue more warnings to push up the sensitivity rate and accept a lower PPV (Table 8.4).

Table 8.4: A selection of classifier accuracy metrics derived from a confusion matrix.

Measure	Formula	Significance
True Positive Rate (TPR), Sensitivity, or Recall	$TPR = \frac{TP}{TP+FN}$	What proportion positive cases are correctly identified?
True Negative Rate (TNR) or Specificity	$TNR = \frac{TN}{TN+FP}$	What proportion negative cases are correctly identified?
False Positive Rate (FPR)	$FPR = \frac{FP}{FP+TN}$	What proportion of negative cases are incorrectly predicted as positive? Also known as Type I error rate.
False Negative Rate (FNR)	$FNR = \frac{FN}{TP+FN}$	Proportion of positive cases that are incorrectly predicted negative. Also known as the false alarm rate or Type II error rate.
Positive Predictive Value (PPV) or Precision	$PPV = \frac{TP}{TP+FP}$	The proportion of predicted positives will actually be positive? This measure is influenced by the prevalence of the outcome.

Despite the underlying accuracy trade-offs, there is often a need to communicate overall accuracy. The F1-score, for example, provides a snapshot of accuracy at a specified threshold by calculating the harmonic mean of precision and recall. One point to note is that the threshold needs to be the "right" one. If all predicted probabilities are less than the threshold, the F1-score simply indicates that none of the values exceeded c.

A more robust measure of accuracy is the *Area Under the Curve* (*AUC*, sometimes referred to as the concordance statistic *C-statistic*) that is derived from a Receiving-Operating Characteristic (*ROC*) Curve. A product of signal detection research during the Second World war, the ROC curve calculates pairs of TPR and FPR at varying classification thresholds c, plotting each pair of values. The area under the resulting curve is a measure of how the classifier performs on average under a wide range of scenarios (Table 8.5).

Table 8.5: Summary measures of classification accuracy.

Measure	Formula	Significance
Accuracy (ACC)	$ACC = \frac{TP+TN}{n}$	The proportion of that records are correctly classified.
F1-Score (F1)	$F1 = \frac{2}{\frac{1}{TPR} \times \frac{1}{PPV}}$	Alternative method of calculating accuracy using a harmonic mean. It balances the TPR with the PPV.
AUC or Concordance (C) statistic	$AUC = \int_0^1 ROC(u)du$	The AUC is the area under the ROC Curve. The ROC is derived by plotting values of TPR (Sensitivity) and FPR for varying probability cutoffs c.

Keep in mind that any measure of accuracy can be distorted by *class imbalance*, when a *minority class*—a class with fewer observations—is far less prevalent than other classes. We can imagine that many potential class variables are imbalanced:

- fire-related fatalities account for only 3,400 fire deaths in 2017 (U.S. Fire Administration (2017)) vs. 2.7 million total deaths in the US in 2016 (Centers for Disease Control (2017))
- approximately 0.17-percent of the US population is homeless (BBC (2017)).
- 8.7% of US population was uninsured in 2017 (U.S. Census Bureau (2017b)).

Class imbalance can arise from any number of sources, whether the minority class is a small part of a population or the sample is biased. Without addressing the class imbalance, summary measures of accuracy can be overstated. Fires, for example, tend to fall into this category as most buildings are not constantly on fire on most days. If 99% of a sample did not experience a fire, then an AUC of 99% may simply indicate a model has learned to patterns associated with fire-free days and has little ability to identify fire risks. Taking a closer look at TPR, TNR, among other measures can paint a truer picture.

Alternatively, a measure like the F1-score may understate its accuracy if the classification threshold is set too high or at the default of $c = 0.5$. The optimal threshold may actually be far below or above 0.5 (see Figure 8.2). Shifting the threshold can improve accuracy, but if adjusted in-sample, we may overstate a classifier's accuracy. Diagnosing a model's accuracy requires clarity of purpose.

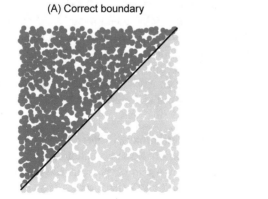
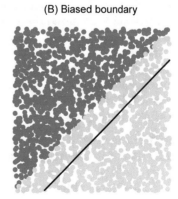

Figure 8.2: Thresholds dictate the placement of decision boundaries, which in turn influence the classification error.

Information criteria

Data scientists tends to focus on accuracy derived from a confusion matrix as the goal is often to predict an absolutely accurate model—in pursuit of the true model. In actuality, we do not know what is the *true* model, but we know that too few variables lead to models with possible bias, whereas too many variables may increase the noise of predictions. A *truer* model exists somewhere in between those two extremes. This idea can be framed in terms of information loss. Farther away from the true model, less a model represents the "true" process, therefore more information is lost. This is the basic idea that underlies *Information Criteria* such as the *Akaike Information Criterion* (AIC) and the *Bayesian Information Criterion* (BIC). Both measures were developed to help approximate the *relative* information lost, which in turn facilitates model selection. The AIC, as developed in Akaike (1973), rates models for how well they approximate the true model, taking the maximized log-likelihood $log(\hat{l})$ that is commonly available in regression models[2], penalized by the number of parameters k.

$$AIC = 2k - 2log(\hat{l})$$

The BIC, as developed in Schwarz (1978), extends the penalty to include the sample size n. BIC has an added quality that it can identify the *true* model from among alternatives. If the *true* model is present, it has been argued that the probability of choosing the right model approaches 1. Then again, the challenge is that *true* is not known—it is theoretical.

$$BIC = log(n)k - 2log(\hat{l})$$

To find the best model relative to alternatives, the analytical goal is to find the model that minimizes the information criterion. For the most part, AIC and BIC tend to agree on which model is best, but occasionally there are differences. In particular, AIC has a tendency of choosing larger, more complex models, while BIC tends toward more parsimonious specifications. Nonetheless, research studies will rely on these measures to illustrate that the optimal model was not selected arbitrarily from a set of alternatives.

Despite their wide use in the social sciences, these measures have limited applicability outside of regression contexts.[3] Furthermore, information criteria are relative measure—they do not provide the sort of closure and confidence of an absolute accuracy measures.

8.3   Logistic regression

8.3.1   The social science workhorse

With the basic ideas behind classification laid out, we can now appreciate a classic technique. Logistic regression, or binomial generalized linear models (GLM), is the go-to classifier for any field concerned with causal inference. Like its cousin OLS, it can parse out how each input variable influences the target variable. Thus, the technique can be helpful in both the pursuit of *interpretability* (focus on $\hat{\beta}$) and *prediction* (focus on $\hat{y}$). A policy analyst can rely on logistic regression to distill the marginal effects of certain variables on social outcomes, while a political advertisement targeting analyst can use the same framework to maximize predictive accuracy. FireCast also made use of logistic regression in its early versions so that buildings could be risk-rated while crafting the narrative to brief firefighters.

The basic gist of the method is as follows: Let's assume the classes of binary target y can be modeled using a linear combination of input variables x_1 and x_2. By setting a classification threshold along the predicted probabilities, we can draw a linear decision boundary that separates one class from the other. Similar to OLS, we can express the modeled relationship between y, x_1, x_2, and any other covariate as a linear model:

$$\hat{y}_i = \beta_0 + \sum_{k=1}^{K} \beta_k x_{ik}$$

---

[2]We will discuss log-likelihood in-depth later in this chapter.

[3]As we will see in the next chapter, methods such as decision trees and Random Forests do not have a set number of parameters k.

where $\hat{y}_i$ is a prediction of a binary outcome and β_k are *coefficients* in the social science literature, or otherwise *weights* w_k in the computer science literature. The problem with binary outcomes in a linear framework is that $\hat{y}$ can exceed the bounds of 0 and 1—the output would make little sense. Imagine if the predicted $\hat{y}$ were -103 or +4. *The magnitude would not have any obvious meaning, so how should one interpret the result?* Fortunately, statisticians have cleverly solved the bounding problem by inserting the model output $\hat{y}_i$ into a logistic function:

$$p(y_i = 1) = \frac{1}{1 + e^{-\hat{y}_i}}$$

thereby bounding the values to a range between 0 and 1 in the shape of a *sigmoid* or "S-Curve".

8.3.2 Telling the story from coefficients

Despite being buried within the logistic function, each coefficient is interpretable. The equation first requires some mathematical tidying. While derivations are generally beyond the scope of this text, understanding these transformations instills intuition for the practical use of logistic regression coefficients. We start by associating the logarithm transformed odds (or *log odds*) of an event with the linear portion of the logistic regression:

$$log(odds) = log(\frac{p(y=1)}{p(y=0)}) = \beta_0 + \sum_{k=1}^{K} \beta_k x_k$$

where the odds are simply the probability of an event divided by no event. To isolate the odds on the left-hand side, we exponentiate both sides of the equation:

$$odds = \frac{p(y=1)}{p(y=0)} = e^{\beta_0 + \sum_{k=1}^{K} \beta_k x_k} = e^{\hat{y}_i}$$

Lastly, we can contextualize odds using an *odds ratio* (OR) in which we compare the odds of an one-unit increase in a variable x_k to a zero-unit change:

$$\text{Odds Ratio } (OR) = \frac{odds(x_k + 1)}{odds(x_k + 0)} = \frac{e^{\hat{y}_i(x_k+1)}}{e^{\hat{y}_i(x_k+0)}}$$

If we recall the quotient rule for exponents, a nifty mathematical identity usually taught in high school, the right-hand side of the equation can be vastly simplified: simply exponentiating the coefficient of interest β_k allows us to interpret the multiplicative effect associated with a one-unit increase in the x_k:

$$OR = e^{\beta_k}$$

Let's put the odds ratio into context. If a logistic regression were trained to associate health insurance coverage with wages ($000) and citizenship (binary), the regression formula may be as follows:

$$y(\text{uninsured}) = 0.468 - 0.048 \times wage + 0.372 \times \text{vnon-citizen}$$

The raw coefficients have little meaning, but a positive value signals that the variable is associated with a higher chance of being uninsured. While not shown, it is also worth noting that each coefficient has a standard error estimate that indicates its statistical significance. By exponentiating the coefficients, we see that a $1000 increase in wages is associated with a 4.69% *lower* chance of being uninsured ($1 - e^{-0.048} = 1 - 0.953 = 0.0469$). Otherwise stated, the higher one's earnings are, the better the chance of having coverage. The non-citizen coefficient ($e^{0.372} = 0.451$) translates to a 45% *higher* chance of being uninsured.

A logistic model's transparency allows an analyst to tell the story behind the science. Because of this prized quality, it has been adopted in a many fields. A 2012 health study using one of the largest lung cancer datasets assembled, for example, was able to link intensity of smoking behavior to specific types of cancer. In addition to quantifying the odds ratios for specific lung cancer types, they found that a smoker's risk of cancer never falls to non-smoker baselines even after quitting for 35 years (Pesch et al. (2012)). In a study in *Geomorphology*, logistic regression played a role in understanding landslide susceptibility in Central Japan, which has implications for civil and geotechnical engineering. Researchers examined 87 landslide events with GIS data containing over a million 10×10 meter resolution grids. By estimating a logistic regression to predict whether each grid cell had experienced a landslide, the model coefficients indicated that areas that are closer to road networks and have steeper slope gradients were at greater risk of landslides. The model was able to achieve a decent level of accuracy ($AUC = 0.8358$) and was applied to produce a risk map (Ayalew and Yamagishi (2005)).

8.3.3   How are coefficients learned?

The secret sauce behind the coefficients is a process known as *Maximum Likelihood Estimation* or *MLE*. While the derivation of the method is beyond the scope of this text, having a high-level understanding of the role of this procedure is essential for using logistic regression, but also for appreciating more sophisticated techniques such as regularized regression (later this chapter). The gist of MLE is as follows: coefficients are incrementally adjusted in order to maximize the likelihood that they represent the process the generated the training data. We rely on a cost function known as the *log-likelihood* ($l(\beta)$), sometimes referred to as *log loss* or *multi-class cross entropy* in data and computer science:

$$l(\beta) = \sum_{i=1}^{n} \log p_g$$

where $p_g = pr(y_i = g|X; \beta)$, or the probability of class g given the values of the inputs x_k and coefficients β. The formula is the cost function for multiple classes. For logistic regression, we are interested in a two-class scenario. We can expand the log-likelihood function into a mathematically convenient form where the probability of $g = 1$ is given as $p_{i,g=1}$ and $g = 0$ as $(1 - p_{i,g=1})$:

$$l(\beta) = \sum_{i=1}^{n} y_i \log(p_{i,g=1}) + (1 - y_i) \log(1 - p_{i,g=1})$$

which in turn can be further simplified :

$$l(\beta) = \sum_{i=1}^{n} [y_i \hat{y}_i - log(1 + e^{\hat{y}})]$$

The idea here is that as we solve for values of β, the prediction $\hat{y}$ will change as well. We take the sum of $\hat{y}_i - log(1 + e^{\hat{y}})$ across all observations in order to track the model's progress. The MLE procedure iteratively adjusts the values of β until $l(\beta)$ is maximized. For a more in-depth treatment on the topic, refer to *The Elements of Statistical Learning* (Hastie, Tibshirani, and Friedman (2009)).

8.3.4   In practice

Logistic regression and MLE have standard implementations in most data analysis software making it widely accessible. It is not uncommon to see program evaluators rely on logistic regression to decompose causal effects. Data scientists also use logistic regression for targeting and prioritization, focusing on accuracy and balance of predictions. In both cases, there are divergent technical standards for how to successfully implement a model. Here, we draw a distinction between *interpretability* and *predictability* and describe common challenges that analysts encounter.

Interpretability. Logistic regression coefficients lend themselves to articulating how $X \rightarrow Y$, thereby facilitating causal inference. For the inference to be valid, the coefficients must follow basic assumptions.

For one thing, how does one choose the best model? In logistic regression, it is wholly dependent on which variables are included. It is customary to test multiple specifications, then compare how each performs in terms of the AIC and BIC, choosing the model with the lowest score.

Arguably, the most powerful quality of logistic regression is the interpretation of the coefficients. They can, however, be distorted by *multicollinearity*—a condition in which two or more input variables are not only correlated with the target variable but also among themselves. MLE then finds it challenging to parse the average effect of x_k on y, holding all else constant. Thus, if two or more variables have identical or similar information, the algorithm is not able to definitively distill each variable's effects during MLE. The consequence is quite serious (statistically, that is): coefficients will behave oddly, sometimes with abnormally large magnitudes and sometimes exhibiting the wrong direction of effect (e.g., positive effect when a negative is expected). One approach to diagnose this condition involves calculating *Variance Inflation Factors* (VIF). Given a specification:

$$Y(employed) = \beta_0 + \beta_1 education + \beta_2 age + ... + \beta_k x_k + \epsilon$$

VIF is a calculated using a two-step process. First, each input variable x_k on all other variables, recording the R-squared R_k^2. The idea is to summarize the correlation between each x_k and all other covariates. Second, the VIF is calculated as

$$VIF_k = \frac{1}{1 - R_k^2}$$

A variable x_k with $VIF = 1$ is *orthogonal* or *uncorrelated* with other input variables, while a $VIF > 1$ indicates some correlation. While there is not a standard cutoff for multicollinearity, it is reasonable to assume that cases where $VIF > 5$ merits more attention.

How can we mitigate the collinearity? For groups of collinear variables, one approach is to select one factor that best represents that group. This places a large degree of faith in the person's ability to identify the right signal There are more computationally intensive *dimensionality reduction* techniques that distill the common signal from a set of variables. *Principal Component Analysis* (PCA) and *Singular Value Decomposition* (SVD) transform k-number of variables into a concise set of uncorrelated component variables, each of which captures unique motifs in the data. The drawback, however, is that each component is a mix of variables, making it subject to one's judgment to define the "theme" of the component.

Prediction. Multicollinearity does not usually adversely impact predictions $\hat{y}_i$, so as long as the relationship among collinear variables are stable. The real challenge lies in how $\hat{y}_i$ will be used in application. One dichotomy consists of ranking versus labeling. Ranking problems make use of the predicted probabilities to sort records, whether to make recommendations (e.g., online recommender systems for products and services) or to prioritize attention (e.g., fire safety inspections). If the use case depends solely on the order of probabilities, then model development is a matter of identifying the specifications that provide the best out-of-sample accuracy.

Labeling problems, in contrast, convert predicted probabilities into predicted classes, requiring attention paid to the sample balance or classification threshold. In a highly imbalanced sample, the predicted probabilities $\hat{y}$ will center around the average probability that an event will occur. This implies that an unadjusted classification threshold c or unaddressed class imbalance will result in poor predictions. There are multiple solutions to the problem, each of which has drawbacks.

- *Adjusted cutoff*: The classification threshold c can be adjusted to optimize accuracy. Setting the threshold to the average probability can achieve a better balance between TPR and TNR, but needs to be done on a cross-validated sample to reduce the incidence of overfitting.
- *Over-sampling*: Over-sampling involves replicating the minority class observations to an equal proportion of the majority class. The inverse is also done: under-sampling the majority class so that it is of equal proportion to the minority. In either case, replicating observations does not change the amount of unique information in the data, but it is a mechanical solution to tricks the classifier into producing

predictions that center around $c = 0.5$. These strategies run the risk of overfitting models. Furthermore, the probabilities no longer reflect reality, but rather the artificially adjusted sample.

- *Synthetic Over-Sampling*: Alternatively, Synthetic Minority Over-Sampling Technique (SMOTE) simulates new observations that look like the minority class that add fresh signal to the sample while minimizing overfitting (Chawla et al. (2002))—see the `DMwR` package.

Ultimately, these sampling challenges require some degree of trial and error to get right.

Hybrid Cases. Data are often collected for a specific purpose, but are re-purposed to support evaluation of programs and policies. The problem, however, is that the data are not collected under experimental conditions. The absence of randomly assigned treatment and control groups means results can be biased— we may be comparing apples to oranges. As there is so much data and random control trials are expensive, methods like Propensity Score Matching (PSM) have been developed. Originally introduced in Rosenbaum and Rubin (1983), PSM brings together prediction and inference in order to re-balance a sample, aiming to force both the treatment group and the control group to be statistically indistinguishable from one another based on observable characteristics. The procedure facilitates an apples to apples comparison. If sufficiently well balanced, any difference in a variable of interest can be attributed to the treatment. This procedure is accomplished by applying a logistic regression to predict assignment (e.g., treatment versus control) using select observable characteristics. The resulting probabilities, or *propensity scores*, are matched between treatment and control groups and tested for any statistical differences.

Since its inception, PSM has been adopted in public health, medicine, and economics, among other fields. For example, Jalan and Ravallion (2003) study the effect of Trabajar, a "workfare" program in Argentina, on income gains. In particular, the authors quantify the effect of participating in the program and which parts of the income distribution benefit the most. The challenge as in any PSM is the lack of a counterfactual—often times social programs will capture data on participants but miss non-participants. Without the counterfactual, it is not possible to infer program effects. By pooling both the program's data with a large national socioeconomic survey, the authors then have an imbalanced sample on which PSM can be applied to produce a balanced sample. Post-matching results suggest that income gains are approximately 50% of the pre-intervention gross wage and that 80% of program beneficiaries are in the bottom 20%.

Non-Starters. Logistic regression cannot be estimated when the number of parameters k outnumbers the sample size n. When faced with ill-posed problems such that $k > n$, it is up to the analyst to choose which variables enter the equation. In policy and strategy environments, there may be a political agenda that could guide the selection of conceptually relevant variables, though it should be emphasized that simply including a variable in a model does not make it "true". Furthermore, each person's life experiences are different, thus we run the risk of introducing an implicit bias when selecting variables. In a high-dimensional problem, manual testing and selection of variables will be prone to bias. As we will see later this chapter, there are more efficient and scalable methods to overcome these big data obstacles (Table 8.6).

Table 8.6: Pros and cons of logistic regression.

Pros	Cons
Interpretable in terms of odds ratios.	Collinear input variables may cause coefficients to return odd, inexplicable results.
Probabilistic properties allow for hypothesis testing that help build statistically-rooted arguments.	Unable to handle cases where k > n
Problems where the distinction between classes is linear	When k is large, finding optimal specification is challenging.
	Class imbalance makes prediction a non-trivial task.

8.3.5   DIY: Expanding health care coverage

DIYs are designed to be hands-on. To follow along, download the DIYs repository from Github (https://github.com/DataScienceForPublicPolicy/diys). The R Markdown file for this example *is labeled `diy-ch08-glm.Rmd`.*

In this DIY, we explore how logistic regression can help describe trends in health care coverage in the United States while supporting a predictive targeting campaign to reach uninsured individuals.

Background. Universal healthcare has become a basic human right in many countries. In the United States, this is not currently a guarantee, shrouded in heated political debate and controversy. Regardless of the politics, there is a lot of useful data on healthcare coverage. According to the American Community Survey (ACS), an annual survey of approximately 3.5% of the US population as conducted by the US Census Bureau, over 22.4% of residents of the U.S. state of Georgia were without healthcare coverage in 2009. That is a fairly sizable proportion of the population—for every ten people, two to three did not have coverage. To close the gap in 2010, a new law was signed into effect to provide affordable healthcare to the uninsured (Stolberg and Pear (2003)).

Imagine that you have been tasked with getting the word out about the new program in the state of Georgia. There is a hiccup, however. While commercial marketing databases are common, there are not many sources of information that indicate whether one has health care. The arguably best data on coverage are survey-based. Thus, we do not know *who* is and is not insured. A brute force marketing campaign *could* reach out to all Georgians though it can easily be seen as a wasted effort as three-quarters of the population are already covered. *How do we reach the remaining quarter of the population that is not already insured?* For marketers and public policy practitioners, this is a classic targeting problem.

Setting up a solution. We operationalize our solution by estimating a logistic regression. Given the label $y(Coverage)$, we can use logistic regression to not only infer which demographics are associated with coverage but also train a model to prioritize who should be contacted about receiving coverage:

$$Y(\text{No Coverage}) = f(\text{Sex, Age, Education, Marital Status, Race, Citizenship})$$

The `glm` function makes it easy to estimate logistic regression in addition to other linear models including ordinary least squares for continuous outcomes, logistic regression for binary outcomes and Poisson regression for count outcomes. At a minimum, three parameters are required:

```
glm(formula, data, family)
```

where:

- `formula` is a formula object. For example $y = \beta_0 + \beta_1 x_1 + \beta_2 x_2 + \epsilon$ can be represented as `y ~ x1 + x2`.
- `data` is a data frame containing the target and inputs.
- `family` indicates the probability distribution used in the model. Distributions typically used for GLMs are *binomial* (binary outcomes), *poisson* (count outcomes), *gaussian* (continuous outcomes - same as OLS), among others.

We use the U.S. Census Bureau's ACS sample for Georgia to train the logistic regression. ACS samples are thoroughly designed survey samples that have sampling weights indicating how many people each response represents. For simplicity, we have curated a sample with pertinent variables, but ignore the sampling weights. Below, we load the pre-processed dataset.

```
load("data/acs_health.Rda")
```

Focusing on interpretation. In public policy, the focus of regression modeling is typically on identifying an effect or an associated relationship that describes the process being studied. To isolate the effects from each variable of interest, analyses involve a *buildup*, which are a series of models that show how sets of conceptually related variables contribute to the phenomenon in question. In our analysis below, we show a build up of four models: personal characteristics (race, age, and sex), economic factors (wage, education, employment), social factors (citizenship, marital status), and "fully loaded" (all factors). The inclusion of certain characteristics, such as race and sex, are arguably necessary for informing models the health care. Demographic variables may capture latent, unobserved characteristics that in turn can improve the predictability of the coverage variable, but *it is advisable to review if the inclusion of these characteristics are equitable and ethical*.

```
glm_pers <- glm("no.coverage ~ log(age) + race + sex",
                data = health, family = binomial)


glm_econ <- glm("no.coverage ~ wage + esr + schl",
                data = health, family = binomial)


glm_soc <- glm("no.coverage ~ cit + mar ",
               data = health, family = binomial)


glm_full <- glm("no.coverage ~ log(age) + wage + schl + esr + cit + mar + race + sex",
                data = health, family = binomial)
```

Each `glm` call returns a regression object that concisely summarizes the inner workings of a logistic regression model. These results are summarized in Table 8.7. Let's start with identifying which variable groups contribute the most to model fit as inferred with the AIC values at the bottom of the regression table. The specification with the lowest AIC indicates that the model that best captures the process being modeled. The *fully* specified model explains the health care story the best; However, the *economic* specification carries the most explanatory power among any one variable group. It is particularly interesting that the combination of economic, social, and personal factors yield a model that performs far better than any single part.

The body of the regression table shows the relationships (e.g., positive or negative coefficients) and their statistical significance (e.g., standard errors in parentheses). Other than race, all variables are statistically significant at the 1% level, albeit some more than others. For example, education has a large effect on the combined model. Of the four levels in education, all coefficients are interpreted relative to the graduate degree reference group. People who did not finish high school and a high school graduate have a *3.7-times* ($e^{1.3} = 3.7$) and *3.3-times* ($e^{1.2} = 3.3$) higher chance of being uninsured, respectively. In contrast, a college graduate is relatively better off than the previous two groups with a *1.7-times* higher chance of being uninsured ($e^{w=0.5} = 1.7$).

These effects seem to be reasonable. For additional surety, we screen for multicollinearity in the full specification by calculating the VIFs making use of the `vif` function in the `car` package. Fortunately, all values are close to $VIF = 1$ (see Table 8.8) suggesting that there is not likely that collinearity has an undue influence on our estimates.

```
# Load package
pacman::p_load(car)


# VIF function
vif(glm_full)
```

Focus on prediction. Let's now approach classifiers from the lens of conducting a micro-targeting campaign to get the word out about new insurance options. Using logistic regression, we can train a model to produce

Table 8.7: Logistic regression coefficient table for four alternative specifications.

	Dependent variable:			
	No Health Care Coverage			
	Personal	Economic	Social	Full
	(1)	(2)	(3)	(4)
log(Age)	−1.2*** (0.02)			−0.7*** (0.03)
Race: Asian	−0.2 (0.2)			−0.2 (0.2)
Race: Black	−0.1 (0.2)			−0.01 (0.2)
Race: Native Hawaiian/Pac. Islander	−0.9* (0.5)			−0.6 (0.5)
Race: Other	1.4*** (0.2)			0.7*** (0.2)
Race: Two or More	−0.4** (0.2)			−0.3 (0.2)
Race: White	−0.6*** (0.2)			−0.3 (0.2)
Sex: Male	0.3*** (0.02)			0.4*** (0.02)
Wage		−0.03*** (0.001)		−0.02*** (0.001)
Employ: Civilian		2.9*** (0.4)		3.3*** (0.4)
Employ: Not in Labor Force		2.0*** (0.4)		2.8*** (0.4)
Employ: Unemployed		3.8*** (0.4)		4.2*** (0.4)
Education: HS Degree		1.3*** (0.1)		1.2*** (0.1)
Education: Less than HS		1.8*** (0.1)		1.3*** (0.1)
Education: At least BA		0.5*** (0.1)		0.5*** (0.1)
Citizen: No			2.0*** (0.03)	1.8*** (0.04)
Married: Yes			−1.1*** (0.03)	−1.0*** (0.03)
Married: Never			0.3*** (0.03)	−0.5*** (0.04)
Married: Separated			0.3*** (0.1)	−0.01 (0.1)
Married: Widowed			−1.6*** (0.1)	−1.4*** (0.1)
Constant	3.2*** (0.2)	−5.0*** (0.4)	−1.2*** (0.03)	−2.3*** (0.4)
Observations	74,805	74,805	74,805	74,805
Log Likelihood	−32,711.6	−31,587.3	−31,840.1	−28,380.2
Akaike Inf. Crit.	65,441.2	63,190.7	63,692.2	56,802.4

Note: *p<0.1; **p<0.05; ***p<0.01

Table 8.8: Generalized Variance Inflation Factors for Full Specification.

Variable	GVIF	Degrees of Freedom	Adjusted GVIF
log(age)	1.96	1	1.40
Wage	1.62	1	1.27
Education	1.24	3	1.04
Employment	1.61	3	1.08
Citizenship	1.27	1	1.13
Marital Status	2.20	4	1.10
Race	1.34	6	1.02
Sex	1.04	1	1.02

probabilities of being uninsured. The trained model could then be applied at scale to a commercial database of consumers that contains the same variables. As marketing databases are proprietary and quite expensive, we simulate the process by splitting our `health` sample into a training set of 70% of observations to build our model and the remaining 30% as a test set.

```
# Set seed for replicability, then randomly assign
set.seed(321)
rand <- runif(nrow(health)) > 0.7

# Create train test sets
train <- health[rand== T, ]
test<- health[rand== F, ]
```

Different practices apply when gaging the accuracy of predictive models. Data scientists often are weary of measuring accuracy in-sample as it will only overstate predictive performance. Instead, we can partition the training sample into k-number of randomly selected partitions. For each partition k, we train a model on $k - 1$ partitions, then score partition k taking note of its predicted accuracy. We then cycle through each k until each partition is predicted once. Then, the accuracy measures across all k partitions are averaged to summarize performance. This *k-folds* cross-validation strategy is a standard procedure for validating model performance. It sounds like an arduous programming task, but fortunately, the `cv.glm` function in the `boot` library makes validation process quite seamless:

```
cv.glm(data, fit, cost, K)
```

where:

- `data` is a data frame or matrix.
- `fit` is a glm model object.
- `cost` specifies the cost function for cross validation.
- `K` is the number of cross validation partitions.

Note that the cost function needs to be specified by the user. To supply a custom cost function, two vectors need to be specified: The observed responses and the predicted probabilities. For example, we can wrap a function around the `ROCR` package to obtain the AUC metric.

```
costAccuracy <- function(y, yhat){
#
# Calculate AUC using ROCR package
#
# Args:
#   y, yhat = binary target and predicted probabilities
#
# Returns: AUC value

pacman::p_load(ROCR)
pred.obj <- prediction(yhat, y)
perf <- performance(pred.obj, measure = "auc")
return(unlist(perf@y.values))
}
```

What's an appropriate number of partitions k? Smaller values of k reduce the number of available observations, which in turn affords fewer opportunities for the model to learn underlying patterns. The result will be noisier predictions that lead to more volatile estimates of accuracy and a lessened ability to distinguish a well-behaved model from one a poor one. In cases where sub-populations in the sample are small (e.g., some demographics tend to be quite small), the entirety of the sub-sample fits into one partition and causes the model to "crash". Choosing a higher value of k, in contrast, will mitigate challenges with sample loss, but requires more time to train k models. This is lengthened when we consider that k-folds cross-validation would need to be performed on *all alternative models* so that we are sure the selected model is in fact the best. We recommend at least tenfold cross-validation, but the value of k should be sufficiently large to lend confidence to the choice of best model.

We test each specification using tenfold cross-validation and obtain the average AUCs. The cross-validated accuracies, stored in the `delta` element of the `cv.glm` object, indicate that each group of variables have comparable contributions to accuracy. When combined, the fully loaded specification performs the best, achieving $AUC = 0.81$—a decent result. Furthermore, the relative model performance confirms the AIC results.

```
# Load boot library
pacman::p_load(boot)


# Train models
glm_pers <- glm("no.coverage ~ log(age) + race + sex", data = train, family = binomial)
glm_econ <- glm("no.coverage ~ wage + esr + schl", data = train, family =binomial)
glm_soc <- glm("no.coverage ~ cit + mar ", data = train, family = binomial)
glm_full <- glm("no.coverage ~ log(age) + wage + schl + esr + cit + mar + race + sex",
                data = train, family = binomial)


# Calculate k-folds
pers <- cv.glm(data = train, glmfit = glm_pers, cost = costAccuracy, K = 10)
econ <- cv.glm(data = train, glmfit = glm_econ, cost =costAccuracy, K = 10)
soc <- cv.glm(data = train, glmfit = glm_soc, cost = costAccuracy, K = 10)
all <- cv.glm(data = train, glmfit = glm_full, cost = costAccuracy, K = 10)
```

Scoring. We can now apply the fully loaded specification to the test set—our simulated real world targeting list—to guide micro-targeting efforts. This is accomplished using the `predict` function, which requires a `glm` model object and a dataset (`newdata`) to score observations.

```
predict(object, newdata, response)
```

where:

- `object` is a GLM model object.
- `newdata` is a data frame. This can be the training dataset or the test set with the same format and variables as the training set.
- `response` indicates the type of value to be returned, whether it is the untransformed "link" or the probability "response".

As a sanity check, we compute the AUCs for each `train` and `test` samples, revealing comparable performance of 81.2 % versus 80.9 %). Generally, we should expect to see that test accuracy is lower than train accuracy. A small to negligible difference in accuracy between the training and test indicates the model only slightly overfits, whereas a large difference should merit more attention.

```
# Predict
yhat_train <- predict(glm_full, train, type = "response")
yhat_test <- predict(glm_full, test, type = "response")


# Obtain AUCs
auc_train <- costAccuracy(train$no.coverage, yhat_train)
auc_test <- costAccuracy(test$no.coverage, yhat_test)
```

Measures for deployment. Model accuracy is one thing, but whether a classifier that is fit for purpose requires other measures. In the case of prioritization, we should ask *what is the expected hit rate for cases rated at pr > 50%?* and *how many uninsured people can we reach the top-ranked 1000 people?*. The answer to these questions can help with operational planning (e.g., how many people to staff) and cost-benefit analysis (e.g., can we reach enough people for the effort to be worth it?).

Let's calculate the estimated hit rate by first dividing the test sample into at intervals of 5% of the predicted probabilities `yhat_test`. Within each interval, we estimate the hit rate of uninsured Georgians along with the number of people. As one would expect, higher probabilities are associated with higher hit rates, but the number of people who have high scores are a relatively small proportion. Much like FireCast, as we move from high to low scores, we increasingly will encounter those who are already insured which in turn reduces the hit rate.

```
# Load
pacman::p_load(dplyr)

# Bin the test probabilities to nearest 5%
test$score_bucket <- round(yhat_test / 0.05) * 0.05

# Calculate hit rates
rates <- test %>%
            group_by(score_bucket) %>%
            summarise(hit.rate = 100*round(mean(no.coverage),2),
                  cell.n = n(),
                  cell.target = sum(no.coverage))
```

If the outreach campaign focused on scores equal to and greater than 50%, we would target a group of $n = 3406$ people (6.5% of the sample)—approximately 63% of which are uninsured. Alternatively, if the top 1000 people were targeted, the hit rate would be 74.5%—markedly higher rates but a small number of people would be reached. In either case, we can maximize targeting resources by targeting highly probable cases recognizing that there is a diminishing return (Figure 8.3).

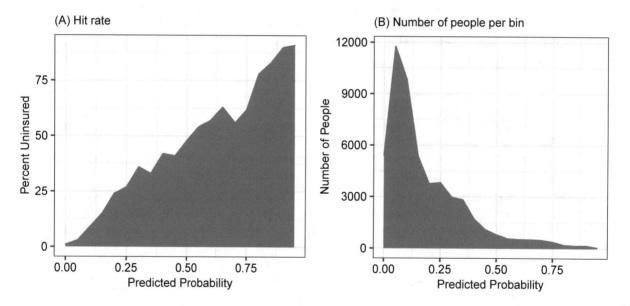

Figure 8.3: The expected hit rate based on the test sample predictions: the percent hit rate (Panel A) and the number of people (Panel B) given the predicted probability of being uninsured.

8.4 Regularized regression

Modeling problems should ideally fit the criteria of a *well-posed problem*. Originally described in Hadamard (1902), a well-posed problem meets three criteria:

- There is a solution.
- There is a unique solution.
- The solution depends continuously on the parameters or input data.

While it would seem that many data projects meet these criteria, there are challenges that arise with modern data. A problem that does not meet one or more of the criteria is considered an *ill-posed problem*. How could these problems manifest themselves in data?

Let's suppose a data scientist is tasked with identifying the drivers of health care insurance coverage. Her sample contains $k = 1000$ and $n = 500$. Conventional methods like logistic regression methods are ineffective for this effort—there is not a feasible solution when $k > n$. The typical workaround in policy environments is to approach the modeling problem from a normative perspective, using only variables that are conceptually relevant. After some trial and error, a data scientist could choose a three-variable specification that not only provides a feasible solution but lends itself to a crisp, compelling narrative. While these results could be promising, it is less so when put into perspective: The chosen specification is just one of 166 million possible three-variable specifications. In other words, simplifying a high-dimensional problem can lead to inaccurate or suboptimal results.

Imagine another case in which an analyst has a near limitless amount of degrees of freedom at his disposal. He finds that adding more variables to a model improves the fit. This, of course, comes with the increased risk of overfitting the model due to added complexity. When applied out-of-sample, individual predictions behave erratically, especially to small changes in the input variables. In other words, the plethora of variables allows a model to appear better than it actual is.

Both scenarios are examples of ill-posed problems that are common in information-rich environments. To overcome these challenges, we need to consider novel modeling techniques.

The regularization option. The creator of *Saturday Night Live* Lorne Michaels once said "To me, there's no creativity without boundaries". For SNL writers, they are given less than a week to write jokes and scripts, prep comedians and actors, hold rehearsals, and put on a professional performance. In this context, the showrunners, cast, writers, and crew can only converge on an idea when given the constraints (e.g., compressed time frame, limited budget).

This idea also relates to ill-posed problems as well. When a logistic regression (and any linear regression for that matter) is overly complex or has too many dimensions, a logistic regression cannot convert on a stable solution. We can introduce a constraint through *regularization*—a broad range of techniques that operationalize ill-posed problems. In this case, placing a penalty on the log-likelihood function will do the trick. This simple change to the log-likelihood function reduces the number of possible feasible regression solutions by forcing coefficients of noisier less correlated variables toward zero. In short, regularization *has the effect of shrinking some coefficients toward and sometimes exactly zero*, but how close to zero depends on the size and type of penalty.

Ridge logistic regression is a breakthrough technique developed in Hoerl and Kennard (1970). A penalty is placed in the log-likelihood function, reducing space for feasible coefficient solutions:

$$l(\beta) = \sum_{i=1}^{n} [y_i \hat{y}_i - log(1 + e^{\hat{y}})] - \lambda \sum_{k=1}^{K} |\beta_k|^2$$

where the tuning parameter λ (also known as the *shrinkage parameter*) adjusts the effect of an l_2 penalty—or the sum of *squared* coefficients. The additional term $(\lambda \sum_{k=1}^{K} |\beta_k|^2)$ is a *bias* that gives the regression less "wiggle room" to find coefficients that satisfy a stable solution. Adjusting λ directly impacts the size of coefficients β_k. Larger values of λ force the model to work with less room, forcing some coefficients toward zero (but not precisely zero). The λ value is determined analytically, meaning that data scientists must test a set of λ values along a grid (range of equally spaced values) through cross-validation. The λ value that minimizes a pre-selected error function corresponds to the model with the optimal balance of coefficients and accuracy.

Another regularized regression is Least Absolute Shrinkage and Selection Operator (LASSO). As developed in Tibshirani (1996), LASSO regression applies an l_1 penalty rather than an l_2 penalty:

$$l(\beta) = \sum_{i=1}^{n} [y_i \hat{y}_i - log(1 + e^{\hat{y}})] - \lambda \sum_{k=1}^{K} |\beta_k|$$

This simple but innovative change in the penalty allows regularization to take a remarkable step forward: *LASSO can force parameters to exactly zero and act as an automated variable selection procedure.* This is a key breakthrough that safely operationalize models, sifting through an insurmountably large variable set to find a unique subset.

While we discuss regularization in the context of logistic regression, we should emphasize that this is only one of many cases where regularization applies. In fact, LASSO and Ridge regressions can be applied to linear regression and survival analysis, and regularization techniques appear in many forms of machine learning techniques that are discussed in later chapters.

8.4.1   From regularization to interpretation

Reguarlized methods are useful for variable selection in high-dimensional datasets. While the models can be applied to predict outcomes, their coefficients cannot be directly interpreted like standard logistic regression coefficients—they are biased due to the penalty within the cost function. Furthermore, the methods do not produce standard errors, making it a challenge to articulate statistical significance. In the worst-case scenario, regularized regression can incorrectly shrink conceptually relevant variables to zero while retaining conceptually irrelevant variables.

Belloni, Chernozhukov, and Wei (2016) developed a strategy to overcome some of these barriers and make use of LASSO's variable selection properties in the pursuit of inference. As implemented in the `hdm` package, the Double LASSO selection procedure conducts variable selection from a large set of potential control variables yet produces an unbiased estimate of a variable in interest. The process unfolds in three steps.

First, a LASSO regression is fit on a set of covariates while omitting a focal variable x_i. Take note of the variables that have non-zero coefficients where λ is optimized.

$$y_i = f(w_{i1}, w_{i2}, ..., w_{ik})$$

A second stage model fits LASSO on the focal variable x_{ik}. Again, we take note of non-zero coefficients. Note that there should not be any additional variables selected if the x_{ik} is truly randomized. The non-zero coefficients from the first and second LASSO models are added to a set of controlling variables A.

$$x_i = f(w_{i1}, w_{i2}, ..., w_{ik})$$

Lastly, we fit a standard logistic regression where the inputs are variables from set A:

$$y_i = f(x_i, w_{ik \in A})$$

This procedure ensures that the variable of interest is retained, correlated controls are retained, and the effects are estimated in an unbiased, generalizable fashion. Along with other tricks from statistical and machine learning, regularization is a game-changer for the social sciences. Clever methods like Double LASSO selection will continue to emerge as strategies to make machine learning techniques possible for inference.

8.4.2   DIY: Re-visiting health care coverage

> *DIYs are designed to be hands-on. To follow along, download the DIYs repository from Github (https://github.com/DataScienceForPublicPolicy/diys). The R Markdown file for this example is labeled* `diy-ch08-lasso.Rmd`*.*

Let's revisit the health care targeting example and focus on the mechanics of applying LASSO regression. While there are many regularized regression packages, we recommend the `glmnet` package for general application of regularized regression and the `hdm` package for inference through LASSO.

```
# Load packages
pacman::p_load(glmnet, hdm)
```

Many ML packages expect input data to be in vector or matrix form. Recall from earlier chapters that matrices do not support mixed data types. A matrix containing both wages (continuous variables) and citizenship (discrete variables), for example, is stored as strings, posing a challenge for prediction. We can solve this format issue by converting discrete variables into a *dummy variable matrix* also referred to as *one hot encoding* in computer science parlance. A variable with *g*-number of classes is replaced with a matrix of *g* variables, each of which contains binary indicators for the class. To avoid the *dummy variable trap*—a situation in which one class can be predicted by all others—always remember that a *g*-class variable should yield a matrix with $(g - 1)$ variables. Below, we load `health_wide`—a re-processed version of the `health` dataset that reflects these format requirements.

```
# Load dataset
load("data/acs_health_expanded.Rda")
```

LASSO-Driven Prediction. We split the sample into train and test sets, creating a pair of vectors for the target y and a pair of matrices for inputs x. Our example below makes the assumption that we would like to consider all available variables in the dataset with the exception of the record identifier.

```
# Randomly assign
set.seed(321)
rand <- runif(nrow(health_wide)) > 0.7

# Subset train/test, drop ID variable, convert to matrix
y_train <- as.vector(health_wide[rand == T, 2])
x_train <- as.matrix(health_wide[rand == T, -c(1:2)])
y_test <- as.vector(health_wide[rand == F, 2])
x_test <- as.matrix(health_wide[rand == F, -c(1:2)])
```

The `glmnet` package is more flexible than any single LASSO or Ridge regression. It is built to estimate an elastic net regression that is a hybrid of *both* LASSO l_1 and Ridge l_2 penalties. In addition to tuning λ, a new tuning parameter α bounded between 0 and 1 controls whether the penalty will resemble more of a Ridge or a LASSO. In our case, we are interested only in variable selection, making our choice of simple: set $\alpha = 1$ to make use of the LASSO.

$$l(\beta) = \sum_{i=1}^{n}[y_i\hat{y}_i - log(1 + e^{\hat{y}})] - \lambda[\alpha \sum_{k=1}^{K}|\beta_k| + (1 - \alpha) \sum_{k=1}^{K}|\beta_k|^2]$$

This still requires us to identify the optimal value of λ through cross-validation. The `cv.glmnet` vastly simplifies the process by conducting the grid search:

```
cv.glmnet(x, y, family, cost, K)
```

where:

- `x` and `y` are the input variable in matrix format and the target variable in vector format, respectively.
- `alpha` is the elastic parameter in which $\alpha = 1$ is a LASSO and $\alpha = 0$ is a Ridge.
- `family` indicates the type of model such as `"binomial"`.
- `type.measure` is the loss function used to evaluate model fitness during cross validation such as `auc` or `deviance`.
- `nfolds` is the number of partitions used for cross validation. Default set to 10.

```
lasso.mod <- cv.glmnet(x = x_train,
                       y = y_train,
```

```
              alpha = 1,
              family = "binomial",
              type.measure = "auc")
```

The model outputs contain rich detail about the influence of *lamba* on the tuning results. As seen in Graph (A) in Fig. 8.4, accuracy is generally higher when the model contains more input variables with an optimum between $-7 < log(\lambda) < -6$. When $log(\lambda)$ is overly restrictive, model performance drops precipitously. The coefficient paths in Graph (B) show how fickle coefficients can be. Some can take on a wide range of values while others are relatively stable, but regardless of the variable LASSO force all coefficients to zero.

```
# (a) Cross Validated AUC
plot(lasso.mod)

# (b) Coefficient Paths
plot(lasso.mod$glmnet.fit, xvar = "lambda")
```

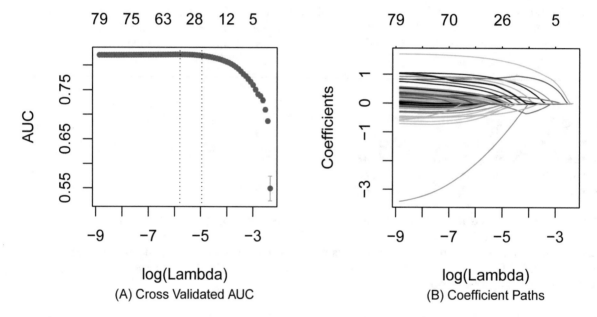

(A) Cross Validated AUC   (B) Coefficient Paths

Figure 8.4: Two graphs that illustrate the effect of tuning the hyperparameter log(lambda). For analytical convenience, log(lambda) is logarithm transformed. Graph (A) shows that as the shrinkage parameter is increased, the AUC falls. (B) Coefficient paths trace how a variable's coefficient evolves for each value of log(lambda), converging to zero at high values of the hyperparameter.

It will be tempting to tell a data-driven story from LASSO coefficients. In fact, it should be a part of any adoption strategy for any prediction problems. But LASSO does not provide generalizable parameter estimates. Nonetheless, a story informed by a cursory review of the LASSO coefficients can go a long way toward trust in the predictions. We can extract the results from the optimized model using `coef.glmnet`, setting the optimal value λ by specifying `s = "lambda.min"`. As the coefficients are biased, we recommend that you focus on identifying whether non-zero coefficients match the normative assumptions held by your stakeholders—*Are the coefficients moving in the right direction? does the model reflect what one might expect? Can you trust how the model discounts certain variables?*

```
opt_coef <- coef.glmnet(lasso.mod, s = "lambda.min")
```

Lastly, producing predicted probabilities is as simple as setting `s = "lambda.min"`. With the results ready, we can once again use the probabilities to guide recommendation engines and prioritization of tasks.

```
# Predict the target
yhat <- predict(lasso.mod, x_test, s = "lambda.min", type = "response")
```

Inference with LASSO regression. Let's now consider an inferential problem using LASSO: *What's the effect of citizenship on health insurance coverage?* To answer this simple question, the Double LASSO selection procedure can be applied to tease out the effect of citizenship on the health care story. The `hdm` package implements the procedure in the function `rlassologitEffect`, seamlessly estimating each stage of the selection process:

```
rlassologitEffect(x, d, y)
```

where:

- x is a matrix of control variables.
- d is a vector containing the focal variable.
- y is the target variable.

The coded example below illustrates that non-citizens are five times more likely (coefficient is $\beta = 1.735$, thus the effect is $e^{1.735}$) to be uninsured than citizens[4]. Unlike the LASSO estimates produced by `glmnet`, the `hdm` estimate is unbiased and generalizable. This result is not much different than that of the plain vanilla logistic regression, but imagine a scenario when the number of parameters k is large. We can see that this procedure could enable generalizable inference at scale.

```
# Set target and focal variable
target <- health_wide$no.coverage
focal <- health_wide$cit.non.citizen

# Set x, removing the id, target, and focal variables
controls <- as.matrix(health_wide[,-c(1,2,5)])

# Estimation
logit.effect <- rlassologitEffect(x = controls,
                                  d = focal,
                                  y = target)
# Summarize results
summary(logit.effect)
```

8.5 Beyond this chapter

Given traditional and newer regularized methods, which should one choose? The answer lies in the use case.

Logistic regression and other traditional methods are heavily used when focusing on *statistical inference* problems. In fact, fields like the social and medical sciences need to able to craft transparent narratives to explain what influences behavior and outcomes, which in turn can serve as context and insights to inform decisions. Traditional techniques are standard tools for inference, but this is not to say there is no room for methodological innovation. Some seasoned practitioners will continue to rely on what they know, but forward-leaning practitioners will drive adoption and acceptance over time.

Prediction problems, on the other hand, are quite different than inference. The focus is to pinpoint and anticipate what will happen to guide decisions on who and what specifically needs help. Whereas inference is concerned with interpretation of the coefficients $\hat{\beta}$, prediction focuses on training an algorithm that can be achieved consistent, generalizable out-of-sample predictions $\hat{y}_i$. While linear techniques could be used, flexible machine learning algorithms can absorb more information and achieve markedly higher accuracy. It

---

[4]To see the model outputs, run the code made available on the DIY site.

is not always clear which technique performs best, however. Data scientists conduct a horse race of multiple algorithms, crowning a winner based on the best predictive performance. Developing predictions is very much part of computer science and certain areas of econometrics, but otherwise are novel to policy, social science, and medical science environment.

Regardless of the use case, logistic regression is a solid first step for tackling any binary variable problem in public policy. It offers a simplified and transparent perspective of a quantitative problem lends itself to telling a narrative, and can support the rudimentary prediction problems. For organizations only starting to work with data, logistic regression is a safe and transparent way to build trust in data before advancing to more cutting-edge options.

Chapter 9

Three Quantitative Perspectives

Decision making in government can be a risky endeavor: policy makers often face complex questions in which constituents lives and taxpayer funding are at stake. Reducing this risk typically requires the decision makers to (1) sufficiently understand the context of the problem space, (2) conduct rigorous analyses that estimate impacts of similar decisions taken in the past, and (3) build models that identify potential paths forward. These fundamental inputs provide a evidence-based framework for decision making.

Let's take the example of issuing emergency approval of a new vaccine during a global pandemic. In such a setting, anxiety and stress can dominate the general sentiment of society. However, society is very heterogeneous. Some people may grasp for any promising solution, while others wait for the worse to pass (potentially a luxury that the first group may not have). Governments face a pivotal decision: *Should a vaccine be approved on an accelerated schedule in order to alleviate societal unrest?* The possible solutions to this simple question have the potential for enormous benefits and risks.

Responsible governance plays a key role. Finding answers to this question requires a degree of trust between the public and private sectors: Will both sides make mutually beneficial decisions that maintain order and enable the conditions for success? But how can we know whether this trust and order are in the interest of the general public? Data science can help decision makers understand the many factors that weigh into a successful response to the pandemic. For example, decision makers may be interested to know the likelihood of success and timing of medical trials, supply chain issues, market-power concerns, the actual demand for the new vaccine, and impacts of pricing on vaccine uptake. Each of these considerations can be addressed through a combination of (1) descriptive analysis, (2) causal inference, (3) and prediction. These three key quantitative perspectives are certainly related; however, each component plays a distinct role in answering the question.

Descriptive analyses rely on summary statistics to quantify the size, extent, and scope of the problem space. These analyses are not causal—they do not prove that A causes B. They do, however, set the stage for more advanced analyses. In the case of a new vaccine that has yet to be tested, descriptive analyses can give context. For example, success rates of similar vaccines, the average cost of making a unit of vaccine, and the likely demand for the vaccine.

Causal inference moves pass summary statistics and asks *What is the effect of X on Y?* These questions focus on detecting and measuring the effect of a *treatment* (or *intervention*). Typically, the causal effect of a treatment is *inferred* by comparing the outcome in a group that received treatment (the *treatment* group) to another group that did not (the *control* group—also called the *counterfactual* for the treatment). In order to estimate a causal *treatment effect* in this manner, it must be the case that the control group provides a valid counterfactual for the treatment group. In other words: *In the absence of a treatment, the treatment group would be indistinguishable from the control group.* When experiments are conducted under the right conditions, researchers are able to unambiguously attribute an observed effect to the treatment.

J. C. Chen et al., *Data Science for Public Policy*, Springer Series in the Data Sciences,
https://doi.org/10.1007/978-3-030-71352-2_9

Imagine a group of individuals who have contracted a virus. In an experiment, the group is randomly divided into two groups. The first group—the treatment group—receives an experimental treatment. The second group—the control group—does not receive anything (or perhaps receives a placebo). If the two groups begin with matching characteristics prior to the experiment and receive identical hospital conditions, then any observed differences in recovery (or side effects) between the treatment group and the control group can be attributed to the experimental treatment. In many situations, the *size* of the treatment effect matters—not just its existence. The next step: Decision makers must weigh the size of the treatment effect against other dimensions like side effects, other risks, and financial costs.

Descriptive analysis and casual inference describe what has already come to pass. *Prediction and forecasting*, in contrast, look forward and ask the question "*What if a pattern were to persist into the future?*". In order for this quantitative crystal ball to be useful, the predictive model should have an ability to mirror real-world phenomena with a high degree of accuracy. Typically this predictive model must prove itself superior to competing approaches and human intuition, though these competing "models" can range from other highly complex statistical models to spreadsheet "look-up tables" to intuition-driven *ad hoc* rules—or even the gut of a veteran decision maker. It is only when a model demonstrates superior predictive performance when decision makers may start to place some trust in the future that it predicts.

Continuing the example from above: using data from the described clinical trial, a predictive model could be trained to predict whether an individual would respond well to the experimental anti-viral treatment—or whether we might expect the individual to develop a reaction. If the model achieves a high degree of accuracy, then the learned patterns could also be applied to patient records in insurance databases in order to anticipate how an untreated population will respond to a vaccine treatment. Or perhaps the predictions could help identify vulnerable population segments or areas for further improvement. The extent to which we can *generalize* the insights of these algorithms onto new populations will depend on how well the algorithm's initial sample represents these new populations. As we discuss later, representation and fairness are important principles in the prediction that are often overlooked.

When used together, these three quantitative perspectives jointly improve the quality of empirical insights. It is helpful to acknowledge that each of these quantitative perspectives evolved under and is guided by a different analytical tradition. In the public sector, however, these perspectives are often treated as a single unit, *e.g.,* data analysis. This lack of precision can sometimes lead to confused applications of methods. In this chapter, we draw distinctions between these three perspectives and introduce vocabulary and concepts that enable their use. We begin with a brief review of descriptive analysis—a sound starting point for any data project.[1]. We then review causal inference—the crown jewel of program evaluation, policy analysis, and the social sciences—from the lens of the potential-outcomes framework and related quasi-experimental methods. The chapter closes with describing principles and best practices of predictive analysis.[2]

9.1 Descriptive analysis

Descriptive analyses are the foundation of any data science pursuit. Much like Exploratory Data Analysis, these analyses characterize the problem space by drawing on simple summary statistics (e.g., the mean, median, correlation, and standard deviation) and plots (e.g., histograms, scatter plots) to fill in the context around a research question. Not only do these analyses help communicate patterns to stakeholders and inform strategies but they also help assess how much trust data scientists can place on the data. For example, a descriptive analysis focused on participation in a welfare program could describe the number of people who rely on the program, the demographic distribution of participants, and the growth of the program throughout time. These summary statistics can be presented to stakeholders to paint a better picture of the problem space and give a clue about which policy strategies to consider. Stakeholders' reactions to descriptive analyses also contain useful information for data scientists—it could validate the insights derived from the data or perhaps signal whether the insights were novel (e.g., news).

---

[1]Refer to Chapter 6 for a refresher on Exploratory Data Analysis (EDA).
[2]We will cover prediction extensively in subsequent chapters.

In general, descriptive analyses focus on a few core analytical tasks, such as:

- *Quantify* the size and extent of the problem space. *E.g.,* how many people use a health care service, which services they use, where individuals tend to access each service.
- Discern *correlated variables*—especially variables that are associated with an outcome of interest. *E.g.,* elderly individuals are more likely to use specific services. Another example: specific services typically appear together (positive correlation) or rarely appear together (negative correlation).
- Determine sources of *oddities* in the data, such as sharp discontinuities, biases, missing values, or other 'strangeness' in the data.

Some of these tasks can be extended through regression analysis, allowing one to attribute the partial effect of each input variable. While a regression model associates an outcome with a set of input variables, the model alone does not determine if a model is *causal* or *predictive*. In fact, classical statistical theory always reminds that *correlation is not causation*—there are well-accepted standards that define the conditions under which causality exists. Likewise, a descriptive analysis should not be mistaken for a prediction.

As we illustrate in later sections, it can be quite challenging for an analysis to meet the conditions that make for a valid causal inference or enable an accurate prediction. Even if an analysis is only descriptive, they can still be quite powerful—many policy stakeholders are often starved for information and some information is already much better than no information. For this reason, many analyses start and end as descriptive analyses.

9.2 Causal inference

Detecting a *causal* relationship is often a tricky task. While humans naturally look for cause-and-effect relationships in our lived experience, it is easy to misattribute the cause to the incorrect effect or even to reverse the two. To combat this tendency of false causal attribution, researchers have devised a well-structured framework for causal inference—essentially a science of inferring the causal effect of an event on an outcome.

In order to detect and measure, a causal effect of some variable X on an outcome Y, certain conditions are necessary:

- *Correlation.* If X causes an effect in Y, then we should observe some correlation between the two variables. Above, we stated that correlation *is not* causation, meaning correlation is not sufficient to establish causality. However, correlation is still necessary for causation.
- *Temporal precedence.* The cause X should precede the effect or outcome in Y.
- *Elimination of extraneous variables* We must be able to eliminate other variables that influence the outcome (akin to ruling out alternative explanations). We want a clear causal pathway between the potential cause X and the outcome Y.

For example, let's return to the experimental anti-viral treatment. Suppose an individual receives the treatment. According to these three conditions, the vaccine is said to *cure* an individual (i.e., *cause* improvement) if (1) the treatment eliminates symptoms (*correlation* between the treatment and the outcome), (2) the symptoms reduce only after administering the drug (*temporal precedence*), and (3) the vaccine is the only known influence (*elimination of extraneous variables*). For any individual, we can easily verify the first two conditions, but the third requirement is far more challenging. How can we rule out all other potential influences? What if some other factor coincided with timing of the drug? Did the drug *cause* the illness to dissipate, or was it some other event correlated with the application of the drug? Perhaps the weather improved around the same time we administered the treatment. Or what if the patient was already beginning to improve when she received the treatment? To be certain that the drug *caused* the patient's improvement, we must either know all other influences of the outcome or be able to observed the same exact individual under identical conditions *except without receiving the treatment*. The first requirement demands God-like knowledge of our patient's biology. It essentially requires knowing all causes to determine the cause. The second requirement necessitates traveling through time—or to parallel universes. While time travel is also implausible, we can actually make some progress here. We effectively have a missing data problem, where the missing datum is the outcome if the individual has not received the drug. Statistics can deal with missing data, but less so with the demands of omniscience.

At its heart, causal inference is a missing data problem. We only see the world as it happened. We do not have the chance to see *counterfactuals* (what the world would have looked like had one thing been different). To resolve this missing data problem—to shed light on the missing counterfactuals—we need an *identification strategy* to tease out (*identify*) the effect of a treatment. Within modern statistical traditions, two main structures are currently used to declare treatment effects as *causally identified*: the *potential outcomes framework* (also called the Neyman-Rubin causal model) (Imbens and Rubin 2015) and *directed acyclical graphs* (DAGs) (Pearl 2009). We will focus exclusively on the former, as it is presently the dominant method in political science, economics, and other social sciences, but we encourage readers to examine the latter as they are quickly becoming common.

9.2.1   Potential outcomes framework

To solve this missing data problem, we need to generalize beyond a single individual's treatment effect—broadening the estimated effect to be an *average* treatment effect—the average within the group of treated individuals. To formally define this treatment effect, it is helpful to introduce some concise mathematical notation.

We will use the variable D to refer to treatment status. Specifically, for the ith individual, $D_i = 1$ denotes that i received the treatment (*e.g.*, our individual received the anti-viral drug). Conversely, $D_i = 0$ implies that individual i was untreated.

Similarly, Y_{1i} gives individual i's outcome *when she receives treatment*[3], and Y_{0i} references individual i's outcome *when she is untreated*.

The difference between individual i's treated outcome (Y_{1i}) and untreated outcome (Y_{0i}) gives the treatment effect for individual i, i.e., $\tau_i = Y_{1i} - Y_{0i}$ (where τ_i is the effect of treatment effect for i). Now we return to the problem of missing data: we cannot simultaneously observe both Y_{1i} and Y_{0i}. If we observe i's outcome when she is treated (Y_{1i}), then we do not get to see i without treatment—Y_{0i} is missing. Similarly, if we observe Y_{0i}, then we cannot observe Y_{1i}.

Identification strategies are essential methods for filling in this missing data. If Y_{0i} is missing (because we observe Y_{1i}), how can we fill it in (or estimate it)?

Part of the solution to this missing data problem is to move to an *average* treatment effect $\overline{\tau}$, rather than the individual-level treatment effect τ_i. As its name implies, the average treatment effect is literally the average of the individual treatment effects for some set of individuals: $\overline{\tau} = \frac{1}{N} \sum_{i=1}^{N} \tau_i$.

Let $\hat{\tau}$ refer to an estimate of the average treatment effect τ. How might we estimate this average treatment effect?

One possibility that naturally comes to mind: Can we just take the difference between the average outcome in the treated group and the untreated group? Let's write this formally:

$$\hat{\tau} = \frac{1}{N_1} \sum_{i=1}^{N_1} (Y_i | D_i = 1) - \frac{1}{N_0} \sum_{i=1}^{N_0} (Y_i | D_i = 0)$$

where the notation $\sum_i (Y_i | D_i = 1)$ refers to the sum of the outcomes for all individuals who received treatment and similarly for $D_i = 0$. Also: Let N_1 and N_0 be the numbers of treated and untreated individuals, respectively.

Now back to our big question: When does $\hat{\tau}$ provide a good estimate for the average causal effect of treatment?

The recipe for success lies in achieving *ignorability*: We want the treatment and control groups to be indistinguishable along dimensions that are relevant to our outcome Y. It would be easy if we could perform the group-wise comparison (the difference in means). However, in "the real world", simple group-wise comparisons, without a legitimate identification strategy rarely result in "good" causal estimates. Why? When

---

[3]The 1 in the subscript again denotes treatment.

individuals are allowed to select into or out of treatment, *selection bias* typically occurs: the two groups will differ on important dimensions. For instance, people who select to receive an experimental drug treatment *may* be fundamentally different from individuals who do not choose to participate in such a treatment. If that is the case, the difference between the treatment and the control groups' averages will include *both* the treatment effect *and* differences between the two groups of people. This latter component biases our estimates of the treatment effect and is often called selection bias. This bias results from confounding non-treatment factors with treatment, which means that we run the risk of misattributing the effect of treatment to underlying differences between the two groups. With selection bias, the control group (individuals for whom $D_i = 0$) does not provide a realistic counterfactual for the individuals in treatment group.

How can the incidence of confounding factors be minimized? Researchers in experimental settings tend to rely on randomized control trials (RCTs), which can achieve ignorability by randomly individuals' assigning treatment statuses—essentially flipping a coin to determine whether an individual is placed into the treatment group or the control group. Through random assignment, we introduce statistical independence between the treatment (D_i) and the other factors that affect our outcome. This independence greatly reduces the risk of confounding factors.[4]

Some public programs *can* be randomized. For example, when a school has more applicants than seats, the school will often conduct a random lottery to select its students. However, it is often the case in public policy that we cannot randomly assign policy changes to individuals—due to ethical or cost issues. Instead, researchers look for *natural experiments* in observational data in which randomization naturally occurs. For example, we might be interested in the effect of a minimum wage change and compare businesses on either side of a state border as in Card and Krueger (1994). Or perhaps we could leverage close election results to evaluate incumbency advantages as in Lee (2008).

A primary feature of these natural experiments is that, due to the assumptions regarding way, the treatment is applied, individuals become "as good as" randomized. The assumptions that we put on the treatment process and the data are part of our *identification strategy*, or the way we effectively turn observational data into a reasonable proxy for experimental data. The term identification strategy first appeared in the economics literature with Angrist and Keueger (1991) and has since become ubiquitous when discussing the estimation of causal effects. We would like to emphasize, what makes an estimate "causal" is not the model specification nor the data, but rather the specific assumptions one places on the process and model— the validity of an identification strategy. Estimates are only causal in so far as the assumptions made are believable.

In this section, we introduce two strategies that enable the pursuit of causal inference: *Regression Discontinuity* (*RD*) and *Difference-in-Differences* (*Diff-in-Diff* or *DID*). Both techniques are frequently used with experimental data and observational data, allowing inference under different situations.

9.2.2 Regression discontinuity

As first described in Thistlethwaite and Campbell (1960), regression discontinuity (RD) design is a framework which, under certain assumptions, mimics an RCT (Cook and Wong 2008) when given observational data. This is a powerful concept as gathering data through an RCT can be cost-prohibitive and invasive. The Thistlethwaite and Campbell (1960) study examined the effect that receiving public recognition through honorary awards had on educational outcomes (e.g., receipt of scholarships). Importantly, they relied upon a qualifier that identified which students would then go on to receive the honorific. In this particular case they used scores from the CEEB Scholarship Qualifying Test with those receiving a sufficiently high score being eligible for the honorific. In short, the honorific was a deterministic outcome of the test score.

If the relationship between the test score and honorific is known, *why do we need anything else to make a statement about causality?* The short answer is *selection bias*. Test scores measure the underlying ability, which is not strictly observable. There are groups of students who always would have received the honorific based on their scores, we call these *always takers*. Conversely, there is a group that lacks the ability to obtain a high score on the test and thus be eligible for the resulting award, we call these *never takers*. It is hard to

---

[4]Significant differences between groups can still happen by chance (i.e., "the luck of the draw").

make the case that we can estimate a causal effect of the honorific if we know the outcome beforehand for a group of students. This is why the focus is on *compliers*, or the group of students around the cutoff used to determine the eligibility for treatment. We would imagine that the probability of receiving an honorific jump at some cutoff along with the test score variable—the score that helps identify the probability jump is known as a *running variable* (or forcing variable).

Why do we focus on compliers?. Suppose the cutoff is a grade of 80%; then it might be reasonable to think that a student which received a 79% is interchangeable with one who received an 81%. More plainly, we can consider them to have identical ability and observable characteristics (on average), and that their presence on either side of the cutoff had to do with pure chance. This is how the RD design approximates the RCT and allows us to draw causal inferences from the estimates.

On a more technical level, RD consists of a treatment which is a monotonic (i.e., either never increases or never decreases), deterministic (i.e., not random), and discontinuous (i.e., involves a sharp break in trend) function of a running variable, Z. Here, we will use Z to denote the a vector of length N of observed forcing variable values, and z_i to denote the value for the ith individual. Additionally, assume the treatment is of the form,

$$q_i \begin{cases} 1 & z_i \geq Z_0 \\ 0 & z_i < Z_0 \end{cases}$$

where Z_0 is some cutoff for treatment eligibility, q_i denotes the treatment status of the ith individual and Q to denote the N-dimensional vector of treatment indicators for the sample. A key feature is that the probability of treatment "jumps" from zero to one at some threshold.[5] The simplicity of this design makes it a common causal inference strategy in public policy.

Simulating a discontinuity. One challenge we face with any RD is that it is unlikely that we know the true treatment effect, but through simulation, we can test to see how close the technique can be to measuring the treatment effect. Visualization plays a key role in identifying a discontinuity—a sharp discontinuity should involve a slope change or step change. Let's simulate what a random dataset with a discontinuity could look like in the wild. The simulated dataset below includes a running variable Z with $N = 10000$ normally distributed observations centered at zero along with a binary variable Q to indicate whether a record is above the cutoff of Z0= 0.

```
# Set random seed for reproducibility
set.seed(3264)

# Let's generate data
N <- 1e4    # Number of observations
Z <- rnorm(N, 0, 5) # Running Variable of mu zero, standard deviation 5
ZO <- 0  # Cut-off point
Q <- (Z > ZO) * 1 # Binary variable

# Construct data frame
example_data <- data.frame(Z = Z,  Q = Q )
```

In Figure 9.1, we plot the simulated series. By design, there is a sharp jump in probability of treatment at $Z = 0$. Furthermore, notice that there is not any overlap between the treatment and control groups at zero—an agent (e.g., observation in this case) is either treated or not based on $z_i = z$. The case where two agents have different treatment statuses with the same value of the running variable is strictly forbidden, otherwise the analysis in invalid.

---

[5] For simplicity, we have confined the conversation to what is called a "sharp" regression discontinuity design. For cases in which compliance around Z_0 is not perfect (or assumed perfect) there is also a "fuzzy" RD design. See Lee and Lemieux (2010) for a discussion.

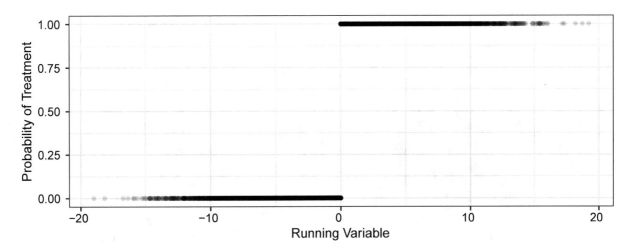

Figure 9.1: Plot of running variable and by probability of treatment.

With the running variable set, we can expand this simulation to add an outcome variable that is dependent upon the running variable and treatment status.[6] We use the following equation to simulate a real-world outcome:

$$Y = \alpha + \zeta Z + \tau Q + \varepsilon$$

where Y is the outcome, Z is the running variable, Q is the treatment indicator, and ε is an identically and independently distributed error term. Our objective is to measure the effect of Q on the outcome Y. In the simulation, we explicitly set the treatment effect as $\tau = 3$, thus our objective when applying the tools of regression discontinuity is to estimate a treatment effect of the same size.

```
# Basic assumptions
epsilon <- rnorm(N) # Random error
tau <- 3 # Size of the treatment
alpha <- 20 # Intercept
zeta <- 0.5 # Slope of the running variable

# Linear formula
Y <-  alpha + zeta*Z + tau*Q + epsilon

# New dataset
simulated_rd_data <- data.frame(Y = Y,
                                Z = Z,
                                color = ifelse(Q == 1,'blue','lightblue'))
```

The simulated treatment effect can be clearly seen in the pair of graphs in Figure 9.2. The histogram of the outcome Y is bi-modal, indicating that there are two sub-populations present—*something* caused the distribution to split. We see further evidence of this hidden influence in the scatter plot of the outcome Y against the running variable Z. The treatment and control groups sharply split at a threshold value. It is worth noting that, in practice, finding the running variable will require careful exploratory data analysis.

Measuring the treatment. With the simulation in place, we turn our attention to measuring the treatment effect, written as the expectation of

$$E[(y_i|q_i = 1) - (y_i|q_i = 0)|z_i = z]$$

---

[6]Note that this assumption of linearity is not required and is only being made here for ease of exposition. While higher order polynomials are discouraged, they often see use in empirical work. See Gelman and Imbens (2018) for more in-depth treatment.

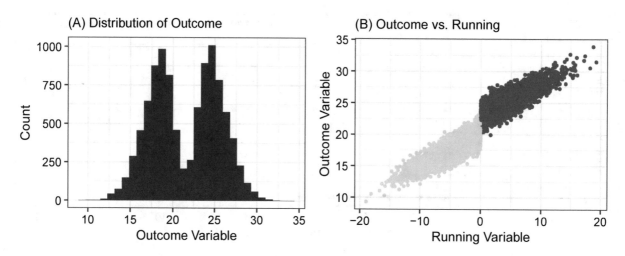

Figure 9.2: (A) A Histogram of the outcome variable, (B) A scatter plot of the running variable and outcome.

where $(y_i | q_i = 1)$ is the observed outcome and $(y_i | q_i = 0)$ is the corresponding counterfactual. More plainly, we would like to know what the effect of treating individual i is on their outcome, y_i, relative to a world in which they were untreated. Of course, we are unable to observe both worlds simultaneously. Adopting notation from Gelman and Imbens (2018), we can write this as

$$\tau = \lim_{z \downarrow Z_0} E(y_i^{\mathrm{obs}} | z_i = z) - \lim_{z \uparrow Z_0} E(y_i^{\mathrm{obs}} | z_i = z),$$

where τ is the local average treatment effect (LATE)—this is our strategy for retrieving the explicit treatment effect in the simulation.

Why do we estimate LATE rather than an average treatment effect over the entire sample? To have confidence in our treatment effect, LATE only applies to *compliers* who immediately feel the effect of the treatment, while *always* and *never takers* are not influenced by the treatment. Agents that are closer to the treatment cutoff are more similar to one another—their relative position to one another is arbitrary. This arbitrariness can be interpreted as random assignment. Yet agents that are far from the threshold are more likely influenced by unobserved factors other than the treatment (e.g., student ability). LATE focuses on finding a subset located slightly above and below the threshold to tease out the treatment effects. These assumptions are what allows us to draw a causal inference from an ordinary regression coefficient.

We have laid out the mathematics and can now estimate the treatment effect using the `rdrobust` package. Remember that we are estimating the effect based on the compliers, requiring us to identify which observations would be considered our always and never takers as well. The `rdrobust` package simplifies an otherwise computationally intensive process.

```
# Load RD Robust package
pacman::p_load(rdrobust)


# Estimate RD
rd_estimation <- rdrobust(y = simulated_rd_data$Y,
                          x = simulated_rd_data$Z,
                          c = 0,
                          all = TRUE)


# Show result
summary(rd_estimation)
```

Table 9.1: Regression estimates of LATE with bandwidth size.

Method	Coefficient	SE	Side	Full n	Effective n
Conventional	2.987	0.061	Below	4984	2316
Bias-Correction	2.973	0.061	Above	5016	2300
Robust	2.973	0.072	Total	10000	4616

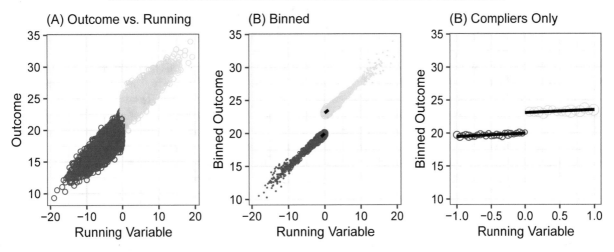

Figure 9.3: Three views of the outcome and the running variable: (A) raw data, (B) binned by discrete interval, (C) compliers only.

As seen in Table 9.1, the regression estimate for τ was 2.987, representing the average vertical distance between the two point clouds at the cutoff point, $Z_0 = 0$. Notice that we combined the regression techniques from the previous chapter, along with some assumptions about the relationship between the running variable, Z, and the treatment status Q to derive a causal estimate of Q on Y.

In order to estimate the treatment effect, the sample size was reduced from $n = 10,000$ observations to $n = 4616$. This effective sample size reflects an automated search for the optimal bandwidth (e.g., window of observations) to ensure the treatment effect reflects the effect on compliers.[7] Indeed, it is a weakness of RD that so much data is both required and discarded in the process, we did not use nearly half our sample!

For good measure, let's plot the results and get an idea of what the regression lines look like and how the measure of τ is reflected in the data. In Figure 9.3, we visualize a standard set of visual robustness checks. Whereas Graph (A) plots the raw data, Graph (B) bins the data to approximates the value of the outcome for small equal intervals. Close to the center of the plot, we see two parallel but staggered lines that mark the effective sample size. In Graph (C), we take a closer look at the compliers and find that the regression line closely aligns with the binned estimates, which lends confidence to the results.

There are many creative applications for RD. In the transport market, for example, Cohen et al. (2016) apply a RD design to a large scale dataset to estimate the entire demand curve for UberX (ride-sharing service). This, in turn, helped to estimate the associated consumer surplus in four US cities—approximately $2.9 billion in 2015. In politics, Lee (2008) found that incumbency has a large influence on Congressional Elections (specifically for the U.S. House of Representatives) that were narrowly won. In fact, incumbency raises the probability of an electoral win by 0.4–0.45 (40–45 percentage points). These findings are exciting and illustrate the power of the technique. And in practice, the treatment effect can directly be used to inform cost benefit analyses and decisions for whether a program or policy is effective or a good investment.

---

[7]We do not go into the selection of the bandwidth and leave it to the reader to examine various options for selection. See Imbens and Kalyanaraman (2012) and Calonico, Cattaneo, and Titiunik (2014).

At the same time, RD estimates are applicable to only a narrow bandwidth around the treatment. The results are internally valid but may not be generalizable beyond those conditions—a great way for telling a story, but may require significant disclosure of the conditions under which the effect can be replicated when implementing a policy.

9.2.3 Difference-in-differences

Difference-in-Differences (DID) combine regression techniques with assumptions about how the data was generated to facilitate causal inference. Like RD, DID relies on features of the data in a different way: *we monitor two otherwise similar groups for a divergence in trend that can be attributable to an outside treatment.*

Perhaps one of the most famous examples of a DID is Card and Krueger (1994), which examines the effect of minimum wage changes on employment in the fast-food industry. In this study, the authors leveraged the proximity of fast-food restaurants in two states, Pennsylvania (PA) and New Jersey (NJ), which differed primarily through the introduction of a minimum wage change in New Jersey. Their outcome of interest, employment, was measured in both states before and after the minimum wage change—a simple pre-post design. They then compared the changes in employment between the two states, making the assumption that fast-food restaurants just across the border in Pennsylvania could be considered as valid *counterfactuals* (what did not happen) for those treated restaurants in New Jersey. We can write this two-period, two-state effect as

$$\underbrace{\Big(E(Emp_{st}|s = NJ, t = 1) - E(Emp_{st}|s = NJ, t = 0) \Big)}_{\text{First difference in employment in NJ}} - \underbrace{\Big(E(Emp_{st}|s = PA, t = 1) - E(Emp_{st}|s = PA, t = 0) \Big)}_{\text{First difference in employment in PA}}$$

Note how in both states, we first take the difference of the *pre- and post-treatment* employment (Emp_{st})—a difference over time. Then, we take a second difference *between* the two states, hence the name Difference-in-Differences. The key assumption of the DID framework is the *parallel trends assumption*. Simply put, this assumes that in the absence of the treatment, the distance between the treated units and their untreated counterparts would have stayed constant over time.

Let's see what that means in practice. Suppose we have two groups of agents: one treated $(N_{d=1})$ and one untreated $(N_{d=0})$. Over time, these two groups tend to follow a similar trend which we can write as

$$Y_{t|d=1} = \alpha_{d=1} + \lambda t + \phi_{d=1} Y_{t-1|d=1} + \varepsilon_t$$
$$Y_{t|d=0} = \alpha_{d=0} + \lambda t + \phi_{d=0} Y_{t-1|d=0} + \varepsilon_t$$

where $Y_{t|d=1}$ is the outcome of interest at time t for the treated group $d = 1$, and $\alpha_{d=1} \neq \alpha_{d=0}$. For simplicity, we will assume in this case that $\phi_{d=1} = \phi_{d=0} = 0$ and the distribution of ε_t is common to both groups.[8] Despite the randomness introduced by ε_t, both groups move in parallel due to the shared λt term, the only difference is contained in their intercepts.

Getting started. Let's import a simulated dataset to illustrate how a DID works in practice. The `did_data.Rda` file contains two data frames:

- `did_data` contains $n = 2000$—$k = 100$ records over $T = 20$ periods—it is a panel dataset. The variables of interest are the outcome `y` in stacked format, a `treated` dummy for each control and treatment, and the `time` variable.
- `projected_data` contains a simulated *counterfactual*—what would the treatment look like if no treatment occurred (a straight projection of the previous trend).

---

[8]Additionally, we have suppressed the subscript i for readability but one could imagine that the units of observation are individuals or firms in which we record the appropriate data in every time period.

While all records could be graphed, we can more easily spot the parallel trend lines by first aggregating the panel data using `dplyr`.

```
# Load packages
pacman::p_load(dplyr, ggplot2)

# Load data
load("data/did_data.Rda")

# Calculate mean by treatment group and time period
did <- did_data  %>%
group_by(time, treated) %>%
summarize(mean.y = mean(y))

# Calculate mean for counterfactual
cf <- projected_data  %>%
group_by(time, treated) %>%
summarize(mean.y = mean(y))
```

Figure 9.4 illustrates the parallel trend assumption in action prior to $time = 10$. Our control group (solid black line) at the bottom runs parallel to the treatment group (light blue and dashed blue lines) prior to $time = 10$. A longer time series in this *pre* period builds the case that treatment and control are indeed comparable. After $time = 10$, the treatment causes a sharp step change, then returns to the parallel trend. This pattern suggests that the effect of interest is the step change—the instantaneous effect of the treatment. In many settings from medicine to technology, the geometry of treatment effects will often follow this pattern.

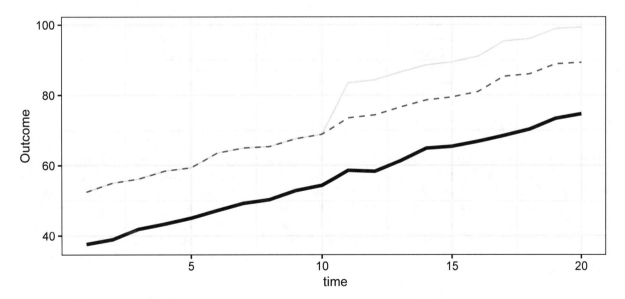

Figure 9.4: Average over time and treatment status.

We can extract the treatment effect through an DID estimator using regression:

$$Y = \beta_0 + \delta_1 treat + \delta_2 post + \delta_3(treat \times post) + \varepsilon$$

where *post* and *treat* are dummy variables for the post-treatment period and treatment series, respectively. Notice that the formula does not include a time trend, which is not necessary when control and treatment groups move in parallel. This regression can easily be implemented:

```r
# Time period treatment was applied
tau <- 10

# Pool the pre and post treatment periods
did_data$post.dumm <- ifelse(did_data$time > tau, 1,0)

# Estimate the model
did_model <- lm(y ~ treated*post.dumm, data = did_data)

# Get regression summary
summary(did_model)
```

What can we infer from the regression results (Table 9.2)? It tells us that the difference between the observed treatment and the assumed counterfactual is approximately 10 units. Because of the parallel trend assumptions, we can safely infer that the treatment at time period 10 *caused* a 10 unit change in the outcome for the treated. Now, this is clearly a very simple example with little noise and only a single treatment at a single time period. Often these designs are done when the noise is significantly greater, requiring careful consideration of adjustment variables and functional form. For further treatment for more complex DID designs, refer to Bertrand, Duflo, and Mullainathan (2004); Goodman-Bacon (2018); and Athey and Imbens (2018).

Table 9.2: Difference-in-Differences regression results.

	Estimate	Std. Error	t value	Pr($>$|t|)
(Intercept)	46.09	0.3376	136.5	0
treated	15.1	0.4775	31.62	3.245e-178
post.dumm	20.15	0.4775	42.2	1.093e-278
treated:post.dumm	9.981	0.6753	14.78	5.403e-47

9.3   Prediction

Causal inference focuses on extracting treatment effects with a high degree of confidence, focusing on answering *why*. Prediction, in contrast, answers questions about *who*, *what*, *when*, and *where*—anticipating what *will happen*. Interestingly, a regression used for causal inference could also be used for prediction. However, what makes for good causal inference is whether the experiment design handily illustrates the validity of the treatment effect (e.g., were the conditions met, are there no other sources of influence). For example, measuring the effect of a new coal plant on housing prices requires evaluating the qualities of a model when applied *in-sample*. As seen in Figure 9.5, a typical estimation flow uses one dataset for both effect estimation and calculating model performance.

Prediction, on the other hand, focuses on making guesses given available information. Models serve as a window into *what can be*, and for them to be useful, they must be accurate when applied *out-of-sample* (i.e., new data). The language of prediction is also different: a model is *trained* to predict a *target* (also known as the outcome or dependent variable) by learning patterns from *input variables* (also known as input features, explanatory variables, independent variables, etc.). When a model has been trained, it can *score* (to be applied) an out-of-sample dataset. This idea of scoring is another way of saying "*Given what the model has seen before and if we assume the new data follows the same paradigm, what will the outcomes be in this new dataset*". It is possible to simulate an out-of-sample dataset, but it is also necessary to understand if fresh data is available at the *cadence* required for a prediction application. In our modern economy, prediction is the name of the game. Self-driving cars need new data multiple times a second so a machine learning system can score road conditions to make driving decisions. E-commerce recommendation systems need user interaction and events so models can learn preferences. There is a whole host of other practical applications when prediction problems are well designed.

To determine if a model will be useful out-of-sample, we need a *model validation* design to tease out its predictive qualities. One basic approach is a train-test design in which data are randomly split into a *train* set to develop the model and a *test* set on which accuracy can be calculated. Partitioning the data allows us to simulate how a model will behave in the wild without having to collect more data. In general, the accuracy of a model will be higher in-sample than out-of-sample. Model validation like train-test among others quantify how much of a drop we should expect when a model is deployed.

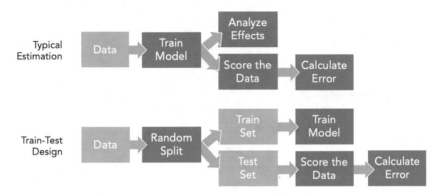

Figure 9.5: Comparison of typical flow for model estimation (for causal and descriptive analyses) versus a train-test design for prediction.

In this section, we introduce the foundational concepts of prediction, starting with understanding are the qualities of accuracy measurement and how it is calculated. As we had briefly described, the trick to successful prediction is to approximate how the model will perform in the wild. We describe a number of common model validation designs that convert ordinary error functions into generalizable accuracy ratings.

9.3.1 Understanding accuracy

Bias-Variance Trade-off. Accuracy and error are two sides of the same coin. Accuracy has a positive connotation—*how close* are we to predicting a value, while error is *how far* are we from the value. What is more important is how we look at error in a prediction context. It turns out that error can be divided into two components: *irreducible* and *reducible*:

$$Error = Reducible + Irreducible$$
$$= (Bias + Variance) + Irreducible$$

Irreducible error is the natural uncertainty that arises from the data collection process. It is irreducible as there is nothing we can do about it other than changing how the data is collected. For example, Chen et al. (2019) applied advanced machine learning techniques to predict the growth of the US services sector. While the results were promising, they illustrated that their prediction error could not achieve an accuracy that is less than or even equal to the sampling error (see Figure 9.6). In other words, if the underlying data is noisy, there is little we can do about it. Another way of viewing irreducible error is how "noisy" the data is—are trends clear or fuzzy? Clear patterns are likely to have less irreducible error, while fuzzy patterns have more irreducible error. *If the target is too noisy, the prediction can be anyone's guess and it is not a good use of time to pursue.*

For simplicity, we consider reducible error in a regression context, which can be further decomposed into *Bias* and *Variance* (Hastie, Tibshirani, and Friedman 2001). The following formula shuffles around many of the same elements (e.g., $\hat{f}(x)$), but the nuances have large implications:

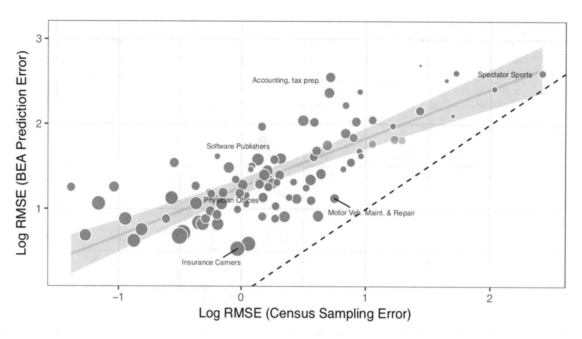

Figure 9.6: Prediction error of machine learning models developed in Chen et al. (2019) and the sampling error of the target variable collected by US Census Bureau surveys. The dashed line indicates the hypothetical scenario when prediction error is equal to sampling error.

$$Reducible = \underbrace{\left(E[\hat{f}(x)] - f(x)\right)^2}_{\text{Squared Bias}} + \underbrace{E\left[(\hat{f}(x) - E[\hat{f}(x)]^2\right]}_{\text{Variance}}$$

Bias is the difference between the expected model $\hat{f}(x)$ and the theoretical true model $f(x)$ when fit on the training data. This quantity can be thought of as the error that arises from erroneous assumptions—the data scientist's operating assumptions do not reflect what is truly happening in the data. If we think back to the cases of nonlinearities in the regression chapter, for example, a straight line is an oversimplified approximation of a curve—it has high bias.

Variance is the error associated with when a model learns random, irrelevant patterns in the data—a fixation on the eccentricities of a specific dataset rather than the prevailing signal. Another way to think of variance is to consider how one model will perform on multiple datasets on the same subject. If a model $\hat{f}(x)$ is trained on datasets $\{D_1, D_2, ..., D_3\}$ drawn from the same process (e.g., housing data, welfare use, etc.), would their predictions be similar when applied to D_n? If the predictions wildly vary from set to set, then the model has high variance. In practice, higher variance occurs when a model is complicated by too many variables.

Data scientists must strike a balance between bias and variance—working with the *Bias-Variance Trade-off*. In Figure 9.7, we illustrate three models that were fit on Bitcoin prices between 2010 and 2017. The first model (A) fits a straight, inflexible line through the time series, missing the fluctuations. The model is considered to be *underfit* as it consistently misses the variability in Bitcoin prices. Underfit models have large *bias* as the model has too few variables to describe the pattern in the data. Model (B) follows the Bitcoin price trend quite well with one hiccup: it is noisy and introduces patterns that are not largely seen in the data. The model is *overfit* as it has high *variance* from having an excess of input variables—it finds correlations between variables and the noise. As a result, when the model is applied out-of-sample, it introduces these artifacts that did not actually exist. In contrast, Model (C) is balanced—it is not too simple and not too complex. A Swedish word provides a simple description of this balance *lagom—just the right amount.*

What does this mean for policy? In pursuit of policy, there will always be pressure to tell the story about how specific X's influence Y. These cases require higher bias in order to tell a clear story. But for producing accurate predictions, we might need to lower the bias (i.e., more complexity) in order to increase accuracy. But models should not be so complex (very little bias) that predictions are noisy (high variance). Therefore, *models for telling the policy narrative will generally be simpler than models for prediction.*

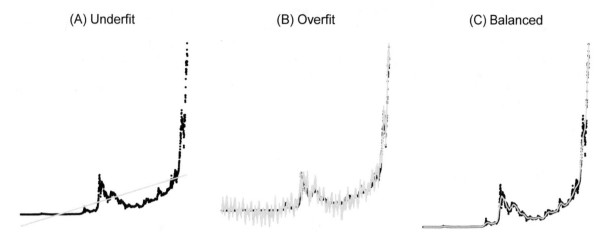

Figure 9.7: Bitcoin prices and how three different models for the data. Data obtained from Coindesk.

Loss Functions. The bias-variance trade-off describes the qualities of accuracy at a conceptual level. We concretize the notion of error through *loss functions* that allow us to directly calculate error. All models are wrong, but the question is by how much? For a set of predictions, we need to apply our beliefs on how accuracy is computed—*what is the cost of being incorrect.* This idea is much less about the cost induced by being wrong, but rather how we treat wrongness.

Let C be the cost of an inaccurate prediction and $\hat{y}$ be the predictions themselves, then we can write the cost function as

$$C = f(\hat{y})$$

The form that f takes should reflect our beliefs about error and help compare the performance between alternative models. Suppose we believe that all errors—regardless how large or small—hold the same weight. Mean Absolute Error (*MAE*, or L_1 loss) operationalizes this belief:

$$MAE = L_1(\hat{y}, \tilde{x}) = \sqrt{\frac{1}{N} \sum_{i=1}^{N} |y_i - \hat{y}_i|}$$

where y_i is the observed outcome and $\hat{y}_i$ is the prediction we generated. The L_1 loss function is strictly positive and a perfect prediction would result in an L_1 value of zero. Lower the value of the L_1 loss function, the better the predictions.

Large errors, however, are hard to ignore. A large miss can cause the trust and faith that your stakeholders have in your work to erode quite quickly. To find models that are more robust to outliers, consider a convex loss function— one that places a greater penalty on large misses. Squaring the errors leads to what is called an L_2 loss function which can be written as

$$RMSE = L_2(\hat{y}, \tilde{x}) = \sqrt{\frac{1}{N}\sum_{i=1}^{N}(y_i - \hat{y}_i)^2}$$

Notice that the magic lies in how we treat the residual $(y_i - \hat{y}_i)$. Like L_1, smaller values of the L_2 error are preferable.

Loss function in action. The value of loss functions can be easily seen when applied to real data. Below, we import a dataset of nighttime lights derived from satellite imagery collected by the US Defense Meteorological Satellite Program (DMSP). For this example, we use a derivative dataset containing the median light intensity for each month over a 20-year period between 1993 and 2013 (National Oceanic and Atmospheric Administration 2016; Gaba et al. 2016).[9] The World Bank, Development Seed among other international development organizations have experimented with this data to measure and predict the impact of electrification in the developing world (Min 2014; Min and Gaba 2014)—it is an intriguing proxy of infrastructure development. While the dataset is quite extensive, we focus on light intensity data for the northern Indian state of Uttarakhand.

```
load("data/ntlights.Rda")
```

Our objective is to produce a regression forecast model using a time trend and monthly dummies, choosing from two alternative specifications:

$$Light_t = \beta_0 + \beta_1 time_t + \beta_2 time_t^2 + \varepsilon_t \qquad Light_t = \beta_0 + \beta_1 time_t + \beta_2 time_t^2 + \delta_1 month_{t,1} + ... + \delta_{11} month_{t,11} + \varepsilon_t$$

The first equation is a quadratic time trend model while the second adds seasonal dummies. Granted, these are simple models that produce deterministic predictions, but are suitable for illustrating loss functions. The imported data can be partitioned into a train set (1993 through 2007) and test set (2008 through 2013). We then train the regressions with the pair of specifications, score the test set, then add the predictions to the test set.

```
# Partition into train and test
train <- lights[lights$year <= 2007, ]
test <- setdiff(lights, train)


# Train two models
model1 <- lm(vis.median ~ poly(date,2), data = train)
model2 <- lm(vis.median ~ factor(month) + poly(date,2), data = train)


# Score test set and add to the test set
test$yhat1 <- predict(model1, test)
test$yhat2 <- predict(model2, test)
```

Before calculating the loss functions, let's inspect the two time series predictions relative to the actual nighttime lights series for Uttarakhand (see Figure 9.8). Model 1 (blue) produces a straight line—it does not anticipate the fluctuations or outliers, indicating that the underlying model is likely underfit. Model 2 (light blue), in contrast, anticipates some of the seasonal patterns in some times of the year while *injects* patterns in others—an indication of being overfit. In short, both models are far from perfect, but *which seems to be more accurate?*

To put the loss functions to the test, we need to first write an `errors` function to calculate the MAE (L_1 loss), $RMSE$ (L_2 loss), as well as the out-of-sample R^2 for a set of predicted values `yhat` and the target `y`. The custom function can then easily construct a data frame of results for each model.

---

[9]The sample contains the median night-time light intensity for each DMSP satellite that was in operation, thus a single month can have multiple values from different satellites.

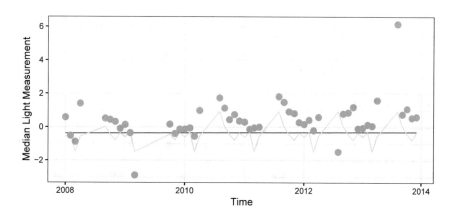

Figure 9.8: Comparison of action median light intensity for Uttarakhand. with two simple forecasts: `model1` (blue) and `model2` (light blue).

```
# Set up loss function
errors <- function(yhat, y){
out <- data.frame(r2 = cor(y, yhat)^2,
                            mae = mean(abs(y - yhat)),
                            rmse = sqrt(mean((y - yhat)^2)))
          return(out)
          }


# Calculate results
results <- rbind(errors(test$yhat1, test$vis.median),
                errors(test$yhat2, test$vis.median))
results <- cbind(model = 1:2,
results)
```

It turns out each function tells a different story (Table 9.3). When judging based on R^2, Model 2 tends to *fluctuate* more closely with target variable than the other model—their shapes are more similar. But closer inspection shows that Model 2's errors can be fairly large. While most peaks are anticipated with seasonal dummies, the troughs tend to miss far more than Model 1. In the presence of large outliers, Model 2's predictions move in the right direction, which makes it the preferred choice when judging predictions using $RMSE$. However, when all errors receive equal weight (MAE), Model 1 is the better choice. There is no discrepancy in these results, just different value judgments at play. In most cases, $RMSE$ and MAE will rank models in the same order, but sometimes, differences manifest themselves as in this case—your value judgments matter.

Table 9.3: RMSE and MAE estimates for Models 1 and 2.

Model	R2	MAE	RMSE
1	0.115	0.95	1.37
2	0.266	1.047	1.321

Pro Tips. You can only compare within a loss function, not between—*do not compare an L_2 to an L_1.* Additionally, the values of the loss function do not give any indication of uncertainty. Our two model examples provide a point *estimate* on the error, but what is the *consistency* of error? Model 2 is slightly better than Model 1 in terms of $RMSE$, but what if we were to draw another sample of nighttime lights data from a different time period—would we get the same results? Uncertainty matters and model validation designs can help.

Finally, while we only covered the L_1 and L_2 loss functions there are many other options to choose from (e.g., Huber Loss, Mean Absolute Percentage Error, etc.). In some cases, we may want to use asymmetric loss functions—the cost of under-predicting the target may be costlier than over-predicting. Choosing the right loss function reflects your policy objectives and managing your risks.

9.3.2　Model validation

Earlier in this chapter, we introduced a Train-Test design to simulate the conditions of applying the data in the wild. This simple design, however, only gives us a single opportunity to check if a model performs well. When two or more models have losses that are close to one another, it could very well be that one model is better than another by chance—that does not inspire confidence in the process of prediction.

To lend confidence to our choice of model, we need to create the possibility to have *multiple* testing opportunities. *Cross-validation* (CV) makes this possible.[10] As illustrated in Figure 9.9, k-folds CV splits a sample into k-number of randomly drawn, non-overlapping partitions (in the diagram, there are $k = 10$ partitions). The objective is to produce a prediction for each fold k using the other $k - 1$ folds, thus each model under consideration is trained k times and produces k sets of predictions. While this sounds complex, it is actually a brilliant approach to see how stable our predictions are. Loss functions can be calculated for each partition, then estimate the average loss and its standard deviation to show the consistency of each model's performance.

As we increase k, we increase the amount of power that is available for a model to learn and reduce the variance of the error estimate. But the trade-off of higher values of k is the computational expensive as more models need to be trained. A special case of k-folds CV sets $k = n$, which means each iteration of the CV process will leave out one observation. This *Leave-One-Out* (*LOO*) trains n models, providing the best possible estimate of a model's performance but can be costly to do.

Time series data are another special case k-folds CV can be implemented by partitioning k equal-length segments of the time series to preserve the order of observations. Models can be trained on $k - 1$ segments, then predict the kth until each segment is predicted out-of-sample.

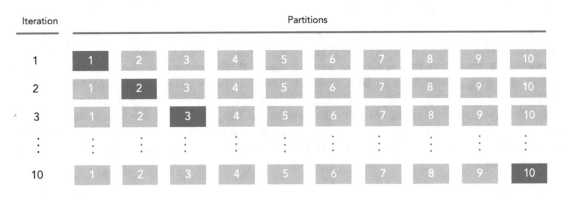

Figure 9.9: The flow of k-folds CV. A sample is randomly partitioned into k folds. Models are trained excluding one fold at a time, then predict on the fold that was held out. Repeat until all folds are predicted once.

CV in practice. Model validation is an essential part of any prediction project, but it may seem unwieldy due to the large number of repetitive steps required. In practice, implementing a basic CV process requires only a few lines of code. Below, we apply 50-fold CV using the `Boston` Census tract-level housing market dataset from Harrison and Rubinfeld (1978). To start, we load the dataset, specify $k = 50$ and create a vector `split_index` that randomly assigns each observation to one of k groups.

---

[10]Note that CV takes on different meanings in sub-domains in data science. In deep learning, for example, CV refers to Computer Vision.

```
# Load MASS dataset for Boston data
pacman::p_load(MASS)

# Set value of k and create random assignment vector
set.seed(1000)
k <-  50
split_index <- sample(1:50, nrow(Boston), replace = TRUE)
```

Using the CV framework, we test four simple regression models to check which variables carry the most predictive power:

$$W = \beta_0 + \beta_1 Rooms + \beta_2 Crime + \beta_3 NOX + \varepsilon$$

$$W = \beta_0 + \beta_1 Crime + \beta_2 NOX + \varepsilon$$

$$W = \beta_0 + \beta_1 Rooms + \beta_2 NOX + \varepsilon$$

$$W = \beta_0 + \beta_1 Rooms + \beta_2 Crime + \varepsilon$$

For each value of $k \in \{1, .., 50\}$, a train sample and test sample is partitioned based on `split_index` and a regression is trained for each specification. The `errors` function is used to construct the CV results for each partition, appending the loss estimates to the `cv_results` data frame. At the end of $k = 50$ iterations, the table contains four sets of 50 loss estimates. For small datasets, this process occurs nearly instantaneously, but larger datasets and complex models may require some patience.

```
# Loop through
cv_results <- data.frame()

  for(i in 1:k){
# Split sample
train <- Boston[split_index != i,]
test <- Boston[split_index == i,]

# Train a pair of models
mod_all <- lm(medv ~ rm + crim + nox, train)
mod_no_rm <- lm(medv ~ crim + nox, train)
mod_no_nox <- lm(medv ~ rm + crim, train)
mod_no_crim <- lm(medv ~ rm +  nox, train)

# Store results
cv_results <- rbind(cv_results,
                cbind(mod = "All", errors(predict(mod_all, test), test$medv)),
                cbind(mod = "No Room", errors(predict(mod_no_rm, test), test$medv)),
                cbind(mod = "No NOX", errors(predict(mod_no_nox, test), test$medv)),
                cbind(mod = "No Crime", errors(predict(mod_no_crim, test), test$medv)))
  }
```

Let's take a look at the RMSE estimates using kernel density plots in Figure 9.10. Three of the four distributions follow an identical bell curve pattern centered at $RMSE = 5$ with a long right tail, while the peak of the "No Room" model is to the right of the other distributions. These patterns provide us a few insights that we would not have seen in a standard Train-Test design. First, the *Room* variable is the most important factor—it carries the most predictive power while *NOX* and *Crime* have modest predictive power. Second, CV illustrates that there is a fair degree of variability from one model to the next. The *All* model has a mean $RMSE = 5.59$ with a standard deviation of 2.5—the predictions are quite variable. We can conclude that while the provides original study found a marginal cost associated with *NOX*, the CV process shows that the variable holds limited power in predicting prices.

As we will see in subsequent chapters, many `R` packages build in CV designs to simplify the problem of choosing the best possible model.

```
# Load package
pacman::p_load(ggridges, ggplot2)

# Graph densities
ggplot(cv_results) +
    geom_density_ridges(aes(x = rmse, y = factor(mod)),
                        fill = "#0070C0", colour = "white", scale = 2, alpha = 0.9) +
    xlab("RMSE") + ylab("Model") +  theme_bw()
```

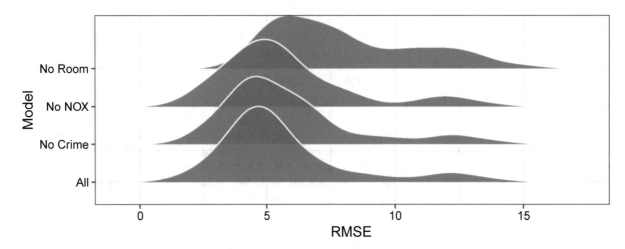

Figure 9.10: Kernel diagram of RMSE distributions by regression specification.

9.4   Beyond this chapter

While the three quantitative perspectives should be used in concert to extract the most value from data, many organizations will typically rely on one more than others. This reliance stems from a focus on what is necessary to accomplish their stated mission. For example, macroeconomic researchers use descriptive analysis to tell the story of the economy while their microeconomics counterparts tend to study treatment effects through natural experiments. Budget officers often monitor spending through descriptive measures; using these they develop forecasts of future financial performance to inform spending adjustments. Tax collection offices have been known to apply predictive techniques to identify potential cases of tax evasion.

There are occasions when the opportunity presents itself to leverage all three perspectives. One such opportunity arises when a government agency must balance scarce resources to accomplish a goal. Imagine a public health agency that is tasked with reducing the incidence of foodborne illness in restaurants. If the agency is understaffed, its inspectors might only be able to conduct inspections at restaurants every few years. Descriptive analysis can inform hypotheses and identify which covariates are indicative of food poisoning. An exploratory data analysis can also help assess if data quality is sufficient to enable data-driven interventions. If the signal quality is adequate, data scientists can develop predictive models which can identify the restaurants that should be inspected. While model accuracy may appear promising, the policy impact can only be measured through a randomized control trial, randomly assigning some health inspectors to visit a random sample of restaurants while assigning others to a list constructed through predictive means. This workflow is common in agencies that conduct field operations and risk management activities.

Another opportunity arises in policy evaluation settings that relies on causal inference using non-experimental data. Recall the role of a counterfactual in causal inference—a group of observations that serves as a reference point. In some cases, a natural counterfactual does not exist, requiring data scientists to simulate a reference group through prediction. For example, an interrupted time series design can quantify an observed treatment effect in time series data. As a reliable counterfactual might not be available, time series models are trained on pre-treatment data, then produce a forecast on the post-treatment period as if the treatment had not occurred. Thus, the counterfactual is a projection of *what if* the past paradigm had continued, placing significant weight on the quality of the counterfactual—how well the predictive model can convincingly mimic pre-treatment conditions. This is just one of many quasi-experimental techniques that benefit from predictive methods. As predictive analysis becomes more common, policy researchers and program evaluators will likely incorporate all three perspectives in their research activities.

There is always room for innovation in public policy. By mapping an agency's demand for insights to these perspectives, we can not only baseline the current state of practice but also identify ways that scientific insights can drive outcomes we observe every day. Whether it is traffic patterns, night light usage, bitcoin prices, or any other manner of the phenomenon, the goal is to draw some axiomatic belief about how the world works. We can then leverage these mechanisms and beliefs into making better predictions about the future which fuel smarter, more objective policy making.

Chapter 10

Prediction

10.1   The role of algorithms

Every day, we rely on high-frequency and high-resolution data to make even the smallest of decisions, and we do so without realizing the role of predictive models. The power of machine learning (ML) lies in its ability to predict, translating raw information of all kinds (e.g., imagery, sound, tabular) into useful information on which decision can be made. The benefits of these advancements have been principally seen in the tech sector, but are increasingly being adopted in the service of the public good. Microsoft, for example, has shown that machine learning algorithms can perform a mapping task that would normally require years for a team of humans to perform. Computer scientists trained a pair of algorithms to identify building footprints. One set of neural network algorithms were trained on five million satellite imagery tiles to identify pixels that belong to buildings, then an additional filter converted pixels into building polygons. (Microsoft 2018) The methodology achieved a precision of 99.3% and recall of 93.5%—accurate enough that one may have confidence that a computer can do the mapping task without human intervention. When set loose on satellite imagery for the entirety of the United States, the result was the first comprehensive national building database containing 125,192,184 building footprints (Wallace, Watkins, and Schwartz 2018). A complete inventory of the state of the built environment has never been available at this level of resolution and coverage, enabling use cases from emergency services to city planning to demographic research.

The Microsoft project illustrates the productivity gains made possible through automation, but ML can also perform hard analytical tasks. The ride share company Uber, for example, optimizes their demand load for their services (e.g., enough drivers to meet) through a forecasting model to anticipate extreme events such as sports matches and holidays. To do so at a global scale with hyper-local detail, Uber data scientists trained a Long Short-Term Memory (LSTM) algorithm—a kind of artificial neural network—to learn the trends, seasonality, and quirks. The forecast model in turn provides the advanced intelligence needed to help Uber make smart business decisions (Laptev, Smyl, and Santhosh Shanmugam 2017).

The public sector has also adopted predictive strategies to improve their services. In any given month, the Bureau of Economic Analysis (BEA) estimates the Gross Domestic Product (GDP)—a mosaic of data and forecasts that gage the performance of the economy. Some key datasets, like consumption of services, are not available in the preliminary GDP estimate, thus analysts produce projections until the "gold copy" of the data is available a short time later. While projections are easy to produce, accurate projections are elusive. If a projection is even marginally incorrect, economic estimates may need to be revised. In recent years, BEA has experimented with machine learning algorithms, such as *Random Forests* and *Regularized Regression*, to sift through thousands of economic variables so that projections are as accurate as possible, reducing revisions by billions of dollars (Chen et al. 2019). These predictive gains, in turn, can mitigate unnecessary market responses to revisions.

Different flavors of ML. This chapter is devoted to supervised learning algorithms—machine learning designed to predict and mimic a target variable as closely as possible. Some algorithms are designed with

© Springer Nature Switzerland AG 2021
J. C. Chen et al., *Data Science for Public Policy*, Springer Series in the Data Sciences,
https://doi.org/10.1007/978-3-030-71352-2_10

structured, tabular data in mind. Others are better for unstructured information like images and sound, while others may occupy a general-purpose role. They trade-off interpretability for raw predictive accuracy. In recent memory, computer scientists have been working hard to make the black box qualities of machine learning more transparent. As we will see later in the chapter, we can look under the hood for some algorithms to understand which variables they fixate on. If a prediction can mimic a phenomenon, prediction makes it possible for us to make decisions with confidence and precision.

What exactly do we mean by focusing on accuracy? Let's take a simple two-class problem as illustrated in Figure 10.1: the decision boundary could be linear, non-linear, or discontinuous. Social scientists and policy analysts tend toward *logistic regression*—it is a well-established part of the policy tradition. While it is *the* champion of parameter estimation, logistic regression is a lossy choice for non-linear, complex patterns as it has trouble mimicking the patterns in the data. Non-parametric techniques, in contrast, are far more flexible. *k-nearest neighbors k*-NN, for example, performs classification for an individual point by borrowing information from other nearby records with known labels. *Decision tree learning* such as *Classification and Regression Trees* (CART) and *Random Forest* learn patterns by subdividing a sample into more homogeneous regions by finding thresholds among the input variables.

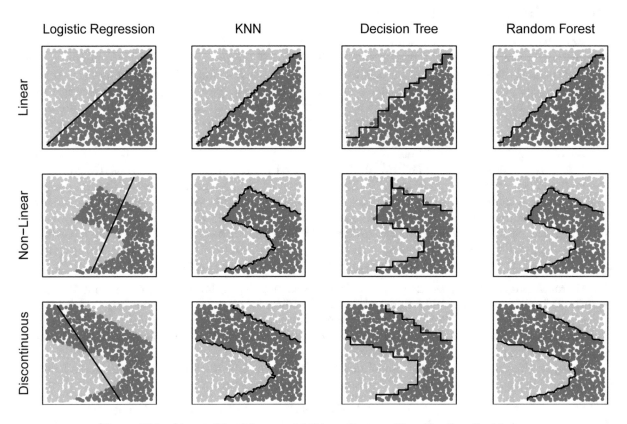

Figure 10.1: Linear, Non-Linear, and Discontinuous Classification Problems.

There are countless types of machine learning algorithms that are part of the modern data science toolbox. In this chapter, we explore a few of these methods, focusing on traditional machine learning methods and their potential role in informing public and social problems. We begin the chapter with the idea of a data science pipeline, treating prediction as an engineering problem. We then introduce three common machine learning techniques (*k*-NN, CART, and Random Forest) that are well suited for the typical tabular data and use cases in the policy space. The state of the art has advanced greatly beyond these basic techniques. To close the chapter, we conduct a high-level survey of advanced methods that are well suited for emerging types of data.

10.2 Data science pipelines

Prior to this chapter, we have covered the essential tasks that underlie data science, such as data cleansing, processing, visualization, and modeling. But these are not tasks that are exclusive to data science. Social and natural scientists, for example, incorporate these skills in the pursuit of an answer to a research question. Because the objective is to answer a specific question, analyses are coded in an ad hoc manner—resembling as a long series of commands with hardcoded assumptions. The data might also be cleaned by hand, making it a challenge to re-apply the same code to other problems or at least quite time-consuming. It goes without saying that ad hoc code does not scale. And if code is not scalable, the intricate detail of an analysis needs to be simplified into less precise rules of thumb for people to use—data can only inform strategy.

Data science, in contrast, brings together research-grade insights with engineering software programs to achieve scale. Rather than answering a specific question once, data science products are designed to continuously answer questions and adapt to the problem space. Raw data can be processed and manipulated. Processed data feeds algorithms to learn patterns. The trained model can score data feeds that in turn support decisions and actions. Each of these steps can be standardized and fit together like a series of modules or interchangeable parts, like an assembly line for transforming data into insight. This is the basic idea behind a *data science pipeline*—standardize the processing steps so that data scientists can focus on the use case. Many industries have adopted the pipeline approach, allowing them to move beyond data-informed strategy to data-enable tactics. Ad hoc tasks are standardized, code is simplified, and development of products becomes far faster to iterate than it might have once been.

Computer scientists have modularized the core tasks of data science by creating programming frameworks. *Apache Spark*, for example, was designed to help data scientists construct pipelines with minimal code and scale to large streaming data source streams. For more general use, the *Scikit-learn* library in the *Python* programming language is designed with data science pipelines in mind. The R-equivalent is the *caret* package, which interfaces with dozens of machine learning and data processing packages.

In this chapter, we focus on the `caret` package to train prediction algorithms. While we acknowledge the package is built with expansive functionality[1], we focus on seven essential functions that allow this single package interface with hundreds of machine learning algorithms with only a few lines of code (see Figure 10.2).

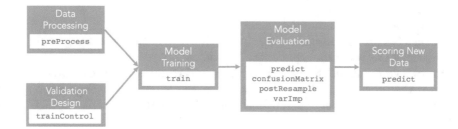

Figure 10.2: A simplified data science pipeline and associated functions.

The `preProcess` function allows you to dictate the types of *data processing* routines that should be applied to the data, populating the metadata object that instructs `caret` what to do. The library can handle basic processing routines such as variable standardization (e.g., mean centered and normalized by the standard deviation), missing value imputation (e.g., k-NN), and dimensionality reduction (e.g., "PCA"). Thus, you still will need to master more advanced processing skills such as working with text and matching. One benefit of using `preProcess` is that it records the processing steps under which a model will be trained. The mean used in mean-centering, for instance, needs to be retained so that new test samples can be adjusted using the

---

[1] See Kuhn and Johnson (2013) for in-depth treatment on `caret`.

same conditions as were used to create the training sample. Without capturing this essential information, the prediction scores for new data samples will not perform as expected.

To illustrate its use, `preProcess` is applied to a hypothetical dataset `wages`. We specify four `methods`: mean-centering ("`center`"), scaling by standard deviation ("`scale`"), imputation of missing values ("`knnImpute`"), then dimensionality reduction ("`PCA`"). The output is stored in the `process_obj`.

```
# Set up processing object
process_obj <- preProcess(wages,
            method = c("center", "scale", "knnImpute", "PCA"))
```

Next, we might consider to think through how to *validate our model*—a critical step for determining if the algorithm will perform out-of-sample. Rather than implementing model validation through a set of complex nested loops, we can instead use the `trainControl` function. `caret` is pre-loaded with useful model validation patterns such as simple k-folds cross-validation (`method = "cv"`), more complex approaches like repeated cross-validation (`method = "repeatedcv"`), or the computationally exhaustive Leave-One-Out CV (`method = "LOOCV"`). Like `preProcess`, the `trainControl` function produces a metadata object that tells `caret` how to set up the training data for model validation.

```
# Example 1: 10 iterations of repeated CV with 20-folds each
val_control <- trainControl(method = "repeatedcv", number = 20, repeats = 10)

# Example 2: Plain 10-folds cross validation
val_control <- trainControl(method = "cv", number = 10)
```

With the processing and model validation ready, we now can `train` a model using:

`train(formula, data, method, trControl, preProcess, ...)`

Where:

- `formula` is a formula object (e.g., "`y ~ x1 + x2`"). For classification problems, the target variable should be a factor.
- `data` is a data frame of training data.
- `preProcess` is a pre-processing object. Alternatively, you can feed a string vector with names of transformations (e.g., `knnImpute`, `scale`, etc.).
- `trControl` sets the model validation approach and requires a `trainControl` object.
- `method` is the type of algorithm, usually taking the name of a package such as `kknn` and `ranger`. For a complete list of algorithms available, visit the `caret` package's Github repository.[2]

ML algorithms often have a set of hyperparameters that need to be tuned in order to optimize for accuracy and generalizability. The number of parameter sets that need to be tune can easily balloon, requiring complex code to achieve a strong result. `caret` makes the process simple by performing an automated grid search—a systematic search for the best parameter set subject to a selected loss function. You can control the grid search by specifying one of the two following arguments:

- `tuneLength` is an integer that indicates how granular the hyperparameter grid search should be. Specifying `tuneLength` implies that you trust that `caret` is searching the right range of values within the hyperparameters, but you would like to control how detailed of a search should be conducted. For example, setting `tuneLength = 10` instructs `caret` to test 10 scenarios while `tuneLength = 1000` tests 1000 scenarios.

---

[2]Visit https://github.com/topepo/caret/tree/master/models/files.

- `tuneGrid`, in contrast, allows you to explicitly dictate the scenarios to be tested. `caret` expects a list of scenarios in the form of a data frame. Each column in the data frame is a hyperparameter specific to the algorithm being tuned, requiring the user to review the instructions for each method.

When we bring together these three functions to tune an algorithm (we use Random Forest from the `ranger` package as an example below), we can condense a large amount of code into only a few arguments:

```
mod_obj <- train(wage ~ .,
        data = wages,
        method = "ranger",
        preProcess = c("center", "scale", "knnImpute", "PCA"),
        trControl = trainControl(method = "cv", number = 10),
        tuneGrid = expand.grid(mtry = 1:10,
                    splitrule = "variance",
                    min.node.size = 2))
```

The output of the `train` function is saved to an object `mod_obj`, which is a sophisticated piece of scaffolding that holds an extraordinary amount of information. Like any other model object, it stores the trained algorithm and model diagnostics. We can retrieve all of the cross-validation results by calling `mod_obj$results` or pull the best parameters by calling `mod_obj$bestTune`. We can also take a peek under the hood at the *variable importance* by passing the model object to the `VarImp` function:

```
varImp(mod_obj)
```

But the most impressive functionality is that this one model object can shepherd data through a full data science pipeline. It holds the instructions not only for the trained algorithm but also all the pre-processing steps that prepare data for use. To apply the predictions to a new wage dataset `new_data`, we can call `predict` and `caret` takes care of all the data processing in the background.

```
predicted_values <- predict(mod_obj, new_data)
```

Now that the algorithm is up and running, we can evaluate the model's performance on out-of-sample data. If the labels (`actual_values`) are available, we can compare the performance of model predictions (`predicted_values`) using two functions. `postResample` computes performance metrics (e.g., Root Mean Squared Error (RMSE), R-squared, and Mean Absolute Error (MAE)) for continuous outcomes, while `confusionMatrix` computes a comprehensive set of confusion matrix statistics for discrete outcomes.

```
# Predicted accuracy for regressions
postResample(predicted_values, actual_values)

# Confusion matrix for classification models
confusionMatrix(predicted_values, actual_values)
```

Mastering these seven functions is enough to get you started with building data science pipelines. But *user beware*. The ease of constructing pipelines also makes it much easier to build models that do the wrong thing very well. Take the time to learn the inner workings of each algorithm that before actually applying it. Ultimately, it is the data scientist's responsibility to ensure that prediction algorithms are built and used appropriately. With the foundational mechanics laid down, we will now explore a few ML algorithms that often see action.

10.3 K-Nearest Neighbors (*k*-NN)

Let's consider the following scenario. You are driving down a country road without cell signal—just like the pre-smartphone era—and realize you are lost. There is an absence of signage, so you decide to stop the first person you see to ask where you are and get some directions. To double-check, you might ask for a second and third opinion from the next closest people. At some point, the amount of information collected is enough and you have an idea of where you are. In other words, people who are closer might have some knowledge of your whereabouts

k-NNs exploits this idea that predictions can be made by asking k-number of neighboring records for their opinions. Neighbors are determined based on distance based on treating input variables as a set of coordinates. k-NN is quite different from other algorithms because it is *instance-based*—it does not learn or retain patterns but produces predictions by comparing against an inventory of records (the training data). Such a simple idea has allowed k-NN to serve as the spackle of prediction models.[3] It is the method of choice for filling holes (imputing) in data while retaining the "shape" of data to which it is applied. In fact, it is also used by `caret` to impute missing values.

10.3.1   Under the hood

As illustrated in Figure 10.3, we can produce a prediction for each record i by following a simple iterative process.

Before starting, the analyst must set two parameters: k and *kernel*. The value of k determines the number of surrounding neighbors that factor into prediction y_i. The *kernel* is the structure under which neighbors are weighted. The true value of k is not known—it is a hyperparameter optimized through tuning. Nonetheless, for any set of hyperparameters, we hold the values fixed as we train the algorithm.

For each record i in the test sample, we follow these steps:

1. Calculate distance. To identify the closest training records around record i, calculate the distance d_{ij} from i to each point j in the training sample. Distance most commonly takes the form of Euclidean distance ($\sqrt{\sum_{i=1}^{n}(x_i - x_0)^2}$), which is appropriate with continuous values. For cases in which the underlying data are discrete, Manhattan distance ($\sqrt{\sum_{i=1}^{n}|x_i - x_0|}$) is more appropriate.

2. Vote! The prediction $\hat{y}_i$ is a vote of the k-nearest points around point i, but it is handled slightly different for each classification and regression problems. In classification, k-NN returns the proportion of neighbors that belong to each of class c, which is treated like a probability of c. The class prediction for each record i is done through *majority voting* in which the most prevalent class wins. In regression contexts, the simplest prediction of $\hat{y}_i$ is simply the average of the k-nearest neighbors.

There are various flavors of voting. Plain vanilla k-NN uses what is known as a *rectangular* kernel in which all neighbors receive the same weight. We could also use a decay function to place more emphasis on closer neighbors such as in the case of *inverse distance*:

$$\hat{y}_i = \frac{\sum_{i=1}^{k}\left(\frac{y_i}{d_i^p}\right)}{\sum_{i=1}^{k}\left(\frac{1}{d_i^p}\right)}$$

where d_i^p is the distance from the ith neighbor to the pth power. By increasing p, we place more emphasis on closer observations. The *Gaussian kernel* and *Tukey Bi-Weight* are alternative weighting schemes to impose distance decay.

Because of its simplicity, k-NN needs data to be treated carefully, in particularly through variable normalization and tuning.

---

[3]In the United States, 'spackle' is a type of plaster used to fill holes in walls.

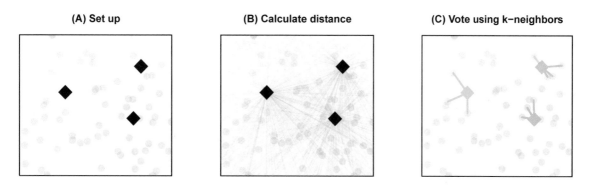

Figure 10.3: The process of training a k-NN algorithm.

Normalization. Because each input variable's values are treated like coordinates, the algorithm treats all inputs equal influence on the distance calculation. To ensure equal treatment means to normalize variables. For example, if an input variable x_1 has a range from 1 to 10,000 and x_2 ranges from 0.1 to 0.3, a k-NN will likely rely more on one variable than another. To normalize the variable set, we can calculate the z-scores for each x_k:

$$scaled = \frac{x_k - \mu_{x_k}}{\sigma_{x_k}}$$

where the transformed variable is mean centered with unit variance. This normalization approach only applies to continuous variables, however.

What if the inputs were of mixed data types? It turns out that there are many normalization strategies, but the common theme is to convert inputs into the same units. Real numbers can be converted into bins of a categorical variable in order to be used with other discrete variables, then a Manhattan distance is applied to identify neighbors. For example, an age variable could be subdivided into discrete age bins (e.g., 18–24, 25–29, etc.) and the distance between each sequential bin is 1. But there is some flexibility in *how* bins are defined. Age does not necessarily need to be defined as 18 to 24 for the lowest bin, but instead could be 18 to any other arbitrary threshold. In government, there are often standard age classification and matching bins to common classes allows for comparing across analyses. The bottom line is normalization is key.

Tuning. The performance of k-NNs is also sensitive to tuning. While we do not know the true value of k or the best kernel, we can conduct a grid search using k-folds cross-validation to hone in on the best parameter set. But given that there may be as many values of k as the sample size n, the goal should be to grid search as efficiently as possible.

One could search from $k = 1$ to $k = \sqrt{n}$ in intervals of one's choosing (e.g., steps of 1, 10, 100, etc.), keeping track of how each k performs in terms of a loss function (e.g., TPR, FPR, F1-statistic). By setting a coarse search grid, we can approximate the accuracy curve for values of k, then search more finely where accuracy looks most promising. But if time is not of the essence, then an exhaustive search at finer increments may be worth the wait.

To illustrate this process, we have drawn a simple random sample of USDA CropScape land cover data (W. Han 2019), a useful dataset for monitoring crops such as corn (yellow) and soybeans (green). As shown in Figure 10.4, while the CropScape data has full coverage in US farmlands, we simulate a survey sample that only covers 10% of land. Using what we know, we can impute the other 90% of the farmland using the k-NN algorithm. In the imputations below, we compare values of k at 1, 5, 10, and 100. The sensitivity to k is apparent as the value increases. The predicted location of cornfields increasingly creeps into areas where soy would be expected. Indeed, the accuracy trade-offs are quite clear.

Other considerations. Despite its simplicity, k-NNs can be computationally taxing due to the large number of pairwise calculations required between training and test samples. In other words, as the sample grows,

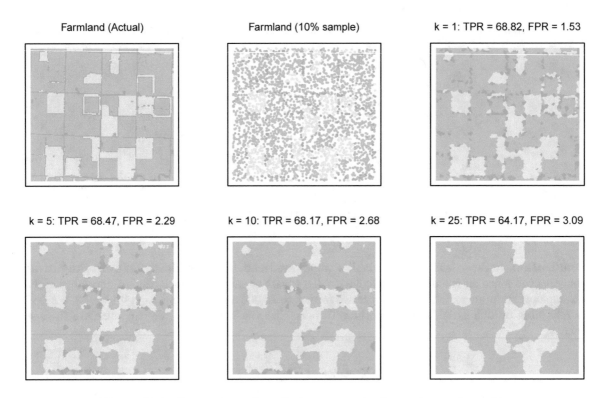

Figure 10.4: Comparison of prediction accuracies for various values of k.

it requires more time to process. Furthermore, as more input variables are incorporated, we also can face challenges with prediction quality as each variable is considered equally. This means that both signal-packed and noisy variables have equal chance of influencing the prediction, thus variable selection matters quite a bit.

In practice, k-NNs are best suited when datasets are smaller and contain fewer variables. They are also well suited when data is recorded along a grid (e.g., spatial, sound, imagery) and there is a lack of a theory of that guides how the inputs relate to the target or prediction. Furthermore, k-NNs are excellent when patterns are non-linear and discontinuous. But like many algorithms that we will encounter, k-NNs lack a clear avenue to interpretation. Most commonly, we will see k-NNs applied to imputation problems when there is not an obvious pattern behind the missing values (Table 10.1).

Table 10.1: The pros and cons of k-NN.

Useful Properties	Challenges
Efficient and timely when there are relatively few variables.	Mixed data types require to convert all data into dummy matrices
Effective in capturing patterns in cases where proximity matters.	Does not offer an interpretation.
Common choice for imputing missing values.	

10.3.2   DIY: Predicting the extent of storm damage

DIYs are designed to be hands-on. To follow along, download the DIYs repository from Github (https://github.com/DataScienceForPublicPolicy/diys). The R Markdown file for this example is labeled diy-ch10-knn.Rmd.

Perhaps one of the most important but under-appreciated responsibilities of city government is to take care of its trees. One might not even notice that trees are well cared for, but they do make their presence known from time to time. When a branch or entire tree falls, it can inflict property damage, bodily harm, and traffic disruptions. In some cities, like NYC, a resident can call the City's services hotline known as 311 to request the tree be removed, then the Parks Department is dispatched. During large storms, the local government's role is even more important, especially after Hurricane Sandy left its mark on New York City's urban landscape.

Let's suppose you are a city official tasked with triaging tree removal after a hurricane. The problem after storms is the chaos—there's only partial information available on the current situation. Some parts of a city have phone access and can still make requests for help, whether it is for downed trees or other localized problems. Other areas may not be as lucky. Thus, the disposition of most areas is largely unknown, which might lead first responders to support areas where calls for help are able to reach them and overlook more heavily impacted areas. In short, responders need to know *where* and *what* the problems are but not *how* or *why* things turned out the way they did—at least not during times of crisis. In this DIY, we rely on k-NN to impute a fuller picture of downed trees after Hurricane Sandy. Using 311 data available from the day Hurricane Sandy hit NYC, we impute downed tree status for large portions of the city, comparing the predictions with what is learned in the 7 days following the storm. We can fill in so that emergency responders can make more informed aid allocations.

Prepping the data. Rather than responding to individual calls for help when impacts are widespread, it may be more effective to allocate resources by grid cells. For this DIY, we have pre-processed NYC 311 call records into a grid of $1000ft \times 1000ft$ cells for a total of $n = 7513$ cells in all. The data contain only a few variables including an `id`, a county label `boro`, geographic coordinates of the grid centroid[4], and a set of binary variables indicating if at least one downed tree was reported on the day of Hurricane Sandy (`tree.sandy`) or in the following seven days (`tree.next7`). The two binary variables were developed under the assumption that residents of a neighborhood would report downed trees if they were spotted. If a call were made, then we could flag a neighborhood as one that experienced tree troubles. Alternatively, if complaints from a neighborhood are devoid of tree-related issues, we can assume that the area was unaffected.

For simplicity, let's focus on the two largest and populous boroughs (`boro`) that share the same land mass, Brooklyn (BK) and Queens (QN), that comprise 59.6% of the city or $n = 4477$ grid cells. We only have 311 information for 43.5% of grid cells at the time of the hurricane, leaving the tree status unknown for 56.5% of the region. If we were at the emergency command center, we have two options. On the one hand, we could wait for data to be collected, but it will likely take some time to gather complete information. Alternatively, we could apply a simple imputation model to approximate the current state of affairs so interim decisions can be made while more information is gathered.

```
# Load data
nyc <- read.csv("data/sandy_trees.csv")
```

```
# Extract Queens and Brooklyn
pacman::p_load(dplyr)
nyc <- filter(nyc, boro %in% c("BK", "QN"))
```

Train. Our training objective is to anticipate where downed trees will be reported over the next seven days by training a model on the first day's reports (target variable `tree.sandy`). Our main set of driving factors are latitude and longitude, which means the predictions are a function of geographic proximity. A quick tabulation shows that 79.7% ($n = 1550$) of grid cells in the training set have at least one downed tree reported – evidence that the storm had widespread impacts. To fill in the remaining 20.3%, we supply a data frame of all grid cells in Brooklyn and Queens that can be scored by the k-NN.

```
# Extract the training and test samples
training <- subset(nyc, !is.na(tree.sandy),
        select = c("ycoord", "xcoord", "tree.sandy"))
testing <- subset(nyc,
        select = c("ycoord", "xcoord", "tree.next7"))
```

---

[4]If latitude and longitude were recorded in decimal degrees, the value of one degree is dependent on where one is on the globe. When working in local regions, *state plane feet* allows each unit of a coordinate to be equal within the region.

```
# Split out
table(training$tree.sandy)
```

```
##
##    0    1
##  396 1550
```

With the data ready, we use `caret` to interface with the `kknn` package, which has the ability to handle both classification and regression problems as well as conduct grid searching to identify the optimal combination of kernel and number of nearest neighbors K.

```
pacman::p_load(caret, kknn)
```

Tuning models. Within `caret`, we specify the algorithm using the `method` argument in the `train` function (`method = "kknn"`). We also have the option to provide `caret` with a list of tuning scenarios, noting that this requires some knowledge of how the `kknn` package was designed. Below are three key parameters that need to be specified:

- `kmax` is the maximum number of neighbors to be tested.
- `kernel` is a string vector indicating the type of distance weighting (e.g., *rectangular* is unweighted, *biweight* places more weight toward closer observations, *gaussian* imposes a normal distribution on distance, *inv* is inverse distance).
- `distance` is a numerical value indicating the type of Minkowski distance. (e.g., 1 = binary, 2 = Euclidean).

In this example, we consider 200 scenarios that are constructed by applying `expand.grid` to find all combinations of the parameter set.

```
scenarios <- expand.grid(kmax = 100,
                distance = 1,
                kernel = c("rectangular", "inv"))
```

Each scenario is then evaluated through tenfold cross-validation. The `train` command does much of the hard work, running the k-NN algorithm 2000 times (10 cross-validation models for each k and *kernel* combination), then returns the optimal parameters that minimize classification error. The results are stored in the `fit_knn` object.

```
# Set seed for replicability, run with 10-folds cross validation
set.seed(100)
 val_control <- trainControl(method = "cv", number = 10)
```

```
# Train model
fit_knn <- train(factor(tree.sandy) ~ .,
        data = training,
        preProcess = c("center","scale"),
        trControl = val_control,
        method = "kknn",
        tuneGrid = scenarios)
```

Evaluate performance. We extract the optimal tuning parameters by calling `summary(fit_knn)`, which indicates that a *rectangular* kernel with $k = 20$ neighbors achieves a misclassification rate of under just 20%. *Is this sufficient for deployment?*

```
# Score, predict probabilities
testing$prob <- predict(fit_knn, testing, type = "prob")[,2]
  testing$yhat <- ifelse(testing$prob >= 0.5, 1, 0)
```

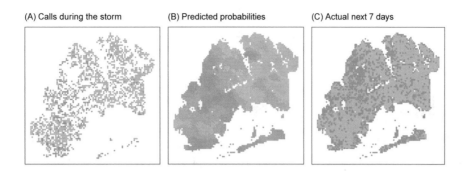

Figure 10.5: Comparison of actual and predicted areas with reported downed trees. Red indicates at least one tree was reported in a given 1000 x 1000 square-mile area.

```
# Fill NAs in test set actuals
testing$tree.next7[is.na(testing$tree.next7)] <- 0
```

A closer visual inspection of the predictions in Figure 10.5 shows that the k-NNs predicted probabilities can adequately approximate the pattern of downed trees that will eventually be reported over the next 7 days after the storm. The algorithm appears to be relatively effective in detecting both the unaffected (blue) and affected (orange) with a blue-orange gradient where the evidence is less clear. These probabilities can come in handy when *prioritizing* aid—send help to where you are more certain it is needed.

In fact, if emergency responders prioritized visits to higher probability areas, we would likely achieve high hit rates in the early days of the emergency response (e.g., finding downed trees when a visit is conducted)—nearly 80% at its peak, then gradually declining with lower probabilities. Thus, while probabilities will not perfectly predict which areas have been impacted, they would allow responders to first focus on worse off areas first.

Despite the promising result, we should be cognizant that k-NN is just one of many prediction algorithms. We naturally should ask: *Is there a better algorithm?*

10.4 Tree-based learning

One of the core challenges of working with data in public sector is the quantitative tradition—most methods are regression-based as the goal has been to estimate and infer relationships. Thus, if a social scientist approached the downed tree problem, it is likely that a logistic regression would be the model of choice. However, prediction problems prioritize accuracy first. We, of course, now know that such a model would draw a linear decision boundary that would invariably ignore the isolated pockets of activity. Furthermore, the coefficients in this case would not be meaningful—downed trees are not distributed monotonically along latitude and longitude.

In contrast, k-NNs can detect pockets of activity. However, they are memoryless techniques—they do not learn patterns, facilitate inference, or identify important variables.

Decision tree learning is a promising alternative that learns the structure present in data while allowing analysts to interpret patterns.

Tree learning recursively splits a sample into smaller, more homogeneous partitions so that affected areas are separated from unaffected areas. When fully grown, a decision tree resembles an inverted tree (see Figure 10.6). The *root node* (the full sample) is first split at a threshold along an input variable, producing a pair of *child nodes*. The optimal threshold is found by an *attribute test*, which compares many candidate thresholds in search of one that maximizes the homogeneity within the child nodes. In our example, 80% of the root node is comprised of areas with downed trees. Each time an attribute test splits a node, we see a pair of

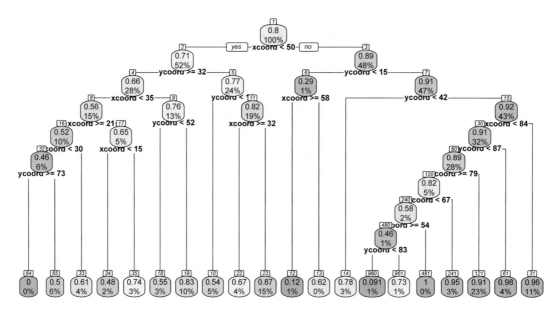

Figure 10.6: A grown decision tree made with the `rpart.plot` package.

child nodes that—when combined—have the same number of records as their parent node. The first split is identified at a threshold ($xcoord < 50$). While the split is imperfect, each node has become slightly more homogeneous: 71% of the child node below $x < 50$ is comprised of downed trees compared with 89% in the other child node.

The algorithm will continue to recursively split the sample, concentrating more affected areas into some nodes and unaffected areas into others. Notice the differences in complexity (number of splits) on the left side of the tree compared with the right side—the splitting activity is more intense for some nodes than others, indicating that the tree is able to identify small pockets of activity. A decision tree's ability to mold to the data makes it possible for it to exploit finer patterns and in turn capture details that a regression-based method would otherwise overlook. As the tree continues to partition the sample into smaller child nodes, it will stop growing according to pre-defined *stopping criteria*. Each node that is without child nodes is referred to a *leaf*—it is the terminal node, the end of the line, where predictions are derived.

While the idea of tree learning may be unfamiliar to policy environments, it does have certain appealing qualities that traditional methods lack. For one, recursive partitioning finds *variable interactions*, or an effect on the target that is dependent on the product of two or more input variables. These interactions allow tree learning to catch discontinuous and non-linear patterns in the data—an ability that would require substantial effort in a linear regression framework. In addition, the search for the optimal split allows tree learning to conduct *automated variable selection* that is not only robust to outliers but even useful when the number of variables outnumbers the number of observations.

10.4.1   Classification and Regression Trees (CART)

The process of growing trees. Now that we have the gist of decision tree learning, let's take a look at how Classification and Regression Trees (CART) work under the hood. As laid out in Breiman et al. (1984), the algorithm follows a three-step process:

1. *Base Cases*. Growing a tree can be computationally costly. To save time, tree learning algorithms cleverly check for "base case" criteria that determine whether it is worth the effort to grow a tree. In particular, base cases check if all values at the root node are of the same class (or below a certain variance threshold), if input variables can achieve any useful splits, among other conditions. When any base case is true, the algorithm will stop. More often than not, your tree will not grow due to a lack of signal. While this can be frustrating, knowing that the data does not lend itself to a prediction problem early on in a project

saves one from the heartache of building a house of cards.

2. *Recursive Partitioning.* If base cases are are not detected, the algorithm proceeds with growing a tree. An attribute tests is applied to each successive partition, which recursively splits the sample into smaller, more homogeneous partitions. More on attribute tests later.

3. *Stopping Criteria versus Pruning.* All trees should stop growing at some point, but question is *when?* One approach is to apply *stopping criteria* that prevent the tree from over-growing. These criteria might include terminating the recursive splitting process if leaf nodes have fewer records than a pre-specified threshold (e.g., do not split if $n < 5$), the nodes can no longer become more homogeneous, etc. While these stopping criteria may sound reasonable, premature termination can prevent the tree from reaching its full potential—it can bias the predictions. Alternatively, a tree can be grown to its fullest, then *pruned* to an optimal level—not too bare and not too shaggy. Since this involves reviewing the tree's performance to make tuning decisions, pruning is conducted on cross validated model performance to avoid overfitting.

Once the tree is grown and pruned, predictions can be generated from the observations in the leaf nodes. Predictions for continuous outcomes are the average of all J observations in leaf i ($\hat{y}_i = \frac{1}{n} \sum_{j=1}^{J} y_j$), whereas predicted probabilities of discrete outcomes are calculated as the proportion of each class c in the leaf ($\hat{y}_{ic} = \frac{1}{n} \sum_{j=1}^{J} (C = c)$). When CART scores a new test sample, observations receive the average value of for the leaf into which it fell. The leaf at the bottom right of Figure 10.6, for example, contains 11% of the training sample, 96% of which have downed trees. When applied to out-of-sample records, any observation that reaches that leaf is assigned a 96% probability of having a downed tree.

The leaf nodes also serve another purpose: *to facilitate interpretation.* Each leaf is defined as a set of binary criteria, essentially *if* statements that define a *profile.* Returning to the bottom right leaf in Figure 10.6, we can trace the path from the root node to the leaf defined by a nested set of binary criteria: *ycoord* ≥ 42 and *xcoord* ≥ 84. When the tree is simple, the profile is easy to communicate—as simple as a checklist. However, complex trees with many splits are far harder to communicate.

Attribute Tests. To understand attributes, let's use a simple analogy. In order to extract metal from mineral impurities, we smelt raw metal ore—or apply some process to separate the two materials. While smelting iron ore applies heat, CART models apply attribute tests to smelt the data, splitting it into smaller partitions with fewer impurities. Attribute tests work by drawing on a concept from *information theory*: the optimal split yields the most *information gain* (*IG*). In other words, the optimal split produces two child nodes that reduce impurity (I) by the largest margin— one child node with more metal, the other with other materials. Mathematically, IG is defined as the difference between the impurity of a parent node (I_{parent}) and the weighted impurity of the resulting child nodes:

$$IG = I_{\text{parent}} - \sum_{j=1}^{J} \frac{n_j}{N} I_j$$

The greater the *IG* value is, the more information is extracted from the split, resulting in purer child nodes.

What is impurity exactly? Impurity boils down to a set of metrics that characterize the information content in a node (see Table 10.2). In classification, impurity is measured through either Gini Impurity or Entropy, both of which are calculated from the proportion of records that belong to each of J classes in the target. In practice, both have similar predictive performance although Gini Impurity is slightly faster to calculate. Regression problems rely on Variance Reduction, prioritizing splits that minimize within-partition variance.

Table 10.2: Common measures used for decision tree learning attribute tests.

Measure	Formula	Description
Gini Impurity	Gini Impurity $= \sum_{i=1}^{k} p_i(1 - p_i) \sum_{i=1}^{k} p_i^2$	Commonly use in CART. The Gini Impurity ranges from 0 to $1 - \frac{1}{k}$.
Entropy	Entropy $= \sum_{i=1}^{k} -p_i log_2(p_i)$	Relied on in other decision tree learning algorithms. Values range from 0 to 1
Variance	$Variance = \frac{\sum_{i=1}^{n}(y_i - \bar{y})^2}{n}$	Focus on reducing the spread in the target within partitions.

To illustrate the power of impurity metrics, let's revisit the downed trees example. To simplify the problem, let's suppose that a single latitudinal boundary line can adequately separate affected from unaffected areas.[5] *Which degree of latitude do we choose?*

Table 10.3 presents two candidate boundaries. Split A is somewhere north and Split B is farther north (North+). Each split results in two sub regions (North and Remainder) that are further subdivided into affected areas (Downed trees) and unaffected areas (Untouched).

Table 10.3: Two sets of candidate splits.

Label	A: North	A: Remainder	B: North+	B: Remainder
Downed Trees	10	3	9	4
Untouched	5	21	0	26

To evaluate which split is better, we calculate the *Gini Gain*—the *IG* value for Gini Impurity. Below, we calculate the IG value for each proposed split (see below). Both calculations require the impurity of the root node (I_{region}) to capture the baseline information content in the data. Then, we obtain the Gini Gain for a pair of candidate child nodes by subtracting their weighted impurity. The larger Gini Gain indicates the split that offers the most information gained.

$$IG_A = I_{region} - \frac{n_{A,north}}{N} I_{north} - \frac{n_{A,rem}}{N} I_{A,rem}$$

$$IG_B = I_{region} - \frac{n_{B,north+}}{N} I_{B,north+} - \frac{n_{B,rem}}{N} I_{B,rem}$$

As we step through each element of the calculation in Table 10.4, it becomes apparent that Split B is more effective for isolating areas with downed trees, achieving an 100% hit rate in the *North+* partition—a marked improvement over Split A's 66% hit rate in the *North* partition. Indeed, the IG calculation confirms that Split B would yield twice as much information as Split A.

Interpretation. While you provide CART with a model specification, the algorithm will choose which of those variables are useful for predicting the target—it conducts variable selection. This may raise concern amongst policy staff: *Is CART a black box?* It is only a black box if we limit our definition of interpretability to the traditions of linear regression. CART facilitates interpretation through *variable importance* (or *feature importance*) that measure how much CART relies on each variable.

---

[5]A tree with a single split is called a *decision stump*.

Table 10.4: Calculating information gain for each Split A and B.

Element	Split A		Split B	
Regional	$I(D_{\text{regional}})$	$= 1 - (p_{\text{trees}}^2 + p_{\text{other}}^2)$	$I(D_{\text{regional}})$	$= 1 - (p_{\text{trees}}^2 + p_{\text{other}}^2)$
		$= 1 - ((\frac{13}{39})^2 + (\frac{26}{39})^2$		$= 1 - ((\frac{13}{39})^2 + (\frac{26}{39})^2$
		$= 0.4\bar{4}$		$= 0.4\bar{4}$
Northern	$I(D_{\text{A,north}})$	$= 1 - (p_{\text{trees}}^2 + p_{\text{other}}^2)$	$I(D_{\text{B,north+}})$	$= 1 - (p_{\text{trees}}^2 + p_{\text{other}}^2)$
		$= 1 - ((\frac{10}{10+5})^2 + \frac{5}{10+5})^2$		$= 1 - ((\frac{9}{9+0})^2 + \frac{0}{9})^2$
		$= 0.4\bar{4}$		$= 0$
Remainder	$I(D_{\text{A,remainder}})$	$= 1 - (p_{\text{trees}}^2 + p_{\text{other}}^2)$	$I(D_{\text{B,rem}})$	$= 1 - (p_{\text{trees}}^2 + p_{\text{other}}^2)$
		$= 1 - ((\frac{3}{3+21})^2 + \frac{21}{21+3})^2$		$= 1 - ((\frac{4}{4+26})^2 + \frac{26}{26+4})^2$
		$= 0.21875$		$= 0.23\bar{1}$
Information Gain	IG_A	$= 0.4\bar{4} - \frac{15}{39}0.4\bar{4} - \frac{24}{39}0.21875$	IG_A	$= 0.4\bar{4} - \frac{9}{39}0 - \frac{30}{39}0.23\bar{1}$
		$= 0.13\bar{8}$		$= 0.2\bar{6}$

Decision tree software typically calculates variable importance metrics based on *goodness of split* measures (e.g., Gini Impurity) aggregated from all splits. Despite widespread adoption across programming languages, each language implements the goodness of split calculation differently. In the `rpart` package in R, variance importance of the kth variable (VI_k) is the sum of information gain from all splits that a variable plays a *primary* or *surrogate* role:

$$VI_k = \sum_{j=1}^{J}(IG_{primary,j}) + \alpha \sum_{m=1}^{M}(IG_{surrogate,m})$$

The distinction between primary and surrogate depends on the completeness of data. A primary variable is used to splits in a CART algorithm. However, missing values in the training or test set present problems for splitting—the algorithm does not know how to handle missing information. In fact, this is a common problem in most modeling approaches, including regression. A clever workaround introduced in Breiman et al. (1984) is to employ a surrogate variable, which is backup splitting variable that mimics the primary variable's split behavior as close as possible.[6] VI_k takes into account the information gained when a variable is primary (J-number of splits) and surrogate (M-number of splits). As the surrogate variable is only an approximation of the primary, the two might not always align in their predictions. The variable importance of the surrogate is adjusted by how often it agrees with the primary (α). The resulting variable importance VI_k can either be reported in raw information gain or normalized ($VI_{normalized} = 100 \times \frac{VI_k}{max(VI)})]$.

Variable importance can be also calculated as *permutation importance*, which measures how much does a given variable contribute to predictive accuracy. Given a trained model with k-number of input variables, the model scores k test sets. All test sets are a copy of the same dataset with one slight modification: each dataset randomly shuffles one of the k variables, which has the effect of preventing the model from having access to that variable's information. If a variable is important and the model relies on it to achieve high accuracy, the model would perform substantially worse because the variable is replaced with noise. When we calculate predictive performance, the test set with the lowest accuracy corresponds to the most important variable.

---

[6]Surrogate variables are beyond the scope of this chapter but are worth the effort when dealing with data streams of varying degrees of completeness. If an observation is missing data, CART will skip that observation. Surrogate variables can improve the chance that a prediction is produced—it is a backup.

While both metrics are useful, represent different concepts of importance. The goodness of split metrics provide insight into the modeling process—what was used to build the tree. Permutation importance, in contrast, grounds the conversation in terms of contribution to accuracy—which variables hold the most predictive power, which may arguably be more useful in predictive contexts.

10.4.1.1   In practice

Decision tree learning can be a remarkable for prediction but *how can we use it in practice?*

CART lends itself to constructing *profiles* from its leaf nodes. Profiles are a convenient way to articulate details that can identify a phenomenon of interest—-providing not only a picture of what the phenomenon looks like but also a way to operationalize CART algorithms in the field. Parts of Northern Queens, for example, were spared from the worst of Hurricane Sandy and can be identified as a region defined by a set of geographic coordinates. The coordinate set can be used to define areas where field crews should and should not allocate aid. It is easy to imagine that an incident commander instructs one team to focus on areas north of a major boulevard and another team takes the remainder. Using profiles, however, is only as manageable as the decision tree is simple. Otherwise, an deep, complex decision tree may prove to be complicated for use—requiring bespoke, specialized software to keep track of hundreds of profiles.

CART's variable importance measures are useful for telling a story about how much variables contribute to the phenomenon of interest. While variable importance does not lend itself to crafting a story of how X causes Y, it can provide audiences a look under the hood and some re-assurance about transparency. The variable importance (goodness of fit) for each latitude and longitude are 36 and 39.5, respectively, indicating that both factors had comparable contributions to the prediction problem. However, if the variable importance for longitude were 240, we can infer that storm damage may have been slightly more distributed along an East-West gradient than North-South (Table 10.5).

Table 10.5: The pros and cons of CART.

Useful Properties	Challenges
Rules can be directly interpreted as "profiles".	If left unpruned, overly complex trees may be too voluminous to interpret.
Variable importance metrics help identify most relied upon inputs.	Importance cannot be interpreted in the same vein as parameter estimates in estimates.
Well-suited to capture interactions and non-linearities in data.	Tend to overfit, especially when the tree is unpruned.
Can accommodate both continuous and discrete variables.	Unable to extrapolate beyond the range of the target variable in the training sample.
	Data-driven variable selection means less human control.

There are, of course, some shortcomings of CART models. The algorithm tends to overfit the data. As a tree grows to its fullest extent, the algorithm might "over-split" partitions, making leaf nodes that are insignificantly small (e.g., $n = 1$). You can imagine that using predictions with a partition of $n = 1$ is a foolhardy endeavor. It is for this reason that trees should be pruned.

Tree learning, like many machine learning methods, will experience difficulty extrapolating beyond the bounds of the continuous target variables. Because CART models are a non-parametric technique, they can only produce predictions within a range of values that they have already observed. For example, if we would like to predict a company's revenue that has historically been fluctuated between \$50 million to \$300 million, the CART model can only produce predictions between \$50 million to \$300 million. Successful companies, however, will likely experience long-run growth and their revenue will grow beyond the range in the training sample. A CART producing predictions will simply fail to predict any higher or lower than the training range.

There is hope for certain types of data. When working with time series (e.g., revenue over time, crimes over time, etc.), for example, we can convert the target series into a percent *first difference*:

$$\Delta y_t = 100 \times \left(\frac{y_t - y_{t-1}}{y_{t-1}} \right)$$

Differencing has the effect of converting data into a *mean-reverting process* in which the outcome's values hover around the mean or zero rather than drifting out of bounds.

In the 1990's and early 2000's, the tree learning literature made large leaps forward in finding ways to overcome CART's shortcomings. In the next section, we introduce one of the most influential models in use in computational research and applied data science.

10.4.2 Random forests

Every quarter, economists around the world forecast the direction of the US economy. One well-regarded set of forecasts is collected by The Wall Street Journal, tapping over 60 economists for their thoughts on next quarter's economic performance. The forecasts are then averaged to produce a *consensus forecast* (Williams 2019) that draws on the wisdom of the crowds. The beauty of this ensemble forecasting approach is that it brings together a broad range of perspectives, each of which has its own bias. But when we average across enough people, their biases equalize and we can produce a more accurate answer.

Like a single economist, the forecast from a single CART model can miss the target (see Plot (A) in Figure 10.7), but what if hundreds of CARTS were combined? It might seem like a farcical idea as all trees might return identical results. However, if we bootstrap each training sample (random sampling with replacement), each tree can learn from a different version of reality. Each tree still predicts the same concept, but its slightly different perspective helps to make a more robust forecast. By aggregating the predictions from all trees (*bootstrap aggregation* or *bagging*), we are essentially developing consensus forecast—averaging out the biases of individual trees (Plot (D) in 10.7).

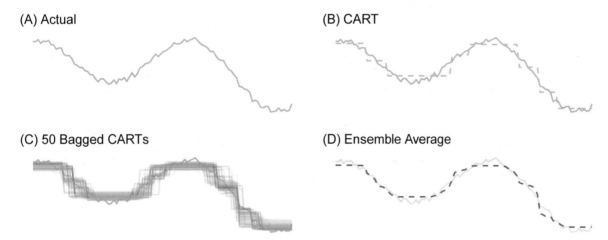

(A) Actual

(B) CART

(C) 50 Bagged CARTs

(D) Ensemble Average

Figure 10.7: Comparison of results of applying a single model to fit a curve versus an ensemble of models.

Although bootstrap aggregation can increase the robustness of predictions, all trees are likely to be correlated because they have access to the same variables—they all may resemble one another. But imagine what if the models within an ensemble are uncorrelated and do not resemble one another, yet they arrive at the same conclusion. Achieving consensus with a wide range of perspectives is similar to having members from opposing political parties agreeing on the same policy action—it arguably has higher potential of being right than a unilateral motion.

Introduced in Breiman (2001), the *Random Forest* (RF) algorithm combines bagging with the *random subspace method*. As illustrated in Figure 10.8, the algorithm bootstraps B-number of samples to grow B-number of trees. What makes Random Forests so robust to overfitting is how each tree is induced. For a training set

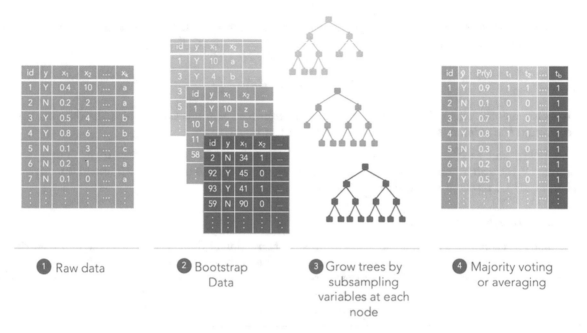

Figure 10.8: Process of growing a Random Forest.

of K variables, a subset of k variables is randomly drawn at each node of each tree. The number of variables is fixed at $k < K$ for all nodes and trees. The algorithm then evaluates the k variables with an attribute test and splits the node based on the one that maximizes information gain. By drawing k, this random subspace method forces trees to look different from one another.

The secret sauce of the random subspace method is the value of k. When k is small relative to K, there is a lower chance that the same combination of variables will appear from one draw to the next, which increases the chance that the trees are distinct and therefore uncorrelated. Forcing trees to be uncorrelated has the effect of asking a panel of experts to look at the same problem with different information. If the majority of trees are able to reach the same conclusion with partial information, then we can have confidence in the prediction.

There is a trade off, however. When we lower k, we run the risk of ignoring the most predictive variables— they might not be sampled in each node. In practice, the size of the variable subsample k needs to be tuned. For a test sample of size n, the ensemble outputs a matrix of predictions that is $n \times B$ large, producing a large distribution of predictions for every observation i.

How do all these predictions get reconciled into a single prediction? For regression problems, the prediction $\hat{y}_i$ for the ith record averages across each tree $f_b(x_i)$:

$$\hat{y}_i = \frac{1}{B} \sum_{b=1}^{B} f_b(x_i)$$

RFs trained for classification problems calculate the probability of each class c, then the predicted class is determined through majority voting. For a given class c, calculate the proportion of trees that predict a class label c:

$$\hat{y}_{ic} = \frac{1}{B} \sum_{b=1}^{B} [f_b(x_i) = c]$$

The result is one of the most robust, standard algorithms in data science.

10.4.3 In practice

Although Random Forests are complex, they are an almost sure-fire solution that delivers reasonable results without significant tuning, making it a general purpose algorithm for tabular data. Nonetheless, there may be marked gains when hyperparameters are tuned with care. Tuning focuses on two parameters:

- *Variable Subsampling*. Each tree is grown on a fixed number of variables k. The default value of k is typically set to $k = \frac{p}{3}$ for regressions and $k = \sqrt{p}$ for classification problems where p is the total number of variables.[7] When samples are large, grid searching for the optimal k can be quite time-consuming. A happy medium involves a simple under-over approximation by testing values of k at the default setting along with values at $2 \times k$ and $\frac{k}{2}$. If the model does not significantly improve at other values of k, then an exhaustive grid search may not be worth the effort.
- *Minimum Node Size*. The minimum node size controls the size of leaf nodes in each tree, which in turn dictates the size and complexity of each tree in the forest. Larger minimum node sizes equate to shallower trees and shorter training elapse times. By default, node size should be set to $n = 5$ for regression problems and $n = 1$ for classification problems.

The *number of trees* in the forest can also be tuned, but the effect of the number of the trees influences the stability of consensus rather than accuracy itself: *how many trees are needed for predictions to converge and become reliable?* Too few trees lead to noisy predictions; However, too many trees might not yield marginal improvement in accuracy. It would be reasonable to forego tuning this parameter (Hastie, Tibshirani, and Friedman 2001) and rely on a default number of trees, typically set at 500.

An aside on cross validation. For most ML algorithms, hyperparameter tuning is coupled with a cross-validation strategy. Random Forests, however, have a clever built-in validation strategy that takes advantage of an artifact of bootstrapping: *approximately two-thirds of the sample's unique observations are captured in each bootstrap sample, leaving one-third of observations untouched.* These untouched observations, or *out-of-bag* sample (*OOB*) are a built-in test set that is useful for gaging model fit. The OOB error is estimated by averaging performance across OOB samples, which has been shown to produce error estimates that are as accurate as a test sample of the same size as the training (Breiman 1996). Nonetheless, algorithm horse races will test all algorithms using a more typical cross-validation strategy.

How about interpretation? Unlike their CART siblings, articulating profiles from individual leaf nodes is not possible due to the bootstrap aggregation and random subspace sampling. Instead, Random Forests can help data scientists under Random Forests do, however, make use of *variable importance* metrics that average across its forest—average of Gini Impurity for classification problems or mean decrease in variance for continuous values. In addition, graphical techniques have been developed to show the relationships contained with the algorithm. Partial Dependence Plots (PDP) show how an input variable correlates with the target in the form of smoothed line. As Random Forests capture non-linearities, PDPs visualize the shape of the relationship. In addition, Break Down Decompositions provide a granular interpretation of how each input variable contributes to a specific predicted instance.[8]

With all the additional moving parts, *are Random Forests really worth the trouble?* This algorithm is the emerging workhorse of data-driven public policy and its use is only accelerated as it is widely available in most programming languages. This leaves one question: Can its superior accuracy outweigh its complexity? As we will see in the DIY example, the gain in accuracy can be quite significant (Table 10.6).

---

[7]While discrete variables appear to be one variable, Random Forests and other ML algorithms treat these variables as a dummy variable matrix (also known as one-hot encoding in computer science). Thus, a discrete variable with 200 categories is treated as 200 variables, which in turn sets $k = 14$ for classification or $k = 66$ for regression problems.

[8]These diagnostics are useful for Random Forests as well as many other machine learning techniques.

Table 10.6: The pros and cons of Random Forests.

Useful Properties	Challenges
Incredibly flexible and produces reasonable results without tuning.	Unable to extrapolate beyond the range of the target variable in the training sample.
Variable importance metrics help identify most relied upon inputs.	Importance cannot be interpreted in the same way as parameter estimates in linear model.
Well-suited to capture interactions and non-linearities in data.	Deeper, more complex trees are computationally costly.
Can accommodate both continuous and discrete variables.	Data-driven variable selection means less human control.
Out-of-bag error estimate has the same approximate effect as a more costly cross-validation approach.	

10.4.4   DIY: Wage prediction with CART and random forests

*DIYs are designed to be hands-on. To follow along, download the DIYs repository from Github (*https://github.com/DataScienceForPublicPolicy/diys*). The R Markdown file for this example is labeled* `diy-ch10-trees.Rmd`.

How much is a fair wage? Societies have pondered this question for ages, but perhaps more so in the modern age. There have been long-standing concerns over the gender, ethnic, and racial pay gaps, which had seen progress at one point but more recently has stagnated (Hess 2019, 2016). To remedy these pay differentials, some US cities such as New York City and Philadelphia as well as states like California have banned employers from asking for applicant salary history (Cain, Pelisson, and Gal 2018). What *should* be considered to be a fair wage? One way we can evaluate a wage is to predict it based on historical data on industry, experience, and education while omitting demographic factors. In fact, decomposing the contributing factors of wage attainment has been a task that policy researchers have long studied with the hope of improving labor policy. For example, Krueger and Summers (1988) examined wage differentials of equally skilled workers across industries, taking advantage of labor quality.

In this DIY, we prototype a tree-based model to *predict* wages based on worker characteristics gathered from a widely used survey. Having an accurate prediction model can be used to evaluate if staff are undervalued in the market, which in turn can be used to pre-empt possible churn by offering pay increases. A wage model could help set expectations on employment costs as well as scoring new positions to support budgetary and human resources use cases.

Data. Drawing on the US Census Bureau's 2016 American Community Survey (ACS), we constructed a training sample (`train`) and test sample (`test`) focused on California. Each sample randomly draws a set of $n = 3000$ records, mainly to reduce the computational overhead for this exercise while preserving the patterns in the data.[9] The data have been filtered to a subset of employed wage earners (> $0 `earned`) who are 18 years of age and older. Each sample contains the essential variables needed for predicting fair wages, namely experience, education, hours worked per week, among others:

- `id` is a unique identification number
- `wage` of the respondent in 2016.
- `exp` is the number of years of experience (approximated from age and education attainment)
- `schl` is the highest level of education attained.
- `wkhp` is the hours worked per week.
- `naics` is a NAICS code used to identify the industry in which the respondent is working.
- `soc` is a description for the Standard Occupation Code (SOC) used for job classification.

---

[9]The ACS is a probability sample with sampling weights. To produce population statistics from the data, we need to account for these weights; However, for this use case, we will treat each observation with equal weight.

- `work.type`. Class of worker indicates whether a respondent works for government, for-profit business, etc.

```
# Load data
load("data/wages.Rda")
```

Let's get to know the data. In Figure 10.9, we plot the relationship of wage against years of experience. We find that wages increase with each additional year of experience up to 20 years, plateaus for 20 years then gradually decline after 40 years. While there is a clear central trend, each person's experience is quite variable—some achieving high salaries even while the age trend declines. The value of education, as seen in Figure 10.10, also has an impact on wages, but it is only realized once enough education has been accumulated. In fact, the box plot suggests that median wage only grows at an accelerated pace once an individual attains an Associate's Degree. There is a large increase in the median wage among Bachelor's to Master's degree holders, although the wages of a Bachelor's rivals the earning potential of graduate degree holders.

Training. Of the two tree learning algorithms that we have covered, namely CART (rpart package) and Random Forest (ranger package), which will produce the best results? We can arrive at an answer by running a horse race in which the average RMSE is estimated for each model through cross validation.[10]

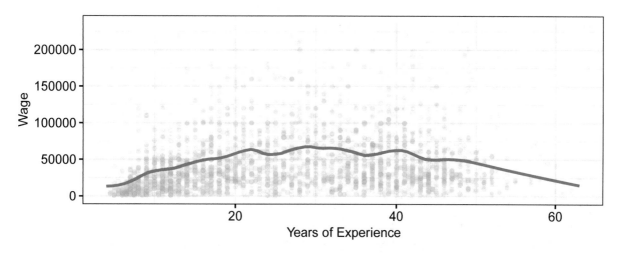

Figure 10.9: Wage by years of experience.

```
# Load packages
pacman::p_load(caret, rpart, ranger)
```

A fair horse race tests all algorithms under the same conditions. In this case, we compare models with the same training and test sets, same input variables and trained using fivefold cross validation.[10] As for the input variables, we could hand-select a set that we believe best represents wages at the risk of biasing the predictions. Alternatively, we *"toss in the kitchen sink"* in which all variables are included in all models, allowing the algorithm to identify the most statistically relevant variables.

```
# Validation control
val_control <- trainControl(method = "cv", number = 5)

# Specification
input_formula <- as.formula("wage ~ exp + schl + wkhp + naics + soc + work.type")
```

---

[10]The number of folds could be increased but with greater time cost. As you will see when we begin training the models, the Random Forest will take some time to run. If the number of folds was increased, we not only would obtain more precise model accuracy estimates but the time required also increases. Thus, we select a smaller value for demonstration purposes.

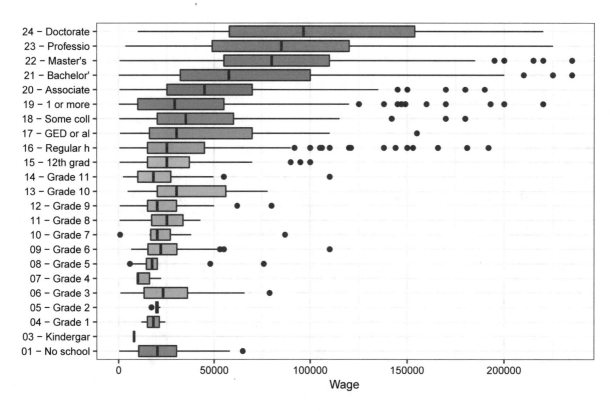

Figure 10.10: Wage by education attainment.

CART. To train CART in `caret` (`method = "rpart"`), most of the effort is spent on tuning the complexity parameter `cp` that controls how deep the CART grows. When $cp = 0$, a CART can grow to its fullest, otherwise, any value $0 \leq cp \leq 1$ will limit the extent to which the tree will grow—essentially a stopping criteria. For a prototype, an approximate solution will do. By setting the `tuneLength = 10` argument, `caret` will test ten different `cp` scenarios, keeping the value that minimizes the loss function. The code below trains a regression tree on all available inputs, excluding the `id` column (column one).

```
# Set seed for reproducibility, then train
set.seed(123)
 fit_rpart <- train(input_formula,
        method = "rpart",
        data = train[, -1],
        tuneLength = 10,
        trControl = val_control)
```

Let's take a look at the results captured in `fit_rpart$results` as shown in Table 10.7. As the `cp` value falls to zero, the model becomes more complex and learns more intricate patterns in the data, which in turn causes the error to fall sharply. It also becomes clear how sensitive CART performance is to tree complexity—if it is not complex enough, the model will produce underwhelming results that miss the mark. Overly complex trees, in contrast, produce noisy results. Finding the Goldilocks value of `cp` is name of the game.

Pro Tip: Although the lowest `cp` value has the best performance (lowest RMSE), it is statistically indistinguishable from other `cp` values that are within one standard deviation of the lowest RMSE. A well-accepted rule of thumb, as proposed in Hastie, Tibshirani, and Friedman (2001), is to *choose the largest `cp` value that is still within one standard deviation of the lowest RMSE.* In effect, this decision-theoretic defaults to the

Table 10.7: Cross-validated performance for each level of tree complexity.

cp	RMSE	R-Squared	MAE	SD(RMSE)	SD(R2)	SD(MAE)
0.008749	34457	0.364	24736	1217	0.035	1143
0.008921	34424	0.365	24793	1085	0.035	1024
0.009219	34516	0.362	24933	1116	0.032	1003
0.01699	35246	0.332	25698	385	0.04	249
0.01799	35569	0.32	25939	813	0.037	317
0.02324	36391	0.289	26532	1116	0.025	512
0.02598	36529	0.283	26670	1048	0.028	463
0.02821	37372	0.249	27392	1114	0.049	808
0.05154	38132	0.216	28012	2161	0.053	1429
0.1622	41751	0.148	31994	3004	0.001	2675

simplest model available that does not lead to a substantial loss in accuracy. `caret` automatically identifies and stores the optimal cp value in `fit_rpart$bestTune`. Despite tuning the CART model, even the best model has relatively modest performance—perhaps a Random Forest can offer an improvement.

Since the ***Random Forest*** algorithm computationally demanding, we need a package that is built to scale. The `ranger` package was developed with this in mind, making it a perfect choice for rapid model training. Through `caret`, we train a Random Forest model by specifying `method = "ranger"` and tune four hyperparameters:

- `mtry` (optional) is the number of variables to be randomly sampled per iteration. Default is $\sqrt{k}$ for classification and $\frac{k}{3}$ for regression. Default set to the square root of the number of variables.
- `ntree` (optional) is the number of trees. Default is 500.
- `min.node.size` (optional) is the minimum size of any leaf node. The default settings vary by modeling problem. For example, the minimum for regression problems is $n = 5$, whereas the minimum for classification is $n = 1$.
- `max.depth` (optional) as the maximum number of splits between a tree stump (one split) and a fully grown tree (unlimited). This is similar to a stopping rule in CART. By default, `max.depth = NULL` indicating unlimited depth.

We could task `caret` to test a large number of automatically selected scenarios by setting `tuneLength`. To save on time, we instead specify sensible default hyperparameters: each tree sub-samples $\sqrt{k}$ variables per tree ($mtry = 24$ in this case).[11]

An aside: why not tune the Random Forest? Random Forests are computationally intensive and conducting a grid search can be time-consuming. As an initial prototype, the objective is to produce *something* that is as accurate but operational as possible to prove that the concept works. *Show the thing* and prioritize speed. A Random Forest trained even with default hyperparameters generally offers improvements over CART and linear regression. Thus, testing a single hyperparameter set might yield clear gains over the alternative CART—a good zero-to-one improvement—a version 1 (v1). If we see little or no improvement, then we can assume that a Random Forest could require more exhaustive effort to optimize—a good candidate for a version 2 effort (v2).

Since `ranger` is built for speed, it foregoes diagnostic calculations like variable importance that slow the training process. Simply specifying `importance = "impurity"` in the model call instructs `ranger` to retrieve the goodness of fit-based measures during training.

```
# Set hyperparameters
scenarios <- expand.grid(mtry = 24,
            splitrule = "variance",
            min.node.size = 5)
```

---

[11] As each level of a categorical variable (e.g., `soc`, `naics`, `schl`) is treated as a dummy variable, we test $mtry = 24$.

```
# Train and cross validate through user-specified grid search
fit_rf <- train(input_formula,
         method = "ranger",
         data = train[, -1],
         trControl = val_control,
         tuneGrid = scenarios,
         importance = "impurity")
```

Evaluating models. Whereas the best CART model has a RMSE of 34,424 and a R-squared of 0.37, our Random Forest scenario performs markedly better with a a RMSE of 30,930 and a R-squared of 0.51.[12] The decision is an easy one: pick on the Random Forest.

Raw accuracy is hard to sell, even if the performance gains are large.[13] What is technical satisfaction for a data scientist might not be for policy makers—there is a need for a policy narrative that maps where an algorithm or program in the normative. There is a need for closure and transparency, especially when decisions are being made.

A natural starting point is to inspect the variable importance metrics contained in the `fit_rf` object to tell that story. As shown in Table 10.8, the Random Forest derives the most information from the number of hours worked (`wkhp`), followed by years of experience (`exp`) and various levels of education attainment (`schl`). Remember, importance does not show direction of correlation, but rather shows how much the algorithm relies on a variable to split the sample.

Let's take the analysis one step farther and visualize the *shape* of the underlying relationships. Partial dependence plots (`pdp` package), for example, render a trained algorithm in a lower dimensional space to expose the average relationship between a model's predictions and its continuous variables. Figure 10.11 plots the partial dependence for years of experience (`exp`), finding that the Random Forest's predictions `yhat` mold to the non-linear trend that we had previously seen in the EDA. The gain in wages with each additional year is relatively modest in the long run, but are quite large early in one's career.

Table 10.8: Top 10 most important variables from the Random Forest model.

Variable	Importance	Variable	Importance
wkhp	100	soccmm-software developers,applications and systems software*	21.11
exp	39.09	naics5415	19.07
schl22 - master's degree	39.06	schl24 - doctorate degre	14
schl21 - bachelor's degr	36.02	schl23 - professional de	13.29
schl16 - regular high sc	22.36	naics722z	12.14

```
# Load packages
pacman::p_load(pdp)

# Partial Dependence Plot using pdp
partial(fit_rf,
    pred.var = "exp", plot = TRUE,
    plot.engine = "ggplot2") +
    ggtitle("Years of Experience") +
    theme_bw() +
    theme(plot.title = element_text(size = 10))
```

---

[12]Note that exact results will differ slightly from one user to the other due to the random sampling in cross-validation.
[13]Contextualized performance gains in terms of dollars and lives saved are a different story, however.

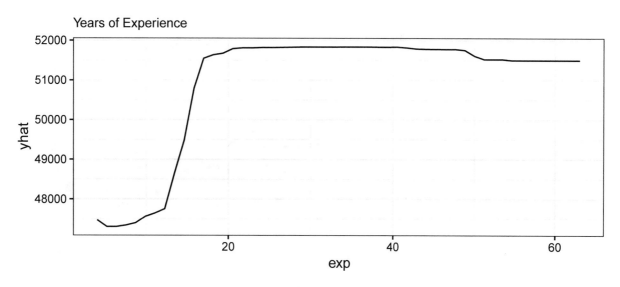

Figure 10.11: Partial dependence plot for years of experience.

For non-parametric algorithms like Random Forest, effect decomposition is not normally possible; However, more recent techniques such as Locally Interpretable Model-Agnostic Explanations (LIME) have been developed to decompose the effect of each input variable on the target (Staniak and Biecek 1988). As implemented in the **broken** function in the **breakDown** package, the LIVE technique approximates input variable effects by creating an artificial dataset in the near proximity around each observation, then identifies input variables that cannot be modified without resulting in a significant impact on prediction for the observation in question. Figure 10.12 maps the input variable impacts on one labor force participant. Whereas the baseline is the average wage in the sub-sampled data (just over 50,000), the prediction is equal to the baseline wage plus the final prognosis (around *45,000*). *How did we arrive at that number?* Starting from the baseline, we first see that working in retail sales has a lower earning potential (red bars), but the market compensates college educated workers more (green bars). 45,000. Keep in mind that the decomposition is controlled to the prediction $\hat{y}_i$—the insights are only useful if the algorithm achieves a high degree of accuracy.

```
# Load packages
pacman::p_load(breakDown)

# Obtain break down for the 100th observation in train
bd_plot <- broken(fit_rf,
         baseline = "intercept",
         new_observation = test[2500, -1],
         data = train[sample(1:nrow(train), 1000), -c(1:2)])
plot(bd_plot, add_contributions = F)
```

Producing predictions. Random Forests produce predictions like any other algorithm by making use of the **predict** function. Below, the trained model scores the **test** set in order to obtain the point estimate. This is suitable for most policy applications.

```
yhat_rf <- predict(fit_rf, test[, -1])
```

In some cases, it can be useful to have access to predictions from each tree in the Random Forest to facilitate an understanding of the uncertainty behind each prediction. These predictions, however, are generally not accessible through caret. In the code snippet below, we illustrate how to train a Random Forest using the **ranger** package, then obtain a matrix in which each tree's prediction is returned by specifying **predict.all = TRUE** in the predict function.

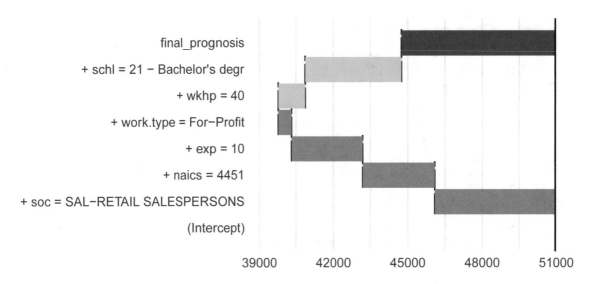

Figure 10.12: Break down decomposition plot for a single observation. As breakdown can require significant computational resources with large datasets, we approximate the decomposition by providing a random subsample of n = 1000 observations.

```
# Train Quantile Random Forest
fit <- ranger(input_formula,
        data = train[, -1])

# Retrieve matrix of all predictions
yhat_all <- predict(fit, test[, -1],
                predict.all = TRUE)
```

Incorporating uncertainty of a prediction can enable compelling use cases. For example, suppose we would identify workers who are compensated adequately and those who could benefit from a raise. In the set of small multiples in Figure 10.13, we can see that exactly 24 randomly sampled individuals' wages fall in relation to the predictions made by the ensemble. For the most part, the 10th and 90th percentile flank the actual wage, which could inform employment decisions. This information could feed into setting, negotiating, and budgeting for wages. Alternatively, by training a Random Forest without the protected demographics of interest. The predicted wages could be useful for identifying systematic biases and help evaluate fairness of pay. For analyzing pay equality, one could use the Random Forest to predict the possibilities for one's wages devoid of demographics, providing a more unbiased way of evaluating fairness in pay. In any case, these models are the backbone for applied use. There are still some last mile tooling that need to be considered so that stakeholders can easily use the ML model's predictions for their day to day. As we will see in Chapt. 13, trained algorithms can be rolled into a *data product*—the last mile that allows data science to be valuable to users.

10.5   An introduction to other algorithms

While machine learning is only starting to find its way into public policy settings, the techniques described in this chapter will likely be sufficient for most use cases. However, the field of machine learning is rapidly evolving and it would be beneficial to periodically review notable advancements. In the *Journal of Machine Learning Research* (JMLR), the number of published peer-review articles increased by 2.5-times from 73 in 2005 to 184 in 2019.[14]. The open source pre-print article archive, *arXiv*, boasted a 23.7-times increase in the number of articles on artificial intelligence from 156 in 2005 to 3,697 in 2018 (Hao 2019). Indeed, the increased availability of inexpensive computational infrastructure and open source software has accelerated progress in this area.

---

[14]These estimates were calculated from articles listed at the JMLR website https://www.jmlr.org/papers/

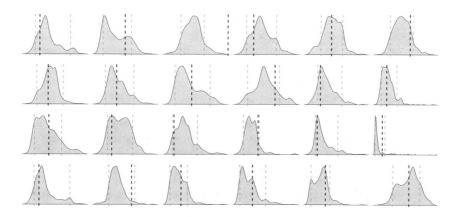

Figure 10.13: Kernel density plots of predictions from 500 trees compared with an individual's actual wage. The black dashed line represents a single wage's wage, the gray dashed lines are the 10th and 90th prediction percentiles, whereas the kernel density contains all the predictions.

In this section, we introduce a pair of advanced machine learning techniques that will no doubt see widespread use in pubic policy: gradient boosting and artificial neural networks.

10.5.1 Gradient boosting

Gradient boosting is a tree-based ensemble algorithm that has gained in popularity due to its use in Kaggle competitions. As developed in Friedman (2001), gradient boosting learns patterns through a stage-wise process by adding M-number of decision stumps.[15] A decision stump is a single split decision tree. It is a a weak learner, which is a model that needs only to be slightly better than random. As each stump is added, the algorithm corrects for previous fitting errors, thereby fine-tuning the predictions. The equation below summarizes how each new stump $f_m(x)$ is combined previously learned stumps to produce the overall model ($F_M(x)$):

$$F_M(x) = \sum_{m=1}^{M} \eta f_m(x)$$

Correcting for errors is akin to compressing an inflated balloon into a desired shape, say a tetrahedron: *As the balloon is squeezed on one side, another part of the balloon expands requiring additional pressure to correct for changes.* Thus, there is a chance that each stump might over-correct for previous errors. To mitigate the chance of overfitting, Friedman (2001) proposed a *shrinkage parameter* (η) to dampen the influence of each stump on the predictions. The shrinkage parameter is bound between 0 and 1, which has the effect of adding only a fraction of each stump is added to the prediction. As η decreases, a smaller fraction of each decision stump is added, allowing the algorithm to take smaller, fine-tuned steps toward convergence. There is a trade-off. When η is small, the number of iterations M needs to be increased to help the algorithm converge to the optimum. Thus, η and M must be tuned together.

Each implementation of gradient boosting has different hyperparameters and features that give the user flexibility to optimize the model for their specific problem. Perhaps the most popular of these implementations is the XGBoost package (Chen and Guestrin 2016). Not only is the package designed for scalable machine learning problems, the package's authors also went to great lengths to add functionality that allows for exhaustive hyperparameter tuning (e.g., shrinkage parameter η, sub-sampling proportion, tree depth among others).

Despite the extensive functionality, the main driver of predictive accuracy is the input variables. Most of the accuracy from a gradient boosting machine will come from feature engineering in which data scientists

---

[15]Gradient boosting technically uses decision trees that can be grown arbitrarily deep. In this introduction, we only discuss a simple decision stump or a one-split tree.

devise new ways to create variables from which the model can learn. The creativity required to produce cleverly engineered input variables comes not from knowledge of machine learning, but rather understanding the context in which it will be applied. For example, patient medical records can be used to engineer a wide range of descriptive variables, such as "time since last visit", "average length of stay", "number of visits in last 30 days", "number of severe diagnoses in the last year", "number of visits for chronic condition", among others.

While Random Forests are able to achieve a reasonable level of accuracy with minimal effort, a gradient boosting model can achieve greater accuracy gains through hyperparameter tuning and feature engineering. For this reason, we recommend that early stage projects should be built with Random Forest algorithms. Then, gradient boosting can be used to further refine predictions in later stages.

For more detail about gradient boosting, consider reviewing the following seminal articles:

- Friedman, Jerome. 2001. "Greedy Function Approximation: A Gradient Boosting Machine". *Annals of Statistics* 29 (5): 1189–1232.
- Chen, Tianqi and Guestrin, Carlos. 2016. "XGBoost: A Scalable Tree Boosting System". *Proceedings of the 22nd ACM SIGKDD International Conference on Knowledge Discovery and Data Mining*: 785–794.

10.5.2   Neural networks

Humans are naturally evolved to interpret on unstructured data such as speech, text, and imagery to convey complex concepts. This unstructured data is intuitive to human eyes. We can examine swaths of pixels and find the higher level concepts embedded within them, whether its people, places, etc. Unfortunately, these abilities are not easily learned by machines. In fact, historically, we would need an army of people to perform a large volume of vision and speech tasks.

While algorithms like logistic regression and tree-based models learn complex patterns from structured data, they produce lackluster results when applied to unstructured data. Traditional techniques simply lack the ability to make sense of pixels or sound waves and associate them with higher level concepts. Fortunately, *artificial neural networks* (ANNs) can fill this gap.

From identifying objects in photos to predicting stock prices, ANNs are a type of machine learning that can achieve feats that were previously thought to be science fiction. ANNs are inspired by networks of neurons—the complex inter-connected mass of cells that makes up the brain. Human memory is stored in the brain's neural network, keeping track of intricate, nuanced information through the seemingly infinite connections between neurons. An artificial neural network, as shown in Figure 10.15, is comprised of *inputs* (or input variables), *neurons* (also referred to as *neural units* or *nodes*), and *outputs*. All inputs flow into neurons, whether their values are discrete or continuous. Neurons sandwiched in-between inputs and outputs are part of a *hidden layer*—a set of formulae that are not directly observable but provide input into other neurons.

A simple neural network with a single hidden layer with a single node is similar to a bivariate regression:

$$h_k = \phi \sum_{i=1}^{I} (w_i x_i + b)$$

where each neuron h_k is a weighted sum across all of its input variables x_i. The w_i is a weight (like a coefficient) learned by the algorithm, x_i is the ith input variable, and b is a bias term that serves as an intercept. The key difference between regression and this simplified neural network is the *activation function* (ϕ), which is inspired by biological neurons. Signals in the brain take the form of electric impulses that are passed from one neuron to the next. Each neuron is not physically connected to the next, thus a neuron must *fire* a signal that is strong enough to jump and register with the next neuron. In an ANN, the activation function represents the gradient that a signal must overcome and how signal will manifest itself with the next neuron.

As illustrated in Figure 10.14, there are a broad variety of activation functions that are suitable for different applications:

- The simplest case is the *identity* function that does not transform values.
- A *rectified linear unit* (ReLU) adjusts values so that values below zero are convert to zero, while values greater than or equal to zero are retained. A favorite of deep learning techniques such as convolutional neural networks, ReLU can help the algorithm capture non-linear patterns but does not preserve information about negative values.
- *Step* and *logistic* functions both map values into a space between 0 and 1. The step function forces values into a binary form: 1 if $x \geq 0$, otherwise 0. The logistic function maps values into a non-linear continuous space between 0 and 1. Large values converge toward $f(x) = 1$ while small values converge toward $f(x) = 0$.

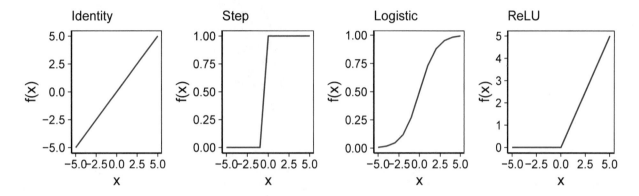

Figure 10.14: Four activation functions.

In this specific case, the single hidden layer ANN is known as a *single-layer perceptron network*. By adding more neurons and hidden layers (Graph (B) in Figure 10.15), we can construct a *multi-layer perceptron network*—a forward flowing system of equations. All input variables flow into all neurons in the first hidden layer, which in turn flow into all neurons in the second layer. The activation functions can be mixed and matched to change the qualities of the signal at each layer and each node. These two ANNs are a case of *feedforward* networks in which the direction of information is one way—from input to output. The tuning process for ANNs is architecture-based, , meaning that the layout of the network impacts model performance (e.g., how many nodes and hidden layers, which activation functions are used and what transformations are applied, how does information move through the ANN, etc.). Each architectural decision changes an ANNs' properties, allowing it to learn non-linear patterns with more hidden layers, accommodate different kinds of data, and distinguish between nuanced targets.

These single and multi-layer perceptrons are among the simplest ANNs. In fact, advanced data science has now turned its attention to *deep learning*—a broad class of neural networks that incorporates *representation learning*. The term "deep" refers to multiple layers. As a deep neural network is trained, each successive layer extracts higher-level information that represents complex concepts. In the case of images of people, the input layer contains pixels. As the signal moves through each successive layer, downstream layers can capture discrete human features such as eyes and ears, which in turn can be combined to identify specific people. Deep learning techniques such as convolutional neural networks have dramatically improved capabilities for object detection in images and video as well as speech recognition (LeCun, Bengio, and Hinton 2015) and expanded the role of algorithms to hard tasks normally reserved for humans.

Some deep learning techniques are more appropriate for sequence and time series problems. LSTM neural networks are optimized for predicting what comes next in time and keeping track of information over time (e.g., what is the demand for Uber rides in all neighborhoods around the world? Which word would be a good auto-completion recommendation?). Convolutional neural networks are well-suited for learning the latent concepts in imagery (e.g., What differentiates a cat from a dog? What looks like urban sprawl versus deforestation?). Interestingly, convolutional neural networks are particularly useful for producing artistic style filters (e.g., style transfer) that are used for photographs posted to social media platforms.

(A) Single node, single hidden layer

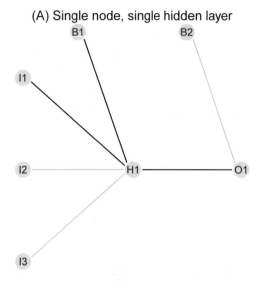

(B) Multiple nodes, two hidden layers

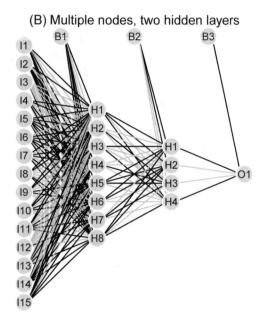

Figure 10.15: Two neural networks. (A) A neural network with a single node with a single hidden layer is functionally the same as a logistic regression. (B) A neural network with multiple nodes and two hidden layers can learn non-linearities in the data. Modern neural networks may include layers upon layers of hidden layers to capture complex, latent patterns in the data.

Neural networks are well-adapted to handle micro-tasks like facial recognition and object detection, which have clear utility for monitoring and operational tasks. However, it is not clear how these techniques can fit into high level decisions. Their opaqueness can obscure the precise rationale behind a prediction, which pose ethical dilemmas (e.g., why a prisoner was not granted parole). This is not to dismiss their potential role, but rather more research is required to open the black box.

Neural networks can instead serve as data augmentation tooling—they can create data that support more transparent use cases. Because the amount of data and infrastructure required to train a new neural network is infeasible for most organizations, it has become commonplace to use pre-trained ANNs. In recent economics research, Amazon's Core AI and Economics team developed price predictions for consumer goods, which is useful for inflation measurement. The innovation of this particular research was the use of pre-trained deep learning models such as Embeddings from Language Models (*ELMo*) for text and *ResNet50* for imagery to produce *embeddings* (i.e., a type of variable) to improve price prediction. In turn, these embeddings were used as inputs into a neural network to directly predict prices (Bajari et al. 2020).

The ecosystem of neural network frameworks is quite expansive. Software frameworks such as Keras, `TensorFlow` and `H2O` are widely used. Furthermore, as deep learning requires large training data to be generalizable, there are pre-trained models that are openly available, such as `AlexNet`, `ResNet-50`, `ResNet-101` among other creatively named implementations. We should note, however, that pre-trained models are known to exhibit biases. The source of such biases will be discussed in.

For a conceptual in-depth treatment of deep learning and neural networks, consider reading: *Neural Networks and Deep Learning: A Textbook* (Aggarwal 2018b). By building a conceptual base, it becomes easier to understand researched published in major conferences such as *Neural Information Processing Systems* (NeurIPS) and publications such as the *Journal of Machine Learning Research*.

For hands-on treatment on deep learning, consider *Deep Learning with R* (Chollet and Allaire 2018). While R can interface with some frameworks, the more common choice for deep learning is `Python`. *Hands-On Machine Learning with Scikit-Learn, Keras, and TensorFlow: Concepts, Tools, and Techniques to Build Intelligent Systems* (Geron 2017) offers a whirlwind overview of a number of the most popular frameworks.

10.6 Beyond this chapter

In this chapter, we provided a broad overview of both traditional ML techniques and recent developments. While these techniques have seen rapid adoption in private industry, the uptake in government has been at a more cautious pace. Beck, Dumpert, and Feuerhake (2018) conducted a 2018 survey of 30 international statistics producing institutions, finding over 130 machine learning projects in development. At the time, only 15% of projects were in production—or in regular use supporting core operations. In contrast, 64% of projects were in the idea or experimental phase, suggesting that the adoption and integration of predictive methods were only in a nascent stage. In other areas of government, namely city agencies with "ground operations", ML techniques have seen more use.

In general, the pace of adoption in government is far slower. This can be explained by a couple of reasons. First, public servants are generally trained in causal inference and estimation rather than prediction, thus many might not be familiar with the distinctions. Second, because the distinctions are not yet clear, many government officials have not had the chance to think through what constitutes a practical but ethical use of the technology. Public sector agencies must naturally be more careful with its actions, weighing ethical considerations and the impacts on constituents. This is stands in stark contrast to private industry where the barriers to using such techniques are governed by the market—business leaders will deploy predictive technologies as long as it has a positive effect for their business. As we will cover in Chap. 14, the ethics of data are indeed quite complex.

Nonetheless, we argue that one should not shy away from these powerful techniques, but rather master them in order to devise equitable ways of using them for the public good. A solid first step is to be able to communicate what is happening in any model and that requires further study. A good place to start is to review *An Introduction to Statistical Learning* (James et al. 2014). The text is written for R users and provides clear explanations of algorithms at an introductory level. For more advanced yet broader coverage of machine learning techniques, we recommend a pair of standard texts. The first is *The Elements of Statistical Learning: Data Mining, Inference, and Prediction* (Hastie, Tibshirani, and Friedman 2001)—a modern classic technical text is written from an applied statistician's perspective. The second is *Pattern Recognition and Machine Learning* (Bishop 2006), which is written from a computer scientist's perspective. Ultimately, data scientists need to build prediction pipelines to brings their algorithms to life and into action. For a hands-on treatment that weaves the `caret` package with theory, consider *Applied Predictive Modeling* (Kuhn and Johnson 2013).

Chapter 11

Cluster Analysis

11.1 Things closer together are more related

Precision *anything* is the domain of data science, since it relies on identifying patterns or similarities within data. Modern marketing strategies, for example, match individuals to advertisements and offers intended to change behavior, like voting for a specific candidate or buying a particular brand of shoes. Presenting an advertisement for bathing suits does not make much sense for a consumer in Antarctica. Similarly, different people along the US political spectrum subscribe to different positions on gun ownership, reproductive rights, among other social issues.

If cost were no concern, each advertisement could be perfectly tailored to each receiving individual. There would be as many narratives or messages as individuals. In reality, however, crafting a high quality message that resonates with the reader takes time and resources, thus it is not a scalable process. The problem of precision marketing is a constrained optimization problem. The objective is to change behavior, while the constraint is cost. *How many different messages should be crafted? Who should see which message?*

To a data scientist, this sounds very much like a job for clustering—when groups of individuals or entities can be identified by similarities in their observable characteristics. Clusters enable cost-effective strategies for applying policy treatments, yet these clusters are not definitions that are set in stone. Instead, they are latently defined. They do not have a label nor do we have a precise empirical definition for them, but we recognize them when we see them. Clustering is especially useful right now in politics, since background data on voters is more readily accessible than ever, whether it's their purchasing preferences or behavior on social media. It is now possible to have more refined and nuanced clusters of voters, which would respond at higher rates to specific messages. These clusters now come with names like "Flag and Family Republicans" and "Tax and Terrorism Moderates"—empirically identified by political campaigns (Koman 2006). The optimal number of groups depends on the amount of campaign resources available, and how well the groups are crafted from data—the rate at which they respond to the produced political messaging (Avirgan 2016). This is the core idea behind *micro-targeting*—campaigns are tailored to distinct groups and even specific people (Brennan 2012).

A rudimentary form of targeting was used as early as Jimmy Carter's 1976 Presidential campaign in which the country was segmented by issue and geography (Cillizza 2007). Surveying played a notable part in understanding which issues matter to which demographics—what emotions are associated with different social and fiscal matters (Curry 2007). In the information age, the possible data-driven strategies are also associated with a "creepy" factor. As it turns out, individual level data such as brand preferences are correlated with which candidate a voter is likely to vote. According to a survey conducted by a brand consultant in 2016, Bernie Sanders supporters are 82% more likely to eat at Chipotle than the average American, and Donald Trump supporters are 111% more likely to eat a Sonic (Higgins 2016). Thus, using data that is common in marketing databases enable campaigns to define hyper-local, finely tuned clusters for highly customized messaging.

© Springer Nature Switzerland AG 2021
J. C. Chen et al., *Data Science for Public Policy*, Springer Series in the Data Sciences,
https://doi.org/10.1007/978-3-030-71352-2_11

The interest and investment in micro-targeting in campaigns has only grown with each successive election cycle. For example, the 2012 Obama campaign's analytics team was five times as large as the 2008 campaign (Scherer 2012). In the 2016 cycle, the campaign of Libertarian candidate Gary Johnson relied on Facebook's political ideology clusters to bolster his vote share (Shinal 2018). Granted, with the revelations of the Cambridge Analytica scandal, criticism is mounting with some calling the microtargeted "dark advertising" on Facebook a threat to democracy.

Nonetheless, clustering is just a statistical technique. Not only it is used for political campaigns and marketing, but it can help extract visual patterns in photographs and detect anomalous activity in web traffic. *It is not a tool that helps to identify "truth", but rather should be viewed as an exploratory technique that brings structure to otherwise unlabeled data.* It is designed to find structure when none is obvious. It is a form of unsupervised learning –when data are devoid of target variables or labels, clustering finds patterns that are embedded in the input variables.

In this chapter, we explore the concepts that underlie clustering algorithms and their applications. As the goal of clustering is quite different from supervised learning, we first introduce foundational ideas on which most clustering algorithms are built —relating records to one another, then finding commonalities between records or exploiting differences. We examine two standard clustering techniques and illustrate how they can be applied to informing economic development targeting as well as structuring household consumption patterns.

11.2 Foundational concepts

We are wired to see clusters of activity, to see patterns. For example, when we are presented with satellite imagery (Figure 11.1), our eyes automatically start to hunt for patterns. Each image shows the intensity of night-time lights in a different U.S. city. Presumably, the population density is correlated with the night-time light intensity, and using light intensity, we can ballpark urban, suburban, and rural areas. This is basic clustering—interpreting activity based on similarities in the data.

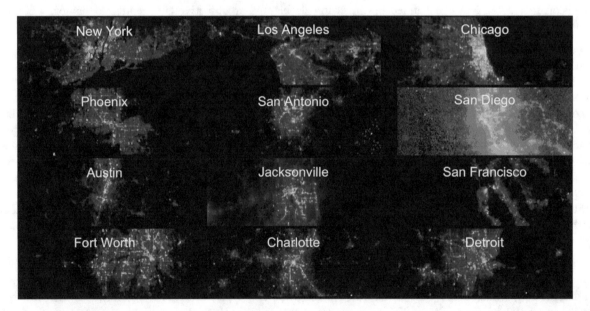

Figure 11.1: Night-time imagery from the NOAA-NASA Suomi NPP Satellite's Visible Infrared Imaging Radiometer Suite (VIIRS).

Harnessing the human eye on a large scale is not an easy or convenient task—it can be labor intensive and costly. How does one do this programmatically, algorithmically, and at scale?

There is not just one definition of a cluster, nor is there a single approach for clustering observations. Is the key to clustering to find points that are close to one another? Or can it be defined in terms of likely membership to a latent group? Or does it have to do with the density of observations? Like all things data science, devising a successful analytical strategy depends on the objective, though we recognize this might not be a satisfying answer in policy circles.

Let's take the case of separating likely populated from underpopulated areas in satellite imagery of the San Francisco Bay Area as seen in Panel (A) in Figure 11.2. In the false color imagery, we might assume human activity is concentrated along two yellow strips of land while the ocean, bay, and rural areas shaded in blue. Each pixel is a data point and its radiance (light intensity) is just one value in a matrix of radiances. In short, imagery is data. The geographic relationships can be also presented in the form of a radiance distribution (Panel (B)) in which there are at least two distinct types of activity in the imagery: the tall peak to the left represent dark pixels where human activity is unlikely, whereas the long tail to the right contain pixels of varying light intensity—the highest of which likely has the greatest population density. How do we separate one group from the other?

A simple clustering approach could identify a radiance threshold that separates light and dark pixels—simply eyeball a line in the radiance distribution and group pixels based on whether they are above or below the cutoff point. If we think through the mechanics of that simple cutoff we are looking for a way to maximize the gap between light and dark. A more rigorous strategy could define clusters based on *centroids*. A centroid is the average value in a cluster and serves as a reference point against which all points are compared to determine their cluster membership. An alternative strategy takes a bottom-up approach, measuring the *connectivity* between pixels—how does each pixel relate to one another based on some concept of distance. One bright pixel is not like to be too "far" from another bright pixel. Those two pixels may be more similar to a slightly brighter pixel but are much less related to a dim pixel. By comparing pixels at a granular level, we can map the hierarchy of relationship from the ground up—which are more associated with others. All of these strategies are valid and are able to bisect the radiance distribution (see Panel (C)), allowing us to quantize the data from a continuous distribution of radiances into two values that represent light and dark. When mapped (see Panel (D)), the area around San Francisco is reduced to two shaded areas—light and dark.

Perhaps most amazing of all, the algorithm was not "aware" of what it was searching—*clusters are formed without any guidance*. It is ultimately up to the data scientist to examine the characteristics of each cluster, determine if the clusters are grounded in reality or are artifacts of noise, and articulate a label for the cluster. The San Francisco example is simple— the landscape can be divided into inhabited versus less uninhabited. More complex social datasets (e.g., politics, economic development) or natural sciences (e.g., genetics) will require more context to inform the cluster label.

There are a many clustering techniques that can be applied to everyday problems. They have the potential to bring structure to everyday amorphous problems and are a fine starting point for exploring data. In this chapter, we explore two of the most commonly used clustering algorithms: *k*-means clustering and hierarchical clustering. Each clustering technique is motivated by different theories of how data should relate to one another.

11.3   *k*-means

Perhaps the simplest of the clustering techniques is *k-means*. The algorithm is designed to identify *k*-number of user-specified clusters, each of which is defined by a centroid that serves as a reference point. It is a quantization technique capable of capturing the gist of data and compressing it into a concise form.

11.3.1   Under the hood

How does one find the centroid? The centroid in cluster c is the mean of each input variable. For example, if wages and age are input variables, the centroid for cluster c would be comprised of two values:

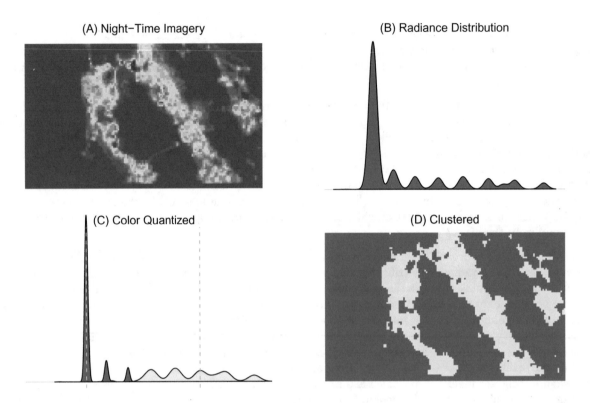

Figure 11.2: Applying clustering to extract likely inhabited areas in the area around San Francisco: (A) False color night-time imagery, (B) Kernel density of radiances, (C) Clustered or color quantized distribution, (D) Imagery classified into light and dark areas.

$$\mu_{c,wages} = \frac{1}{n_c} \sum_{i=1}^{n_c} x_{i,wages}$$

$$\mu_{c,age} = \frac{1}{n_c} \sum_{i=1}^{n_c} x_{i,age}$$

This presents a chicken-and-the egg problem: do the points define where the cluster is, or does the cluster determine the cluster label of each point? In actuality, the optimal location of each centroid is not known and may not even truly exist. Nonetheless finding clusters in a k-means framework can be solved by starting with a random guess, then iteratively refining the centroid coordinates. Before starting the algorithm, the analyst must specify how many clusters k are believed to be in the data. While context and knowledge of the problem space can inform the initial value of k, the value can be tuned as will be described later in this section. The algorithm progresses as follows:

1. *Initialization.* To begin, the algorithm randomly generates k-centroids, each of which defined within the space of the input variables. In addition, each centroid is assigned an identifying number that serves as the label for its cluster c.
2. *Assignment.* The distance between each point i and each centroid is calculated. For continuous values like radiances, wages, employment, among others, Euclidean Distance ($d(x_1, x_2) = \sqrt{\sum_{i=1}^{n} |z_1 - z_2|^2}$) is a suitable choice for measuring distance or dissimilarity, while Manhattan distance is reasonable choice for discrete variables (e.g., demographics characteristics). Each point is assigned to a cluster based on whichever centroid it is closest. The cluster assignment after this first iteration is unlikely to be optimal – it is random.
3. *Update.* The algorithm proceeds to update the centroid by calculating the mean value of each input variable for points in each cluster c. This procedure shifts the center mass of the cluster. With the new centroids in place, repeat Step 2.

This *assignment-update* procedure (Steps 2 and 3) is repeated so that the algorithm can converge on an optimal solution. The algorithm has converged when cluster assignments no longer change.

From a statistical lens, we can frame k-means as an optimization problem. The objective is to find the cluster assignments that minimizes the total cluster variance:

$$argmin \sum_{c=1}^{k} \sum_{i=1}^{n} (X_{c,i} - \mu_c)^2$$

in which variance is the sum of squared distances between each point i in cluster c to its centroid μ_c.

11.3.2   In Practice

While k-means is a simple algorithm, there are a number of quirks that one must keep in mind in order to use it responsibly.

Stability of clusters. While randomization kick starts the algorithm, it runs the risk that the same starting conditions can be run twice and yield different results. In fact, the cluster label for each point can be different from one model run to the next! A centroid can straddle two clusters that the human eye may otherwise deem as separate and distinct (see Figure 11.3).

How can this be? Is this a design flaw?

One source of instability is the number of variables and their scale. Similar to KNN in supervised learning, all input variables should have equal influence on how the algorithm converges on its results. For example, the effect of an age variable on a range of 20 through 50 will have a different weight than a wage variable distributed from 100,000 to 1 million. If each variable has different ranges of values and dispersion, the algorithm will lean on one variable more than others.

High dimensional datasets pose challenges as well. When faced with a large number of variables, both signal-packed and noisy variables are considered equally, allowing irrelevant information to play a disproportionate role in cluster determination. Though this is not a problem specific to k-means, it can cause centroids to fall into local optima—there may be more nooks and crannies on the optimization surface into which the model may fall. To avoid biasing the results toward one variable, consider scaling variables using a standard z-score transformation, which ensures that all variables have unit variance. Also, favor parsimony—limit the number of variables to reduce the effect of confounding factors.

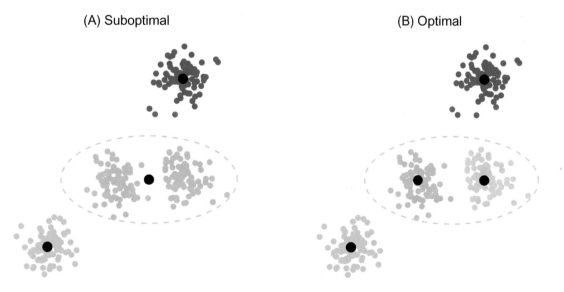

Figure 11.3: Comparison of a suboptimal and optimal results from a k-means algorithm.

Tuning. As we can see in Figure 11.3, the human eye can identify $k = 4$ clusters. If $k = 3$ were specified, k-means would be forced to group two or more clusters together. Conversely, $k = 5$ would force the algorithm to split clusters into fractions of a cluster—a suboptimal result. While it is necessary to find the optimal k, there is a lack of consensus on how to correctly tune k-means algorithms. In some cases, the data scientist may choose a small value of k that is small enough to articulate to audiences (e.g., there are five types of people in the world. . .). While this is convenient, it is arbitrary.

The *Elbow method* is a more data-informed approach that compares the total cluster variance across multiple values of k. As seen in Figure 11.4, the curve that emerges from plotting the variance over values of k will be downward sloping. In an optimization problem, the natural tendency is to select the smallest value of k; However, doing so in this case does not provide any useful information. In fact, the total cluster variance should approach zero as k approaches the sample size. The curve should ideally "bend" like an arm (Panel (A)), indicating that beyond some value of k, there are a diminishing returns to each additional k. The inflection point is that value of k. It is the "Goldilocks" value. There is one problem with the Elbow method: a clear inflection point is not always present (Panel (B)).

A more computationally intensive strategy optimizes the *Silhouette value*, which compares the similarity of a given observation i to observations within and outside its cluster. The silhouette $s(i)$ is defined as

$$s(i) = \frac{b_i - a_i}{max(a_i, b_i)}$$

where a_i is the Euclidean distance between a point i and other points in the same cluster, b_i is the minimum distance between i and any other cluster of the sample. The silhouette takes on values of $-1 \leq s(i) \leq 1$, where 1 indicates that an observation is well matched with its cluster and -1 indicates that fewer or more clusters may be required to achieve a better match. Put into action, for each of k that is tested, the silhouette value is calculated for every observation and summarized as the *mean silhouette*. The largest mean silhouette marks the optimal k.[1]

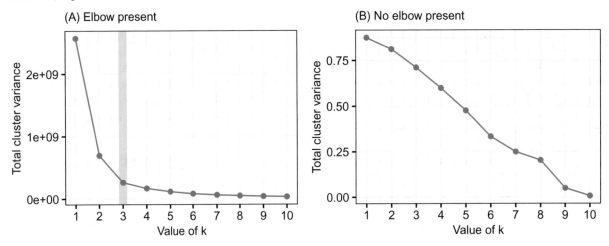

Figure 11.4: Elbow method: Choose k at the inflection point. (A) Inflection point identified at k = 3, (B) No inflection point identified.

Scoring. Suppose a k-means algorithm has converged and clusters have been defined. If presented with new data, how can new observations be mapped to the pre-defined clusters? Much in the same way as defining a cluster, new observations simply need to be assigned to the nearest cluster centroid. One practical way to *score* the new dataset is to treat it as a nearest neighbors problem (kNN) in which the goal is to identify which centroid is the nearest neighbor. By applying kNN setting $k = 1$, each new observation is scored by treating the cluster labels as the target variable and the centroids as the input variables.

Other considerations. Cluster instability and sensitivity to high dimensionality can limit the usefulness and reliability of k-means. But are there other limitations? For one, k-means is based on minimizing total cluster variance, which works best when the shape of clusters is spherical and symmetrical. In fact, k-means generally have difficulty in neatly identifying clusters when the cluster shape is anything other than spherical,

---

[1]We illustrate the silhouette method in the following DIY section.

has outliers that skew the cluster's shape, or contains noise in the data that reduces the distance between clusters. In practice, however, some data simply do not lend themselves to producing neat clusters—there is not clear pattern. Instead, applying k-means in these cases has the effect of ranking groups of observations. For example, applying clustering on highly correlated variables such as age and wage has the approximate effect of segmenting observations by quantile (e.g., older high earners, young low earners), thereby producing grouped rankings.

Nonetheless, k-means is a good first step for many clustering problems. It is remarkably fast and efficient, and the method exposes latent structure in the data. However, like any technique, it has its shortcomings. As we will see later in this chapter, there are more sophisticated algorithms with useful features, such as hierarchical clustering and density-based clustering.

11.3.3 DIY: Clustering for economic development

> *DIYs are designed to be hands-on. To follow along, download the DIYs repository from Github* (https://github.com/DataScienceForPublicPolicy/diys). *The R Markdown file for this example is labeled* `diy-ch11-kmeans.Rmd`.

Economic development corporations and chambers of commerce support local communities by attracting jobs and investment. Given the need for more jobs around the country grows, economic development initiatives are fierce affairs, sometimes pitting one community against another in bidding wars over tax benefits. In 2018, Amazon.com announced new secondary headquarters in New York City and Arlington, VA after an exhaustive 20 city search (Amazon 2018).[2] The global manufacturer Foxxconn announced it will bring high-tech manufacturing to Racine, WS (Romell and Taschler 2018). And a long-standing "border war" between Kansas City, MO and Kansas City, KS has seen a number of high profile companies like AMC Theaters move headquarters a mere miles, chasing economic benefits (The Economist 2014).

Beyond the bidding war spectacle, there are other issues that factor into these siting decisions. Furthermore, not all companies are as high profile as the ones described above but are nonetheless important due to their contributions to the economy. For one thing, the prospective host region of new jobs should have the right economic conditions to sustain and foster the new opportunity. Suppose a tech executive in Santa Clara or Alameda in the Bay Area in California wanted to find another county with similar socioeconomic conditions. *Based on available data, how would one find a list of comparables?* The same question could be asked in reverse for economic developers: *what are other areas that are in direct competition?*

A competitive analysis could first consider which observable characteristics of a city or county are selling points for prospective businesses. *Is it the size of the labor force? Is it the relative size of the target industry? Or perhaps it is related to education of the labor force or the local cost of employment?* In any of these cases, publicly available economic data can be clustered using the k-means technique to surface insights on the market landscape. Below, we illustrate a simple process of finding clusters of comparable economic activity, focusing on finding clusters associated with tech industries.[3]

Set up. We start by loading the `cluster` library that has utilities for evaluating clustering results, then import a county-level dataset that contains information for over 3,100 US counties.

```
# Load library
pacman::p_load(cluster)
```

```
# Load data
load("data/county_compare.Rda")
```

The underlying data is constructed from a variety of U.S. Census Bureau databases, in particular the American Community Survey, County Business Patterns, and the Small Area Income & Poverty Estimates.

- `fips`: Federal Information Processing System code that assigns a unique ID to each county.

---

[2] This later was revised to only Arlington, VA due to local politics in New York.

[3] For simplicity, we define online tech industries using NAICS codes 5182, 5112, 5179, 5415, 5417, and 454111 although we recognize this may exclude sub-industries that are rapidly growing in importance in tech.

- `state` and `name`: The US state abbreviations and the name of the county.
- `all.emp`: total employment (U.S. Census Bureau 2018b).
- `pct.tech`: percent of the employed population in tech industry (U.S. Census Bureau 2018b).
- `est`: percent of company establishments in that industry (U.S. Census Bureau 2018b).
- `pov`: the poverty rate (U.S. Census Bureau 2017a).
- `inc`: median household income (U.S. Census Bureau 2017a).
- `ba`: percent that is college educated (U.S. Census Bureau 2017a).

Clustering. Before we apply k-means, the data should be mean-centered and standardized so that no single input has disproportionate influence on clustering (the equal weights assumption). The `scale` function is applied to the numeric fields (column 4 onward), then the output is assigned to a new data frame `inputs`.

```
inputs <- scale(cty[,5:ncol(cty)])
```

Let's get comfortable with the clustering process. As a dry run, we apply the `kmeans` function to `inputs`, specifying $k = 5$ for five clusters, and setting the seed to a constant value so the analysis is replicable. The resulting object `cl` not only contains diagnostics about the clustering process, but also the coordinates of the centroids and the cluster assignment for each county (`cl$cluster`). Digging into the `cl` object, we can calculate how many counties fall into each of the five clusters by tabulating `cl$cluster` as well as retrieve the loss metric—the total cluster variance (`cl$tot.withinss`). *Is this a good result? Why choose five clusters? Why not two or 50?*

```
# Set seed
set.seed(123)

# Apply clustering
cl <- kmeans(inputs, centers = 5)

# Tabulate number of counties per cluster
table(cl$cluster)

# Retrieve the Total Within Sum of Squares (TotalCluster Variance)
cl$tot.withinss
```

To ensure clusters are identified with a practical degree of rigor, we will search for the optimal value of k by comparing mean silhouette widths as calculated using the `silhouette` function in the `cluster` library. `silhouette` requires two inputs: the cluster assignment and a dissimilarity matrix. The former is an output of `kmeans` whereas the latter is a distance matrix between all observations that can be calculated by applying the `dist` function on the `input` variables.

The `silhouette` function calculates the *silhouette width* for each observation. To make comparisons between each value of k, we instead need the mean silhouette width. Below, we illustrate how this calculation should flow: (1) calculate the dissimilarity matrix, (2) calculate the silhouette and store to object `sil`, (3) calculate the mean of silhouette values contained in the third column of `sil`.

```
# Calculate dissimilarity matrix
dis <- dist(inputs)

# Calculate silhouette widths
sil <- silhouette(cl$cluster, dis)

# Calculate mean silhouette width
mean(sil[,3])
```

Optimizing k. To test dozens of values of k, it would be prudent to wrap the clustering and silhouette procedure into a cohesive function. Not only would this keep the code concise and tidy, but it also allows the optimization procedure to be applied and re-applied to other problems. We combine the k-means and silhouette functions into a single function km. The function requires input variables x and the number of clusters k. In addition, it requires a dissimilarity matrix d. As d is the same for all scenarios of k, we only need to calculate the dissimilarity matrix once and apply the same object to each iteration to save compute time.

```
km <- function(x, k, d){
  #
  # Calculate mean silhouette for a value of k
  #
  # Args:
  #   x = data frame of input variables
  #   k = number of clusters
  #   d = a dissimilarity matrix derived from x

  cl <- kmeans(x, centers = k)
  sil <- cluster::silhouette(cl$cluster, d)

  # Return result
  return(data.frame(k = k, sil = mean(sil[,3])))
}
```

With km ready to go let's test values of $k \in \{2, 30\}$, storing the mean silhouette for each k in the placeholder data frame opt. Notice how the code is relatively compact and itself could be wrapped into an optimization function as well.

```
# Placeholder
opt <- data.frame()

# Loop through scenarios
for(k in 2:30){
  opt <- rbind(opt, km(inputs, k, dis))
}
```

By plotting the contents of opt, we reveal the shape of the silhouette curve for all tested values of k (see Figure 11.5). Ideally, the curve will have a global maximum and is downward sloping as k increases. The global maximum is $k = 2$, indicating that the US' can be divided into two clusters that have distinctive levels of tech activity, one with $n = 526$ and the other with $n = 2611$.

Are these clusters meaningful? Suppose one wanted to create a Buzzfeed list of the most educated counties in the US. It is as easy as taking the proportion of people with a college degree in each county, then sorting from most to least. What if the list were expanded to the most educated and populous? Or if income was added in addition? Clustering can be treated as a way to bucket observations with similar values together, effectively producing a rank. This is evident in Figure 11.5.

The scatter plots suggest that larger employment centers tend to also have higher concentration of tech employment and greater incomes as well—not surprising given the trend toward urbanization. These resource rich areas tend to bubble across all measures. Granted, these results are not causal, but are certainly suggestive. In effect, the k-means algorithm provided an efficient strategy to partition better resourced communities from ones that are less well-off.

Making use of clusters. How should these results be used? It depends on the audience. An economic development authority does not need to compete with every county—it should focus on comparables. By clustering all counties on their observable characteristics, economic developers can map their county to a pool of similar counties. If the authority can articulate which other counties are competitors in its market segment, then it can also craft offerings for prospective companies that distinguish itself from the rest.

A tech company searching for the site of its future headquarters may not want to pay for being in the most expensive cities yet would want access to a reasonably sized, skilled tech labor force. k-means condenses the data of competitiveness into a simple list of where tech is concentrated. The smaller of the two clusters

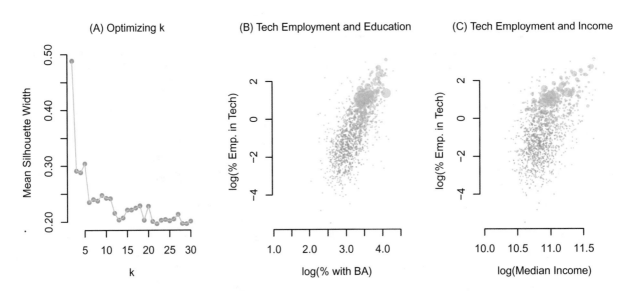

Figure 11.5: (A) Grid search for optimal k with respect to mean silhouette width, (B) and (C) Clusters for select input variables bivariate plots-scaled by total employment.

contains $n = 526$ counties, which is comprised of iconic high-tech areas such as San Francisco and Santa Clara in California as well as large cities such as New York City (New York City) and Seattle (King County, WA). The same cluster finds less expensive and less densely populated alternatives like Durham, NC, and Arlington, VA—both of which are growing centers of technical excellence. In essence, clustering can identify smart alternatives to inform the selection process.

Despite these promising use cases, keep in mind there are limitations. From an ethical perspective, cluster analysis based on the past performance of regions might be a self-fulfilling prophecy. Well-resourced areas will continue to draw more attention and resources, ignoring smaller promising areas. Furthermore, clustering on past data does not provide any indication of future performance of counties. For k-means, in particular, the outputs are "flat"—cluster labels are one dimensional and ignores the rich context that shows how some observations are more like some than others. Alternative methods might be more useful in capturing that context. As we will see in the following section, hierarchical clustering is a step in that direction.

11.4   Hierarchical clustering

There is more to clustering than producing a cluster label. k-means assumes that an entire cluster can be represented by the location of its centroid, but in so doing, we ignore one of the richest sources of information: how points are related to one another. The linkages *between* points capture the relationship between people, places, and events. It is within those links that we tell nuanced stories behind clusters, and how one observation is related to any other. To capture the relationships among points, we turn to a more sophisticated alternative known as *hierarchical clustering.*

Hierarchical clustering relies on the similarity between each observation and all other observations. Hierarchical Agglomerative Clustering (HAC)—a common hierarchical clustering approach—groups together observations in a greedy fashion, doing so iteratively until all observations are part of one mega cluster.[4] By agglomerating the data, one observation at a time, the algorithm catalogs how all points are related to one another. These relationships are the scaffolding of the cluster analysis and can be visualized in the form of a *dendrogram* (see Figure 11.6), a tree diagram that plots the relationship between all points scaled by a measure of dissimilarity. The same information can be rendered in other formats borrowed from evolutionary

---

[4]Hierarchical clustering is technically comprised divisive and agglomerative clustering. The former is a top-down approach, splitting a sample into smaller clusters until each observation is a singleton—reminiscent of decision tree learning. Agglomerative clustering is a bottom-up approach, grouping together observations. Both algorithms are *greedy*, meaning they make the locally optimal splitting or grouping decision in each iteration.

biology such as phylogenetic trees, cladograms, and fan-style trees.

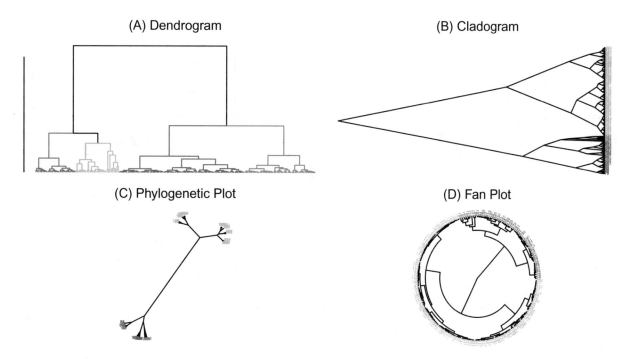

Figure 11.6: Hierchical clustering can expose the underlying structure of data by measuring how every point is connected to all others.

11.4.1 Under the hood

The HAC algorithm is a simple but exhaustive procedure. Like most algorithms, it is an iterative process:

1. Calculate distance d between all points. All points are first treated as their own cluster known as *singletons*.
2. Group records until there is only one cluster:

   – Calculate similarity s based on distance d between all clusters.
   – Find the most similar pair of clusters.
   – Merge the pair into a single cluster.
   – Repeat process.

To illustrate HAC in action, let's take a simple six-observation dataset as illustrated in Figure 11.7. In iteration 0, there are a total of $k = 6$ clusters. In iteration 1, the two closest singletons are linked together, resulting $k = 5$ clusters. The algorithm proceeds to iteration 2 in which the next closest pair of singletons are linked, leaving $k = 4$ clusters. The process continues until all points are part of a single cluster. Since the algorithm captures the linkages in each iteration, the total number of possible clusters is $1 \leq k \leq n$.

There is a twist to the HAC algorithm: we need to define what constitutes the *similarity* between a pair of clusters. Of course, Euclidean and Manhattan distance are favorites in measuring relatedness, but in HAC, it is also necessary to define the linkage method that guides how the similarity is calculated.

Linkage methods. HAC relies on a family of *linkage methods* that dictate which points are used in similarity calculations. The choice of linkage method not only changes how two clusters are merged, but influences the computation complexity and time requirements. Linkage methods fall into two classes: distance-based measures or variance-based measures.

Single linkage, or nearest neighbor linkage, calculates similarity s based on the closest pair of points, i and

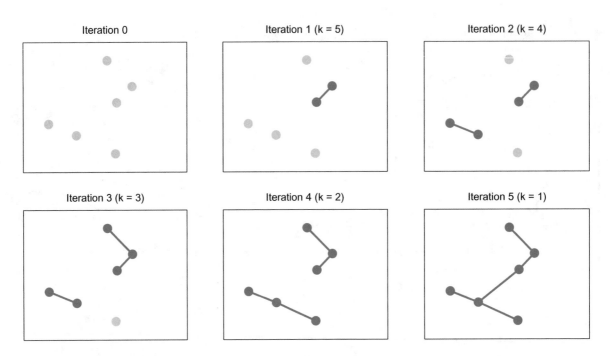

Figure 11.7: Each iteration of a modestly sized hierarchical clustering problem.

j, from a pair of clusters. In other words, first calculate the distance in terms of input variables X between all pairs of points, then choose the pair of clusters with the shortest distance.

$$s_{ij} = min[d(X_i, X_j)]$$

The opposite of single linkage is *complete linkage*, which searches across all points in a pair of clusters to represent distance in terms of the most dissimilar points (maximum distance).

$$s_{ij} = max[d(X_i, X_j)]$$

Single and complete linkage both rely on the extremities of clusters where outliers tend to be which at times can create odd-looking, suboptimal results. Alternatively, linkages can focus on the center mass of clusters where the most representative information is located. *Average linkage* considers more information, distributing its weight across all points by averaging across all distance between all points in a pair of clusters:

$$s_{ij} = \sum_{i=1}^{I} \sum_{j=1}^{J} d(X_i, X_j)$$

where I and J are the number of points in a pair of clusters. Another way to incorporate more information from clusters is *centroid linkage* in which distance is calculated from the cluster centroids, $\bar{X}_i$ and $\bar{X}_i$:

$$s_{ij} = d(\bar{X}_i, \bar{X}_j)$$

Distance, however, does not indicate that combining two clusters would be statistically meaningful. *Ward's Method* approaches clustering from the lens of an analysis of variance (ANOVA). From among all the prospective merges, two clusters are merged if it minimizes the increase in the sum of squares. This strategy is equivalent to finding two clusters that are least statistically different from one another. Through a greedy search, the *merge cost* is for all candidate merges are calculated, keeping the smallest value:

$$s_{ij} = \Delta(X_i, X_j) = \frac{n_i n_j}{n_i + n_j} ||\bar{X}_i - \bar{X}_j||^2$$

where n_i and n_j are the sample sizes of a pair of clusters i and j. $\bar{X}_i$ and $\bar{X}_j$ are the cluster centers.

Each of these linkage strategies has different effects on the shape of the dendrograms —some resulting more distinct clusters than others (see Figure 11.8). Using a random subset of the economic development data, we find that single, centroid, and average linkages are inadequate for separating observations into clean clusters at $k = 4$—one cluster contains the majority of observations. In contrast, Ward's method and complete linkage are able to break the sample into more equally sized clusters. Ultimately, the choice of linkage is a matter of trial and error, but Ward's will generally provide the more robust, statistically grounded solution.

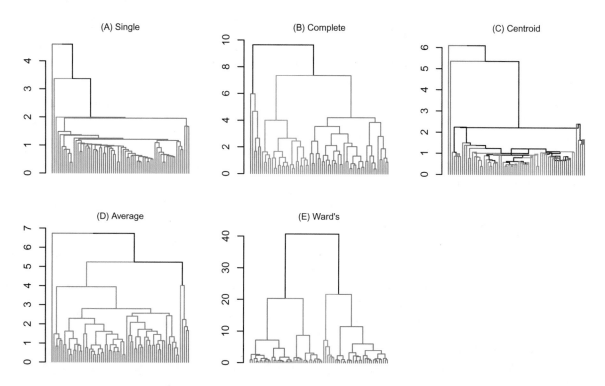

Figure 11.8: Effect of Linkage Methods on Clustering Results. Number of clusters set to k = 4.

11.4.2 In Practice

While dendrograms provide an intimate look inside the relationships within a dataset, HAC is computationally expensive in that as the sample increases, the amount of time required to complete the task grows much faster. In computer science, *Big O* notation describes the time and memory complexity of an algorithm, which in turn has a direct impact on whether a technique is feasible for a use case. The O refers to the "order of the function" indicating the worst case for an algorithm's computational performance. For example, $O(1)$ indicates an algorithm is able to complete a task regardless of the number of observations or inputs. $O(n)$ indicates that time required is a linear to the size of the data and inputs. $O(n^2)$ means the time requirement grows faster then the number of observations in a dataset. Certain implementations of HAC have a time complexity of $O(n^3)$, meaning that the algorithm will require a large amount of time for even a modest sized dataset. Relying on HAC in a policy and strategy thus depends on the decision latency. Most datasets used in policy are relatively small, but occasionally a larger set may require more time to process. If the audience can wait for the outputs, it is well-worth the wait.

11.4.3   DIY: Clustering time series

DIYs are designed to be hands-on. To follow along, download the DIYs repository from Github (https://github.com/DataScienceForPublicPolicy/diys). The R Markdown file for this example is labeled `diy-ch11-hca.Rmd`.

How is the economy doing? Where should we focus our efforts? While these are vague questions, they are common questions that policy makers struggle with. The questions imply the need for a comparison of the economy against its historical performance in a time series—a fairly straightforward problem. But when it is expanded to the *where* dimension, we suddenly must incorporate a geographic component that will easily increase the complexity of the problem. In the United States, for example, a state-level analysis expands the number of time series to 50, whereas a county-level analysis balloons to 3,100 time series. The challenge when moving from a single time series to thousands of time series is the sheer amount of information that needs to be concisely summarized in an informative, relatable manner. But maybe every county is not a snowflake. Maybe each county can be viewed as part of an economic motif. Each county can be compared with all other counties using the qualities of its time series (e.g., seasonal cycles, increasing or declining trends, and irregularities), then counties that move similarly can be part of the same economic motif — essentially identifying clusters. These clusters reduce the complexity of the data so that only a few distinct profiles can form the basis of a concise economic analysis.

In this DIY, we illustrate how to apply hierarchical clustering to time series data to identify and articulate behavioral motifs. It is an idea that can be widely applied to economic data. Clustering can help policy makers understand that there is not just one type of economic growth—it is heterogeneous and some regions may be more resilient than others. The approach can also be applied beyond economics to cyber-security to identify common types of web traffic in order to monitor for unusual activity. Virtually any set of time series can benefit from clustering.

We rely on the Quarterly Census of Employment of Wages (QCEW), a quarterly dataset of employment and wages that represents more than 95% of US jobs. Collected by the US Bureau of Labor Statistics, the data is one of the core data sources that tells the economic story of the United States, showing every level of economic activity from the county level to the national top-line.[5] While approximately 3,200 counties are published in the QCEW, we illustrate the clustering on a subset of data, namely, mean quarterly employment for California's 58 counties.[6] The dataset contains 100 quarterly observations from the first quarter of 1992 through the fourth quarter of 2016. The time series are provided in wide format, storing the date and time index in the first two columns and the remaining 58 columns contain the county-level employment time series.

```
# Load QCEW data
load("data/qcew_cali_sa.Rda")
```

The hierarchical clustering process starts with constructing a distance matrix that relates all points to one another. Rather than using the individual time series observations themselves, we can represent a time series in terms of its relationship to all other time series using a Pearson correlation. Using the `cor` function, we transform the 58-time series in the `cali` data frame into a 58×58 matrix of Pearson correlations. The result is assigned to the matrix `rho`.

The HAC algorithm requires a distance matrix on which linkage methods can be applied—the current form of the `rho` matrix does not fit the bill. Each county's set of correlations can be viewed as coordinates that indicate the location of one series relative to another series—not quite distance. The correlations need to be rationalized as a measure of absolute distance, which can be accomplished by passing the `rho` matrix to the `dist` function to construct the distance matrix.

```
# Calculate correlation matrix
rho <- cor(cali[, -c(1:2)])
```

---

[5]The BLS does not consider QCEW to be a time series, but it contains useful information if treated as a time series.

[6]For ease of analysis, the authors have pre-processed the data. First, the data aggregate monthly records into average quarterly records. Secondly, the data were also seasonally adjusted (SA), meaning that normal year-to-year cycles have been extracted from the data leaving only trend and noise.

```
# Convert into distance matrix (default is Euclidean)
d <- dist(rho)
```

The HAC procedure is neatly packaged into the `hclust` function and can accommodate a number of linkage methods. At a minimum, two arguments need to be specified:

- *d* is a distance matrix (or dissimilarity matrix) from the `dist` function
- *method* is the type of linkage method that guides agglomeration, such as "single", "complete", "average", "centroid", "ward.D", among others. In this example, we apply the *ward.D* method.

The resulting tree is stored in the object `hc`.

```
# Create HAC tree
hc <-hclust(d, method = "ward.D")
```

With the tree grown, we can render a dendrogram. The HCA object can be directly plotted using `plot(hc)`, but given the large number of singletons, consider stylizing the dendrogram for ease of interpretation. Using the `dendextend` package, the `hc` object is converted into a special dendrogram object to which styles can be applied: the tree is cut into $k = 8$ branches that are color coded. The font size ("label_cex") is also adjusted for aesthetics. The resulting `dend` object is shown in Figure 11.9.

```
# Load package
pacman::p_load(dendextend)

# Set as dendrogram for plotting, find 8 clusters, label andcolor code each cluster,
# Then resize the singleton labels
dend <-as.dendrogram(hc) %>%
        color_branches(k = 8, groupLabels = TRUE) %>%
        color_labels %>%
        set("labels_cex", 0.5)

# Plot dendrogram
plot(dend, font.cex = 1)
```

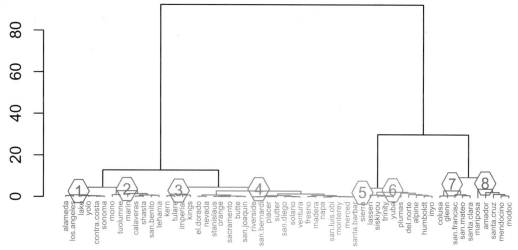

Figure 11.9: Dendrogram of county employment patterns in California.

Extract and inspect clusters. How does one interpret the dendrogram in practice? The vertical axis indicates how dissimilar two records are. A pair of singletons that are linked together near the bottom of the plot are considered to be most similar, whereas clusters linked toward the top are farther apart. To retrieve

k-number of clusters, we scan along the dissimilarity axis to identify a value at which only k links cross. For example, $k = 4$ is associated with a dissimilarity score of approximately 10, and $k = 8$ is approximately 5. Similar to k-means the optimal value of k can be found through the Elbow method or maximizing the mean silhouette width.

For simplicity, let's examine the case of $k = 8$ and compare employment patterns within each cluster. We retrieve a vector of cluster assignments by applying the `cutree` function setting $k = 8$. So that the cluster analysis can be easily tuned to other values of k without substantial editing, we soft-code the desired number clusters to the object `num_k`. We then write the code so that the assumption will propagate to all subsequent steps when the value of k is changed.

```
# Number of clusters
num_k <- 8
```

```
# Define groups
groups <- cutree(dend, num_k)
```

Small multiples are an effective way to communicate patterns when presented with a large amount of data. The format presents a grid of small thumbnail-sized graphs that provide just enough information for the viewer to make quick comparisons. All plots are packed relatively close together so the eyes do not need to move far to detect patterns. If the HAC algorithm were successful at identifying clusters, all series within a cluster should have visually similar trends. Otherwise, the analyst could tune the value of k until each cluster is visually homogeneous. Our small multiples graph provide a visual summary of each cluster in the form of a 8×11 matrix: eight rows of time series line plots—one row for each of $k = 8$ clusters, each containing at most 10 plots with one additional column to label the cluster.

Setting a color palette. In theory, each cluster should be distinct from all other clusters. Choosing a visually appealing color palette can emphasize distinctions and make it easier for the viewer to interpret results. Using the `brewer.pal` function in the `RColorBrewer` package, we customize a color palette ("Set2") with n color steps. The color palette is stored in the object `pal`, which is a string vector of $n = 8$ hex-codes.

```
# Set color palette
pacman::p_load(RColorBrewer)
pal <- brewer.pal(n = num_k, name = "Set2")
```

Laying out the small multiples. While `ggplot2` is the typical visualization analyst's preferred tool, we instead use base R graphics for the small multiples. Before rendering the plots, we define the layout of the plot using `par`: set the `mfrow` argument to contain 8 rows (`num_k`) and 11 columns (one for the cluster label and 10 for graphs). For all graphs to neatly fit, plot margins are adjusted such that only the top margin is given one-inch of space.

With the canvas set, the time series graphs can fill each cell in the 8×11 grid. When dealing with multiple plots on the same canvas, graphs are inserted into the grid from left to right, then top to bottom. The graphs are laid out under five simple guidelines:

- Each row is a cluster.
- The first cell is a blank plot in which the cluster label and sample size are placed.
- The next ten cells in the row (second through eleventh cells) should contain plots or blank plots.
- The first ten counties are selected to represent a cluster, each of which is rendered as a minimalist line graph in which axes and axis labels are suppressed.
- If a county has less than ten counties, the remainder of the row is filled with blank plots.

This process is repeated for all eight clusters and is efficiently plotted through a set of two loops and an if-else statement.

```
# (1) Set plot grid
par(mfrow = c(num_k, 11), mar = c(0,0,1,0))
```

```
# Loop through all groups
for(i in 1:num_k){

    # Get columns where elements in 'groups' matches i
    # Add +2 to retrieve column index in 'cali' data frame
    series <- which(groups == i) + 2

    # Create blank plot, then fill with the cluster label
    # And cluster sample size
    plot.new()
    text(0.5, 0.55, i, cex = 1.5, col = pal[i])
    text(0.5, 0.05, paste0("(n = ", length(series), ")"),
        col = pal[i])

    # Loop through the first 10 counties in each cluster
    for(j in series[1:10]){

        # Some clusters have less than 10 counties
        # If j is NA (< 10 counties), fill with a blank plot
        # If j has an ID, plot the time series
        if(is.na(j)){
          plot.new()
        } else{
          plot(cali[,j],
              col = pal[i], main = colnames(cali)[j],
              type = "l", axes = FALSE,
              font.main = 1, cex.main = 0.8)
        }
    }
}
```

What do these plots tell us? As shown in Fig. 11.10, these clusters illustrate economic motifs. Each cluster has a sort of economic rhythm that distinguishes its growth trajectory likely due to the industries that operate within. Scanning across each row and between rows, it is apparent that each clusters are different from one another. Some appear to continuously grow while some fluctuate. Cluster 6 experienced booms and busts, possibly due to the influence of the booms and busts of technology companies concentrated in San Francisco and Santa Clara. Cluster 8 is comprised of Sierra County that has experienced a long steady decline in employment, likely driven by population loss. In stark contrast is Cluster 7's steady upward growth throughout the dataset regardless of contractionary periods. In between are clusters 2 and 3 that have experienced volatility in employment, most of which have somewhat rebounded in recent years.

Making use of clusters. From a policy perspective, these clusters can serve as a baseline to inform interventions tailored to each cluster's profile. A declining cluster (Cluster 8) should not receive the same policy treatments as a well-resourced cluster (Cluster 6) or a continuously growing cluster (7). At the same time, we do not need to craft 58 distinct policy interventions. With clustering, we can move beyond a one-size-fits-all and be more responsive to constituents' needs.

From a research perspective, these clusters can improve the quality of models. Economists and financial researchers often estimate econometric models using panel time series data. A common strategy employs a fixed effects regression that allows each panel (e.g., county, state) to have its own intercept but assumes that all panels have the same relationship with input variables. For example, imagine if employment in Sierra County was assumed to have the same relationship with inputs as San Francisco County despite differences in growth trajectories. The model may in turn provide misleading coefficient estimates, but also biased predictions. HAC can help identify statistically defensible ways of splitting a panel into smaller groups that exhibit similar behavior and improve the quality of research.

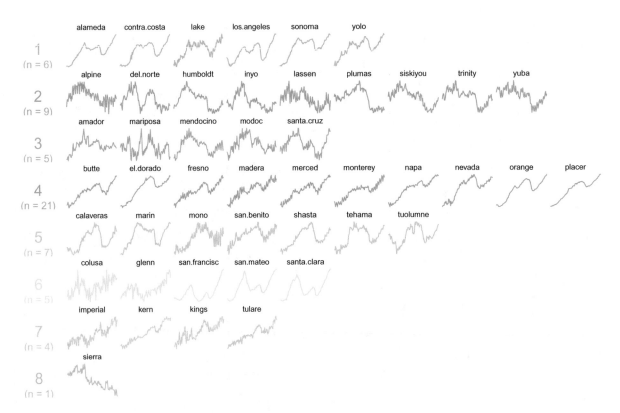

Figure 11.10: Comparison of time series by employment cluster in California.

11.5   Beyond this chapter

In this chapter, we illustrated how two clustering techniques can exposure the structure of otherwise amorphous data. k-means and hierarchical clustering are simple and intuitive but are best suited for exploratory applications to inform hypotheses and narratives. Indeed, these techniques are only the tip of the unsupervised learning iceberg. We close this chapter with a discussion of further areas of study.

The choice of method. When working with clustering, a common mistake is to choose a technique out of convenience. Both k-means and HAC are *hard clustering* techniques in which each point is placed into one and only one cluster. In policy settings where insights need to be clear cut, hard clustering is a safe choice for the sake of transparency. The simplicity of k-means excels when an application requires a short computation time or when the dataset has a large number of observations; However, the technique falls short in its ability to produce replicable, consistent results from one model run to the next. In contrast, the more computationally exhaustive HAC approach can achieve more stable results; However this comes at the cost of longer computation time.

Hard clustering could oversimplify the patterns in the data. For example, a Supreme Court justice can be a swing vote—sometimes siding with more liberal opinions and sometimes with conservative opinions. Placing the swing vote justice in one cluster over an another may be unreasonable. *Soft clustering* takes an alternative perspective: any data point has some probability of belonging to two or more clusters simultaneously. These techniques, such as *Density-based spatial clustering of applications with noise* (DBSCAN) and *Gaussian Mixture Models*, provide greater flexibility, but their complexity make them less popular in pragmatic policy settings. Nonetheless, soft clustering can offer more realistic representations of latent clusters. For further reading on these alternative clustering techniques, consider the following literature:

- Ester, Martin, Hans-Peter Kriegel, Jörg Sander, and Xiaowei Xu. 1996. "A density-based algorithm for discovering clusters in large spatial databases with noise". *Proceedings of the Second International Conference on Knowledge Discovery and Data Mining (KDD-96)*: 226–231.

- Frühwirth-Schnatter, Sylvia. 2006. "Finite Mixture and Markov Switching Models". Springer-Verlag New York.

Dimensionality. High dimensionality poses a challenge for clustering techniques. Because these techniques do not target a known pattern as in supervised learning, an algorithm cannot easily identify the most important variables and thus places equal weight on all inputs. As the number of variables increases, the influence of each variable is diluted and can lead to unstable or suboptimal results.

To manage high dimensionality requires a dimensionality reduction strategy. Recall the "Flag and Family" and "Tax and Terrorism" voters from this chapter's opening story. These conceptual labels are derived from interpreting summary statistics for each identified cluster in a dataset. In these contexts, a data scientist could employ a pragmatic dimensionality reduction strategy: *use only variables that have some intuitive connection to the phenomenon of interest*. This, of course, introduces a degree of subjectivity but can be explained away given reasonable intuition.

A more data-driven strategy involves principal components analysis (PCA). When two or more variables are highly correlated, they contain similar and sometimes redundant information. PCA offers the possibility of summarizing a set of input variables as a smaller set of *principal components*—a set of index variables that each represent a distinct pattern in the data. For example, a high dimensional economic dataset might find that certain employment variables will move together while monetary policy has a separate distinct theme. In applied settings, PCA is a convenient strategy to engineer new concise, features for prediction projects, serve as indexes in dashboards that summarize a complex concept, among others. The quality of the principal components, however, is still wholly dependent on the quality of data and the variety of signal—there is no guarantee that PCA and any other clustering technique will be effective. Nonetheless, consider reading the following article:

- Lever, Jake, Martin Krzywinski, and Naomi Altman (2017). "Principal component analysis". *Nature Methods* 14(7). pp. 641-642. https://doi.org/10.1038/nmeth.4346.

For a more hands-on treatment, *An Introduction to Statistical Learning* (James et al. 2014) offers a concise introduction to PCA using R.

Chapter 12

Spatial Data

12.1 Anticipating climate impacts

Climate change has become an issue at the forefront of society's collective consciousness. Foreseeing the challenges that lie ahead, the United States passed the Global Change Research Act of 1990 that requires the U.S. Global Change Research Program (USGCRP) to deliver a report to Congress and the President to provide a comprehensive, integrated scientific assessment of the effects of global change every 4 years. The concept of global change is expansive, dealing with changes to the Earth's system, such as oceans, atmosphere, oceans, among others. Changes in these foundational elements of life have profound impacts on all things in society such as agriculture, energy production and use, land and water resources, transportation, human health, and biological diversity (U.S. Global Change Research Program 1990). The principal output of the USGCRP is the National Climate Assessment—a report that brings together the best scientific minds and policy makers to make sense of the state of current research on climate.

One of the key areas of climate change is the effect on sea levels. According to the Fourth National Climate Assessment released in 2018,

> *Global average sea level has risen by about 7–8 inches (about 16–21 cm) since 1900, with almost half this rise occurring since 1993 as oceans have warmed and land-based ice has melted. Relative to the year 2000, sea level is very likely to rise 1 to 4 feet (0.3 to 1.3 m) by the end of the century. Emerging science regarding Antarctic ice sheet stability suggests that, for higher scenarios, a rise exceeding 8 feet (2.4 m) by 2100 is physically possible, although the probability of such an extreme outcome cannot currently be assessed.* (U.S. Global Change Research Program 2018)

While the trends are worrying, the implications of these sea-level changes are hard to fathom. How can scientists make these trends more real? Suppose we had more certainty—what would these trends mean for society? Digital Elevation Models (DEMs) can be used to approximate the height of the Earth's surface relative to the ocean. Then, the surface of the Earth can be "filtered" to areas with the greatest risk of inundation given different levels of sea-level rise. This is exactly what the Centers for the Remote Sensing of Ice Sheets (CReSIS) researched to make sense of the potential sea-level rise crisis.[1]

The results are quite striking. As seen in Figure 12.1, even a meter increase would result in a large land loss of the Mississippi Delta, significant land loss in the Netherlands and in the Amazon Delta (Rowley et al. (2007), Li et al. (2009)). To put the starkest scenario (+6 meters) into context, Aschwanden et al. (2019) finds that the entire sheet would likely melt by the year 3000, causing the sea level to rise by 5.1m to 7.0m.

Incorporating the *spatial* dimension of climate change transforms abstract scientific knowledge by giving it context. This is among the many benefits of Geographic Information Systems (GIS)—a field specialized

---

[1]To access maps from CReSIS, visit https://cresis.ku.edu/content/research/maps.

© Springer Nature Switzerland AG 2021

J. C. Chen et al., *Data Science for Public Policy*, Springer Series in the Data Sciences, https://doi.org/10.1007/978-3-030-71352-2_12

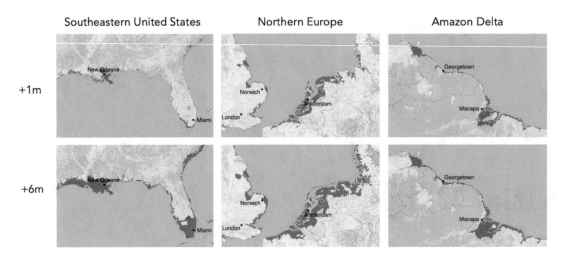

Figure 12.1: Estimates of land loss under one meter and six-meter sea-level rise scenarios. Courtesy of the Centers for the Remote Sensing of Ice Sheets (CReSIS).

in the capture, storage, processing, and retrieval of spatial data (i.e., data about geography and spatial relationships). The inundation maps can be used for much more than shading in portions of the Earth. In fact, emergency services use the maps to identify potentially vulnerable populations and develop response plans. Real estate developers use the maps to prioritize areas for development.

Maps convey *geographic relationships* between people, policies, places, and other objects. By incorporating *spatial data*, a well-made and esthetically pleasing map can be extraordinarily helpful in communicating policy implications. However, maps are only one tool in the toolbox of *spatial data analysis*. Consider these five problems in public policy:

- Find the ten neighborhoods in a city with the longest fire department response times.
- Use real-time satellite imagery to detect early warnings of forest fires.
- Detect changes in high-resolution satellite imagery that suggest building additions that do not match official building permits.
- Estimate the impact of bar openings on nearby crime, business patterns, and property values.
- Determine whether low-income households face a higher risk of sea-level rise.

The applications of data with a spatial dimension are profound and afford decision makers with a tangible dimension on which to direct resources—we can infer and identify where we should act to affect change.

What goes into *spatial data*? In general, the data should include attributes that place an observation (or objects) in space —somewhere on the Earth or elsewhere. Often, these attributes are two variables that place the observations on a grid, for example, latitude and longitude (though any x and y will work). Other times, elevation (or altitude) of the observation adds another dimension. The spatial dimension shows how observations are connected to one another in space or even a k-dimensional space.

As spatial data availability continues to grow in public sector, data scientists will develop methods to better harness, represent, and understand the complex spatial relationships inherent to and underlying public policy. In this chapter, we provide a primer on spatial data with an emphasis on wrangling and shaping. We start with an introduction to two types of spatial data, then discuss how each type of data is structured and how to work with it. With the foundations laid, we show how to conduct geo-processing to extract meaning from spatial data.

12.2 Classes of spatial data

Spatial data broadly fit into two categories: *objects* and *fields*.

Objects are variables that are discrete, disconnected items like cities, building footprints, property lines, rivers, or road networks. Governments need to operate, manage, and maintain their physical assets and often rely on object data to keep track of their operations. In Figure 12.2, for example, we see two sets of objects. Map (A) plots the extensive roadway network that is maintained by Highways England. As roads can be thought of as sets of connected line segments, the roadway network is recorded as object data in Highway England's Pavement Management System (Highways England 2020). Map (B) shows a selection of building footprints in Washington, D.C.—a useful dataset for any city that is interested in city planning. Objects require sets of points to define vertices that, when combined, create geometric shapes such as roads and buildings.

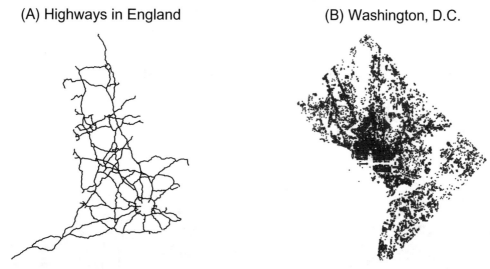

Figure 12.2: Two examples of object data: (A) Highway networks in England and (B) Building footprints for Washington, D.C.

Fields represent spatially continuous concepts like temperature, humidity, air quality, or soil quality— concepts that take on values everywhere throughout space. Earth science data such as climate data are recorded as fields. For example, Figure 12.3 plots maximum daily temperature in the United States on 24 July 2018. The gradation in the colors represents temperatures at equally spaced locations on the Earth's surface. Indeed, the southwestern border of the United States is particularly hot during the summer months while the mountain ranges tend to be quite cool.

As continuous fields and discrete objects are conceptually different, we use different classes of spatial data to analyze them. To quantify continuous fields, we use *rasters*—a format that resembles a grid, while discrete fields are represented as *vectors* to store geometric shapes.

12.3 Rasters

Rasters are grids. Each grid cell of a raster is associated with a value that summarizes the portion of the field contained in that grid (raster) cell. We can think of rasters as discretized or pixelated versions of a continuous field.

Rasters are defined by their *extent* (the swath of space that they cover) as well as their *resolution* (the size of the grid cells). When a raster's cells are small, it is considered to be *high-resolution*, which should contain enough quality information to characterize any field. The drawback, however, is that higher resolution requires larger file sizes, which in turn increase the time needed to process and render. *Low-resolution rasters* (large raster cells) can provide a decent approximation of fields whose values do not change much at smaller scales. However, as illustrated in Figure 12.4, low-resolution rasters *can* miss potentially important variation

Figure 12.3: Maximum daily temperature in the United States on 24 July 2018.

when that variation occurs at a very fine, local scale. Depending upon the use case, missing out on the fine, hyper-local variation may not matter to your project—or it could mean certain failure for the entire project.

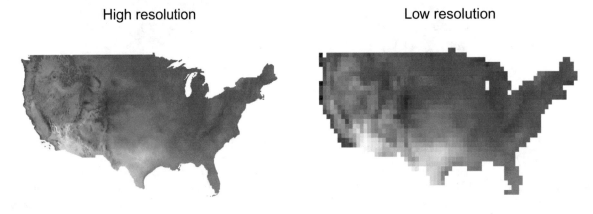

Figure 12.4: Comparing high- and low-resolution rasters. The raster on the right is approximately 1 km resolution; the raster on the left is approximately 200 km resolution.

At the heart of it, rasters boil down to a pair of *coordinates* with an attached *value*. The value denotes the raster's representation of the field within the grid cell centered at the coordinates. If we think back to the temperature map of the United States, each pixel in the photograph contains a coordinate pair for a location on the Earth plus a temperature value. For example, consider a simple raster with 15 total cells. We can write it as a data frame with 15 rows and 3 columns—two columns for coordinates and one column for the value.

```
simple_df <- data.frame(x = rep(1:5, 3),
                        y = c(rep(1, 5), rep(2, 5), rep(3, 5)),
                        value = c(1:5, 2:6, 3:7))
```

The three-column data frame can be treated like a raster—think of it as reshaping the x as rows and y as columns in a spreadsheet, filling in the intersections with the value. The value in $(x = 5, y = 3)$ is 7, which is greater than the value in the next row below $(x = 5, y = 2)$, which is 6. When tabular data is placed in a spatial grid, it can be visualized as a heat map as illustrated in Figure 12.5.

In R, the conversion of data frames to rasters relies on the `rasterFromXYZ` function (`raster` package). In the

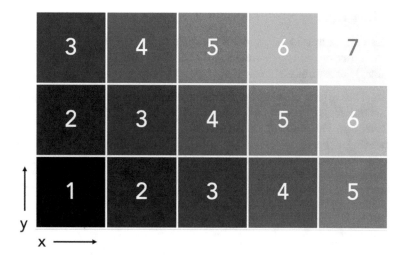

Figure 12.5: Visualizing the data frame when converted into raster format.

snippet below, we apply the function to the data frame `simple_df` using a forward-pipe operator (`magrittr` package). As the object `simple_raster` has been converted to a `"RasterLayer"` class, R can plot the values in `simple_raster` as a grid, using their attached coordinates.

```
# Load raster package, a color palette, and magrittr
pacman::p_load(raster, viridis, magrittr)

# Convert from data frame to raster, applied via forward-pipe operator
simple_raster <- simple_df %>% rasterFromXYZ()

# Render raster using magma color palette
plot(simple_raster,  col = magma(1e3),  box = F, axes = F, )
```

12.3.1 Raster files

Rasters can be stored in any format that can represent space as a grid. A *CSV* can do the trick for simple problems, but lacks the ability to store context and requires specialized knowledge to transform tabular values into a grid. For imagery (e.g., photographs) and remote sensing data (e.g., radiometer data), photographic formats like TIFFs and JPEGs are quite effective for preserving the relationship of grid cells, although the latter can lose precision of data if compressed. Ordinary imagery files also lack context—where was the photograph taken and what part of the Earth does it represent?

For this reason, geospatial raster files tend to have specialized metadata to help retain context. For example, GeoTIFF format combines *TIFF* imagery with metadata that indicates what part of the Earth's surface the photograph covers, along with other geographic data.

Raster formats tend also to be developed with specific domains in mind. In meteorology, data on concepts such as precipitable water can be stored as GRIdded Binary or General Regularly distributed Information in Binary form (*GRIB*).[2] For storing complex data about the Earth system, a favorite of earth scientists is *NetCDF* (Network Common Data Form)—the common choice for climate data. Whatever the raster file format may be, all formats require some specialized logic to open and make sense of the data.

---

[2]The earth and planetary science communities tend to fashion creative acronyms from arbitrary letters. For example, one of NASA's asteroid satellite missions was called OSIRIS-REx (Origins, Spectral Interpretation, Resource Identification, Security, Regolith Explorer) while a specialized algorithm for astronomy was named GANDALF (Gas AND Absorption Line Fitting algorithm).

12.3.2 Rasters and math

Mathematical operators can be applied to rasters, treating them like mathematical matrices. We can think of any field in a raster as any ordinary variable x_k. Thus, if we would like to apply a scalar (e.g., $a \times x_k$), square a variable (e.g., x_k^2), or combine identical size rasters (e.g., $x_1 + x_2$), R can make it happen.

For example, to square the values of `simple_raster`, we can apply R's exponential operator (`^2`), *i.e.*, `simple_raster^2`. This particular type of raster math can prove useful when converting between units (*e.g.*, changing between Celsius and Fahrenheit). In other cases, raster math can identify cells that are above (or below) a cutoff. By creating a *mask*, we can filter data based on their value. Other times, we may need to combine multiple rasters—perhaps to calculate an average, sum rasters to find a total, or combine a measure and a population raster to get *per-capita* measure.

```
# Square
plot(simple_raster^2,  col = magma(1e3))


# Create a mask
plot(simple_raster > 4,  col = magma(1e3))


# Apply the mask to extract cells that meet criterion
plot(simple_raster * (simple_raster > 4),  col = magma(1e3))
```

When working with raster math, we should keep three caveats in mind:

1. To compute summaries of rasters in R, we need to use the `cellStats()` function from the `raster` package. For example, if we want to find the minimum value in the raster `simple_raster`, then we should use `cellStats(x = simple_raster, stat = min)`.
2. To compare a grid cell's value *across equally sized rasters*, first create a stack of rasters using the `stack()` function in the `raster` package. A raster stack is exactly what its name applies—a stack of raster files, where each layer is an individual raster. Once doing so, apply `min`, `max` or other summary statistics functions to the raster stack. The results will be cell-level summaries in raster form.
3. To apply mathematical and statistical operations to rasters and stacks, the rasters need to "conform" (match up), meaning that they need to have identical extents and identical resolutions.

While raster math can produce useful results, how do we convert back into data frame for general data science use cases? Simply apply `as.data.frame`. By default, the function only returns the values of the raster without the coordinates. However, when enabling the optional argument `xy = T`, the resulting data frame will be furnished with both values *and* their coordinates.

```
# Converting raster to data frame ('xy = F' by default)
simple_raster %>% as.data.frame()


# Converting raster to data frame ('xy = T')
simple_raster %>% as.data.frame(xy = T)
```

With these basic raster operations, we can unlock the signal in satellite imagery, photographs, and other gridded data.

12.3.3 DIY: Working with raster math

> *DIYs are designed to be hands-on. To follow along, download the DIYs repository from Github (https://github.com/DataScienceForPublicPolicy/diys). The R Markdown file for this example is labeled* `diy-ch12-raster-math.Rmd`.

Let's work through some examples using some simulated data.[3] In the `raster_simulate.Rda` file, we have simulated four rasters of 50×50 cells: `r1`, `r2`, `r3`, and `pop_r`. We have plotted all four rasters in Figure 12.6.

---

[3]The data used to simulate these rasters is available on Github (https://github.com/DataScienceForPublicPolicy/build-simulation-data) under `raster-simulation.Rmd`.

```
load("data/raster_simulate.Rda")
```

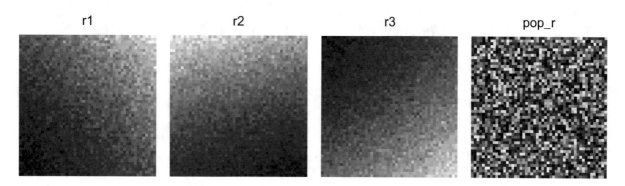

Figure 12.6: Four simulated raster files: r1, r2, r3, and pop_r.

If we imagine that r1, r2, and r3 are the same variable over time for a fixed geographic area, we may be interested to know the mean value in each cell. To find the mean value of each cell across the three rasters, we have a two options:

1. Sum the rasters and then divide by three: $(r1 + r2 + r3)/3$.
2. Create a stack and then take the mean: stack(r1, r2, r3) %>% mean().

When we implement both options, we can test if these yield the same result using all.equal—they do.

```
# Method 1: Sum then divide
r_mean <- (r1 + r2 + r3) / 3

# Method 2: Stack and mean
r_mean_alt <- stack(r1, r2, r3) %>% mean()

# Compare
all.equal(r_mean, r_mean_alt)
```

We can elaborate Method 2 to find other summary statistics for each cell such as the maximum and minimum. Below, we apply max and min to the raster stack in order to find the range.

```
# Find each cell's maximum across the rasters
r_max <- stack(r1, r2, r3) %>% max()

# Find each cell's minimum across the rasters
r_min <- stack(r1, r2, r3) %>% min()

# Range
r_range <- r_max - r_min
```

It may also be necessary to normalize a raster by another raster. For example, divide r1 by pop_r (perhaps r1 contains income and pop_r contains population—then r1 divided by pop_r would yield income per-capita). We can perform division by using R's arithmetic operator for division (/).

```
r_div <- r1 / pop_r
```

When we render these results in Figure 12.7, it is clear that raster math is quite powerful—we were successful in extracting patterns that would be otherwise locked away in a data frame.

Mean   Maximum   Division   Range

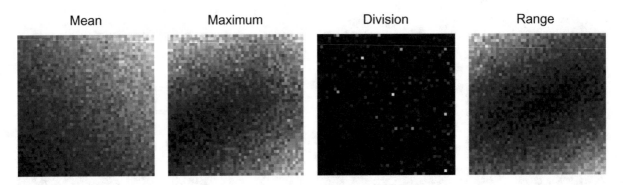

Figure 12.7: Four outputs from raster math calculations.

12.4 Vectors

Just as we represent and analyze fields using raster data, we portray discrete objects with *vectors*. At their core, vector data are simply groups of ordered points—each point has a coordinate. We can divide vector data into three broad categories:

1. *Points:* each group has exactly one point.
2. *Lines:* each group has multiple, ordered points.
3. *Polygons:* each group has multiple, ordered points, and the first and last points are the same (closing the polygon).

In Figure 12.8, we plot examples of the three types of vectors. Each of the three types of vector data—point, line, and polygon—is made up of underlying points. For lines and polygons, the ordering of the points matters: by connecting the dots in different orders, we would clearly find different lines or polygons. Polygons are lines that have been closed off by joining the first and last points. This closure leads the polygon to have an enclosed area, whereas a line does not have an area—just a length. To summarize once more: *Polygons and lines boil down to collections of points.*

(A) Point   (B) Line   (C) Polygon

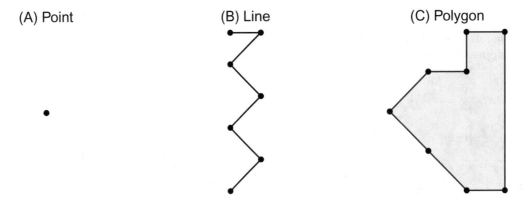

Figure 12.8: Example of point, line, and polygon vector objects.

12.4.1 Vector files

A vector file must store the geometric data of an object, have a framework to make sense of the geometric shapes, and have an easy way to affix attributes to each object. While there are a variety of vector data formats in use, three stand out:

- *CSV.* A single CSV file is well suited to carry the geographic coordinates about *points* along with their attributes. While polygon and line data *could* be stored in a CSV, a non-trivial amount of processing would be required to accurately string together coordinates making the .format only ideal for point data. However, attributes relating to a polygon or line could be stored in a CSV without the geometry data.

- *Shapefile*. A shapefile is a geospatial vector data format that can support points, lines, and polygons. It is a collection of three types of files that contain the geometry (`.shp`) and attributes (`.dbf`) as well as an index file (`.shx`) to make navigation between vector objects as seamless as possible. If the collection of files is missing one piece, the shapefile is inoperable. While it is a common format, shapefiles tend to be bulky and are often used in conjunction with specialized software such as ArcGIS (proprietary) and QGIS (open source)—both require some specialized knowledge of GIS. In R, shapefiles can be loaded using the `sf` package.
- *GeoJSON*. An alternative to shapefiles is GeoJSONs. Like shapefiles, GeoJSONs can support multiple types of vectors but have been developed to be more flexible and open.[4] By taking advantage of JSON format—the data format of the internet—GeoJSONs make it easy to transmit and use vector data in web applications and any data-driven software.

As public policy organizations often rely on CSVs and shapefiles, we primarily focus on these data formats.

12.4.2 Converting points to spatial objects

Because of their importance in spatial data analysis, one must understand how to transform a set of points—perhaps from a CSV—into a spatial object. We demonstrate this process using the example of Active Fire Data derived from readings from NASA's MODIS instrument for the period of February 10th to 17th, 2020.[5] Every time NASA's Terra and Aqua satellites fly overhead, the radiometer data can be used to detect fires—certain pixels will exhibit a certain level of brightness and color. Earth scientists have developed algorithms to automatically detect active fires, which are useful for monitoring situations across the globe. The `modis-fire.csv` file contains `latitude`, `longitude`, `brightness` of fire, among other measures (NASA Earth Science 2019).

```
# Read file
modis <- read.csv("data/modis-fire.csv")

# Check that data was imported
modis
```

To work with vector data, we use the `sf` package.[6] For the task of converting this data frame to an object that R recognizes as a spatial object, we use the function `st_as_sf()`, calling the `coords` argument to indicate which column is the horizontal dimension (`longitude`, otherwise the `x` coordinate) and which column contains that vertical dimension (latitude, otherwise the `y` coordinate).

Pro Tip: The order of the coordinates in a vector file is typical source of error. The `x` coordinate usually is first and should correspond to longitude while the `y` coordinate comes second and corresponds to latitude. If we pass coordinates to the `coords` argument in the wrong order, the data will be mapped with a rather odd rotation. A straight forward method of checking if the data was imported correctly is to plot it.

```
# Load sf package
pacman::p_load(sf)

# Convert modis to 'sf' object
modis <- modis %>% st_as_sf(coords = c("longitude", "latitude"))

# Inspect the new 'sf' object
```

---

[4]Shapefiles were originally developed for use with a proprietary software. GeoJSON was developed to be an open-by-default format.

[5]Data can be obtained from NASA Earthdata: https://earthdata.nasa.gov/earth-observation-data/near-real-time/firms/active-fire-data.

[6]`sf` stands for *simple features*. See the `sf` project's main page (https://r-spatial.github.io/sf/index.html) for more information.

```
class(modis)
  modis
```

A couple of observations. The class of the `modis` object contains both its previous class (*i.e.*, `data.frame`) and a new one, `sf`. This double identity is quite useful: we can perform many spatial operations using `sf` objects while being able to perform standard `data.frame` operations (*e.g.*, all of the friendly functions from the `tidyverse`). In addition, when inspecting the new `modis` object, the columns that contained our coordinates have "disappeared" and have been replaced with a new column named `geometry`. Any time we create an `sf` object, we should expect a `geometry` column that contains the coordinates for each entry in your dataset. In our current example, each fire detection is represented by a single point, so the `geometry` is a single point. But, as we will see in future examples, when each row of our dataset is a polygon, each entry in `geometry` will be a set of points.

While base `R` is equipped to `plot` a map of `modis`, we can leverage `ggplot2` for more a customizable, visually satisfying experience. In the case of vector files, `geom_sf` can handle `sf` objects to render a map as shown in Figure 12.9. The result suggests a world on fire, but not to worry—the points overstate the size of the blazes.

```
ggplot(data = modis, aes(color = brightness)) +
      geom_sf(alpha = 0.3, size = 0.3) +
      scale_color_viridis_c("Brightness",  option = "plasma") +
      guides(color = guide_colourbar(barwidth = 20, barheight = 1)) +
      theme_void() + theme(legend.position = "bottom")
```

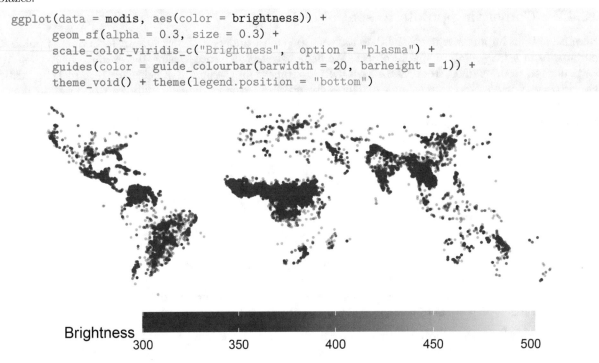

Figure 12.9: Brightness of active fire detections from MODIS instrument (Feb 10 to 17, 2020).

12.4.3   Coordinate Reference Systems

The MODIS fire detections are recorded in decimal degrees—expressing location as *degrees* from the equator and prime meridian. In particular, MODIS coordinates are recorded in World Geodetic System 1984 (WGS84)—just one of many *coordinate reference systems* (CRS) that allow all positions on Earth to be recorded in a standardized way. As the name suggests, the coordinate reference system indicates how to interpret and treat the coordinates of a point. While standard latitude and longitude are common, other CRS represent location in terms of distances (e.g., meters, feet) from some central point. Furthermore, the CRS indicates whether the coordinates use some sphere-like object as the surface or whether they are projected onto a flat surface.

For example, New York City's spatial data are often recorded in:

NAD83_StatePlane_NewYork_LongIsland_FIPS_3104_USFeet

Let's break this down. *NAD83* is the North American Datum of 1983—a set of geographic reference points that identify the precise position of locations on the Earth. State Plane refers to one of 124 geographic zones in which coordinates from a spherical model (NAD83) are projected and represented in terms of a Cartesian coordinate system. In effect, a curved surface is flattened onto a plane. New York-Long Island is a specific State Plane zone. The coordinates are not represented in decimal degrees, but instead as feet relative to a point in the State Plane zone.

The choice of the projection can lead to quite dramatic changes to the spatial data. In Figure 12.10, we render three maps of the US using different projections, illustrating the importance of the CRS. In short, geographic data can be quite complicated and having good metadata plays an important role in making the data useful.

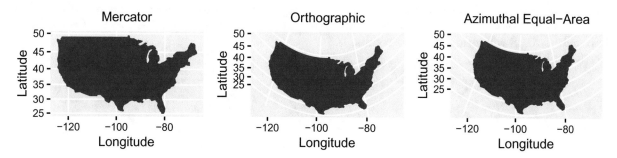

Figure 12.10: Various examples of map projections applied to the United States.

Setting CRS. When importing data from a CSV or other generic file format, the data will not contain the CRS. If the data hygiene practices are strong, then we can simply assign a CRS to a dataset. In the case of MODIS fire detections, NASA data managers are quite good and we know that the appropriate CRS is the *WGS84* projection. In fact, many decimal degree coordinates will be *WGS84* and is a reasonable default assumption.[7]

What do we do with that information? We need to assign it to the data to give it the context for R to make use of it. In R, there are two ways to specify a CRS:

- The EPSG: A numeric code that indicates the CRS.
- The proj4 string: An alphanumeric code that describes the coordinates, projection, and spherical model associated with the coordinates. For example proj4string: +proj=longlat +ellps=WGS84 +no_defs.

The st_set_crs function attaches the desired CRS, ensuring that both EPSG and proj4 appear. Below, we specify only the EPSG and R takes care of the rest.

```
modis <-st_set_crs(x =modis, value = 4326)
```

As illustrated in Fig. 12.10, each CRS makes vastly different assumptions about how to treat spatial data. When we compare objects that use different CRS', then they all must be transformed into the same CRS. Otherwise, the coordinates will not align or even appear on the same mapping space and calculations will return nonsensical values.

The transform process is known as *re-projection*, which requires computing the geographic transformations from one CRS to another. *Do not* simply set a different CRS using st_set_crs. Instead, a special transformation is required.

Changing CRS. Suppose we would like to calculate distances between fires. The challenge is that our current CRS (WGS84) is measured in degrees, and degrees are not constant throughout the globe. For example, at the equator, when you move 1 degree west, you move approximately 111 km. However, 49 degrees north of the equator (the border between the United States and Canada from approximately Minnesota/Manitoba

---

[7]Before assuming *WGS84*, check the range of values. Decimal degrees will vary between specific ranges: latitude in degrees is -90 and +90 while longitude is between -180 and +180.

to Washington/British Columbia), moving 1 degree west is only 72.9 km. On the other hand, the distance between latitudes stays a constant 111 kilometers.[8]

Why does any of this matter? Because when we calculate distances in longitude and latitude, we assume that one degree in the horizontal direction (one degree longitude) equals one degree in the vertical direction (one degree latitude), which simply is not true (unless we are exactly on the Equator).

So what can we do? For relatively small areas, we can project the Earth onto a flat surface and then use a coordinate reference system in meters (or another constant distance unit). The Universal Transverse Mercator (UTM) offers one such system. Specifically, UTM divides the earth in 60 zones—6 degrees wide split above or below the Equator. Each zone receives its own coordinate system. Chicago, for example, sits in UTM zone *16N*, which we can specify via EPSG code *32616* (with WGS84 datum). To change from one CRS to another, we need to *transform* the spatial object—re-project into the new model of Earth and coordinate system. For this task, we use `st_transform`. As a reminder, simply changing the CRS without re-projecting into another CRS will yield disastrous results.

```
# Re-project
modis2 <- st_transform(x = modis, crs = 32616)

# Check CRS for both objects
modis2 %>% st_crs()
```

12.4.4   DIY: Converting coordinates into point vectors

DIYs are designed to be hands-on. To follow along, download the DIYs repository from Github (https://github.com/DataScienceForPublicPolicy/diys). The R Markdown file for this example is labeled `diy-ch12-create-vectors.Rmd`.

In this DIY, we put into practice steps for working with point vectors, importing two CSVs of point data and preparing them as spatial vectors. Both datasets are drawn from the domain of public safety in the US city of Chicago:

`data/chicago-police-stations.csv`

- Locations of $n = 23$ police stations
- Includes district, latitude, longitude, and several other variables

`data/chicago-crime-2018.csv`

- Reported crime incidents ($n = 267687$) by the Chicago Police Department in 2018[9]
- Includes date, latitude, and longitude (at block level), type/description of crime, whether an arrest was made, and many other variables

Below, we begin with importing the two `CSV` files, then clean the headers with the `clean_names` function (`janitor` package).

```
# Load packages
pacman::p_load(dplyr, janitor, sf, magrittr)

# Load the datasets and clean names
station_df <- read_csv("data/chicago-police-stations.csv") %>% clean_names()
crime_df <- read_csv("data/chicago-crime-2018.csv") %>% clean_names()
```

While converting a points data frame into a vector file is as easy as using the `st_as_sf` function (from `sf`), we need to pay attention to the data quality of the geographic coordinates. In real-world data, the location of every event or incident is not likely to be known, sometimes limiting a data scientist's ability to see the whole picture. Furthermore, missing coordinates can prevent `R` from processing data frames into

---

[8]To calculate the distance between degrees longitude: $\cosine(\pi \times \text{latitude}/180)$ km.
[9]For the Chicago PD data, visit: https://data.cityofchicago.org/Public-Safety/Crimes-2018/3i3m-jwuy/data.

vectors. Before converting into spatial data, be sure to remove records that are missing coordinates. While `station_df` has complete information, `crime_df` is missing $n = 4254$ or 1.5% of reported crime incidents. In the code snippet below, we directly convert `station_df` into a spatial file and add a missing values filtering step before converting `crime_df`.

```
# Convert into station_df to sf
station_sf <- station_df %>%
            st_as_sf(coords = c("longitude", "latitude"))

# Remove NAs from crime_df and convert to sf
crime_sf <- crime_df %>%
            filter(!is.na(longitude) & !is.na(latitude)) %>%
            st_as_sf(coords = c("longitude", "latitude"))
```

With the data in spatial form, we need to set the CRS, then transform it into a useful form. First, we set the standard EPSG 4326 (WGS84) as the CRS, then re-project the files into a computationally useful CRS. In this case, we choose UTM zone 16N, which is EPSG 32616. This CRS flattens the Earth's surface and reports coordinates in meters, making it possible to calculate accurate distance between locations as long as data are confined to the Zone 16N. We will revisit these more advanced processing steps later in this chapter.

```
# Set CRS
station_sf <- st_set_crs(x = station_sf, value = 4326)
crime_sf <- st_set_crs(x = crime_sf, value = 4326)

# Transform
station_sf <- st_transform(station_sf, 32616)
crime_sf <- st_transform(crime_sf, 32616)
```

12.4.5 Reading shapefiles

When working with vector data containing lines and polygons, we will need a shapefile. The boundary of geopolitical units (e.g., counties, countries, cities) are often maintained in spatial files. In addition, public resources such as the location of police stations and parks are often stored in spatial information. Data input/output is often the largest barrier to starting a project—shapefiles are no exception. In this section, we build on the previously imported pair of Chicago datasets by loading a polygon shapefile of police districts in Chicago.

To get started, let's load a polygon shapefile for police districts in Chicago using the `st_read` function.[10] The only argument `st_read` needs to load the police district shapefile (`chicago-police-districts.shp`) is the file's location.

```
# Load the police district shapefile
district_sf <- st_read("data/chicago-police-districts")

# Set to same CRS
district_sf <- st_transform(district_sf, st_crs(crime_sf))

# Inspect the shapefile
district_sf
```

Extracting a single variable. Upon loading the shapefile, we find that it contains a collection of $n = 25$ polygons—each row represents a single police district containing its label and number along with its polygon (geometry). If we would like only one variable, then apply `select` (`dplyr` package) to extract the variable of interest. Alternatively, `transmute` can be used to select and transform the variable. However, if we are only interested in the geometry, then we can extract the `geometry` column using `st_geometry`.

---

[10]The function is also called `read_sf`.

```
# Extract police district, which can be plotted if passed to 'plot'
district_sf %>% transmute(dist_num %>% as.numeric())

# Extract the geometry
district_sf %>% st_geometry()
```

12.4.6 Spatial joins

One of the most common spatial analysis tasks is to aggregate the number of points in a polygon. For example, in our crime data, we could be interested in counting the number of incidents that occur in each police district. These *point-in-polygon* summaries require a *spatial join*. A cousin of the tabular joins introduced in Chapter 5, a spatial join looks for spatial intersections between a point vector and a polygon vector. Some spatial joins are analogous to inner joins (*intersects*) while others are like full joins (*unions*). Spatial joins can be computationally intensive as each point needs to be compared with all other polygons to find a match. Fortunately there are simple, easy-to-use geo-processing functions such as `st_join` to facilitate joins and `st_interects` to identify overlapping vector objects (`sf` package). In this section, we describe the process of finding a spatial intersection, then use the resulting joins to summarize and visualize crime patterns by Chicago police district.

Point-in-Polygon Join. In the code snippet below, we conduct a spatial join to determine to which police district (`district_sf`) does each incident belong (`crime_sf`). The goal is to construct a data frame of points with new attributes (variables) added from overlapping police districts. Before starting, it is a good idea to check if both layers have the same CRS. If not, we would need to re-project one vector based on the CRS of the other using `st_transform`. In this case, we apply the CRS from the incidents data to the district file. With the data ready, we apply `st_join` to the point data (`x`) and polygons (`y`) specifying that we are interested in overlaps (`join = st_intersects`).

The resulting file has the same number of points as the original crime file with two additional columns containing information from the overlapping police district polygon: `dist_label` and `dist_num` (the label and number of the matched police district).

```
# Check if CRS match
st_crs(crime_sf) == st_crs(district_sf)

# Conform CRS
district_sf <- st_transform(district_sf, st_crs(crime_sf))

# Join district to crime data
crime_sf <- st_join(x = crime_sf,
            y = district_sf,
            join = st_intersects)
```

Aggregating points from shapefiles. To aggregate the data, we can treat the spatial join results as data frames. One reasonably efficient approach involves applying `table` to the district column of `crime_sf`, returning the total number of incidents per district. Alternatively, `dplyr`'s `group_by` and `summarize` can facilitate more complex summaries.[11]

```
# Count incidents by district (using 'dist_num'), rename table
crime_count <- crime_sf$dist_num %>% table() %>%
                as.data.frame()
names(crime_count) <- c("dist_num", "n_incidents")
```

Choropleth map. A thematic map with shading or patterns proportional to some metric is known as a choropleth map. In the case of crime, the number of incidents can be used to shade police districts in

---

[11]Because our spatial join was not able to assign a district to some incidents, we may need to remove the non-joined observations with `NA` for `dist_num` as well as remove the geometry column from the output.

order to find crime hot spots. However, in the example above, the counts are captured in a separate table `crime_count`. When calculating any aggregates in R, the results will first need to be joined to a shapefile in order to be visualized. Fortunately, `sf` files can be joined to data frames using the `left_join` function as long as there is a common identifying column. In the example below, the common column is `dist_num`.

```
# Join crime counts to district
district_sf <- left_join(x = district_sf,
                  y = crime_count,
                  by = "dist_num")
```

With the data ready to go, a choropleth map can be rendered from the augmented shapefile. Incidents appear to be clustered in certain areas, indicating that there are hot spots.

```
# Plot
ggplot(data = district_sf, aes(fill = n_incidents / 1000)) +
       geom_sf(color = "black", size = 0.05) +
       scale_fill_viridis_c("Number of Incidents ('000)", option = "magma") +
       guides(fill = guide_colourbar(barwidth = 15, barheight = 1)) +
       theme_void() +  theme(legend.position = "bottom")
```

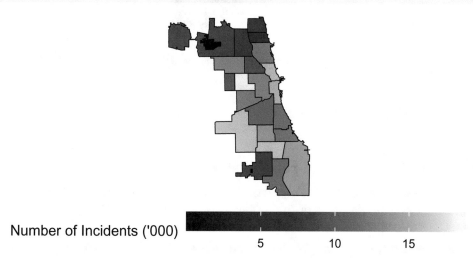

Figure 12.11: Number of incidents by Chicago police district.

More complex summaries. The crimes are quite widespread, but what proportion of crimes resulted in an arrest? Once again, we follow the same paradigm: compute the summary statistic, then join to the shapefile. In this case, we compute the arrest percentage using `dplyr` functions, then join to the `district_sf` shapefile (Figure 12.11).

```
# Calculate arrest status and district number for each crime
district_arrests <-  crime_sf[, c("dist_num", "arrest")] %>%
                as.data.frame() %>%
                group_by(dist_num) %>%
                summarise(share_arrests = 100 * mean(arrest, na.rm = T))


# Join to district data
district_sf <- left_join(x = district_sf,
                y = district_arrests,
                by = "dist_num")
```

In Figure 12.12, the share of incidents that result in an arrest is once again plotted using `ggplot`. *What do we see?* The share of police-involving incidents that result in arrest varies from district to district. The contrast is even more striking when comparing the pair of choropleths—arrest rates are more concentrated in some neighborhoods than others. There are many potential explanations for this difference, but this simple map raises some interesting and important questions about policing, crime, equity, and policy.

```
ggplot(data = district_sf, aes(fill = share_arrests)) +
        geom_sf(color = "black", size = 0.05) +
        scale_fill_viridis_c("Percent with Arrest", option = "magma") +
        guides(fill = guide_colourbar(barwidth = 15, barheight = 1)) +
        theme_void() +  theme(legend.position = "bottom")
```

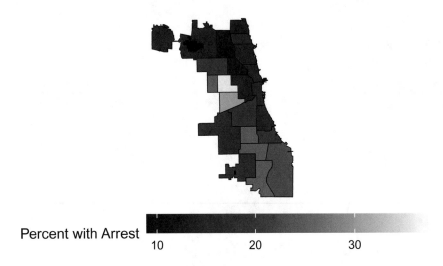

Figure 12.12: Percent of incidents that resulted in an arrest.

12.4.7   DIY: Analyzing spatial relationships

*DIYs are designed to be hands-on. To follow along, download the DIYs repository from Github (*https://github.com/DataScienceForPublicPolicy/diys*). The R Markdown file for this example is labeled* `diy-ch12-spatial-analysis.Rmd`*.*

One of the most useful features of spatial data is proximity. The distance between two locations might hold a clue about a spatial concept. In our housing examples from earlier chapters, proximity to sources of pollution and central business districts might be contributors to *or* detractors from the sales price. Likewise, the distance to police stations might hold some clue about crime patterns—perhaps crimes tend to occur farther away from stations? Furthermore, two locations that are closer together might be more related and distance is the medium that can help quantify these relationships. To answer any questions concerning distance between locations, we need to compute a distance matrix.

Similar to the matrices used for Hierarchical Agglomerative Clustering in Chapter 11, a distance matrix quantifies the relationship between all points in a dataset. Suppose we would like to compute the distance between three Chicago police stations: $1, 2, 3$. The first 3×3 matrix below shows all of the combinations between the three stations and the second matrix fills in some distances in meters.

$$\text{Distance matrix} = \begin{bmatrix} D_{1,1} & D_{1,2} & D_{1,3} \\ D_{2,1} & D_{2,2} & D_{2,3} \\ D_{3,1} & D_{3,2} & D_{3,3} \end{bmatrix} = \begin{bmatrix} 0 & 3090 & 8896 \\ 3090 & 0 & 11882 \\ 8896 & 11882 & 0 \end{bmatrix}$$

where $D_{i,j}$ gives the distance between station i and station j.

The top-left entry (entry $D_{1,1}$ above or entry [1,1] in matrix notation) is the distance between the first station and itself. Of course, the distance between an object and itself is zero. Therefore, the diagonal of the matrix is all zeros. The second entry in the first row (entry $D_{1,2}$ above or entry [1,2] in matrix notation) tells us the distance between the first station and the second station. This number is the same as the first entry in the second row ($D_{2,1}$), because they both give the distance between the first and second police station. In other words, the matrix is symmetric: $D_{i,j} = D_{j,i}$.

Computing a distance matrix between all objects in a shapefile requires only one function: st_distance() (sf package). The function calculates the distance in the units of the specified CRS (meters in this case). Because there are 25 police districts in the district_sf vector layer, the distance array will have 25 rows and 25 columns, representing the distance (in meters) between each of the 25 districts.

```
district_dist <- st_distance(district_sf)
```

This distance matrix can help answer challenging questions that would otherwise be overlooked. In this section, we lay out steps to computing distances and use them to analyze complex spatial problems in the criminal justice system.

Distance between objects in different sets of data. Another way to pose distance-related questions is in terms of distances between objects in two distinct sets of vectors. For example, we may be interested to know how far each crime in crime_sf is from the nearest station in station_sf.

The function st_distance can help with this application as well, requiring two arguments (x and y). The output is a distance matrix where the objects in x are the rows and the objects in y are the columns, *i.e.*,

$$\text{Distance matrix} = \begin{bmatrix} D_{x_1,y_1} & D_{x_1,y_2} & \cdots & D_{x_1,y_n} \\ D_{x_2,y_1} & D_{x_2,y_2} & \cdots & D_{x_2,y_n} \\ \vdots & \vdots & \ddots & \vdots \\ D_{x_n,y_1} & D_{x_n,y_2} & \cdots & D_{x_n,y_n} \end{bmatrix}$$

where D_{x_i,y_j} gives the distance between the ith element of x (*e.g.*, the ith crime) and the jth element of y (*e.g.*, the jth police station). The matrix dimensions match our understanding of the function: we have 263,423 crimes (as rows) and 23 police stations (as columns).

```
# Distances between crimes and police stations
dist_mat <- st_distance(x = crime_sf,
                        y = station_sf)
```

If we want to know the distance to the nearest police station for each crime, then we can employ the same strategy as before—using apply to find the minimum distance along the rows of dist_mat. We do not need to worry about 0 distances now, because we calculated distances between two different sets of objects. Let's find the distance to the nearest police station for each crime and then add it to the crime_sf dataset as a new variable.

```
# Minimum distance to station for each crime
crime_min_dist <- apply(X = dist_mat,
                        MARGIN = 1,
                        FUN = min)

# Add distance as variable in crime_sf
crime_sf <- crime_sf %>% mutate(dist_station = crime_min_dist)
```

Let's plot the kernel density of these distances.[12] Based upon these data, 17.7% of crimes in Chicago happen within 1 kilometer of a police station.

One might wonder how this distribution of crimes' distances to the nearest station compares to the overall distribution of distances to the nearest station for all of Chicago. If police station locations are chosen to be in *high-crime* areas (or if they attract crime), then we would expect the distances between crimes and stations to be concentrated on smaller distances relative to the distances to stations for all of Chicago. If crimes avoid police stations, then we would expect their distribution to be pushed farther out relative to greater Chicago (Figure 12.13).

---

[12]Given the size of the code snippets, this section's geo-processing and visualization code is made available at the Github

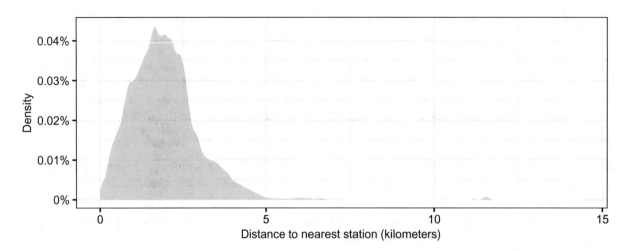

Figure 12.13: Kernel density plot of distance between crime incidents and nearest police station (meters).

Constructing a benchmark. If we believe police station location has some relationship with location of crime, we need to construct a benchmark—some comparison group that gives context. One strategy involves constructing a point vector containing a random or regular set of points placed throughout Chicago. This benchmark makes the assumption that if distance did not matter, then there is an equal chance of crime at every location in Chicago. When we compare the equal chance distribution to the actual crime-distance distribution, we can infer if distance to police station has any relationship with crime.

To construct this equal chance distribution, we first create a single polygon for Chicago by merging the police district polygons (`district_sf`) through the `st_union` function. In effect, the borders between all polygons are removed and merged into a single large city limits boundary. From this new city polygon, we draw a hexagonal grid of points ($n = 10000$ to be exact) using `st_sample`. Then, the same distance calculations are applied to find the minimum distance between each of the $n = 10000$ points to police stations.

```r
# Re-project district shapefile to UTM 16N
district_sf <- st_transform(district_sf, crs = 32616)

# Take union
outline_chicago <- district_sf %>% st_geometry() %>% st_union()

# Draw sample of 'hexagonal' points from Chicago
points_chicago <- st_sample(x = outline_chicago, size = 10000, type = "hexagonal")

# Distances between points and police stations
dist_points_station <- st_distance(x = points_chicago,  y = station_sf)

# Find distance to nearest station for each point
points_min_dist <-  apply(X = dist_points_station, MARGIN = 1, FUN = min)

# Convert points_chicago to sf data frame
points_chicago <- points_chicago %>%
            st_coordinates() %>% as.data.frame() %>%
            mutate(dist_station = points_min_dist) %>%
            st_as_sf(coords = 1:2) %>%
            st_set_crs(32616)
```

In Figure 12.14, we compare the police districts, the grid points, and the minimum distance to police station.

---

project under *spatial-visualizations*.

To an extent, the minimum distance to police station loosely follows the boundaries of the police districts—there are both areas that are well-covered and others that are far less so. We can see that most of Chicago is within 5 km of a police station—with the exception of the airport in the northwestern corner.

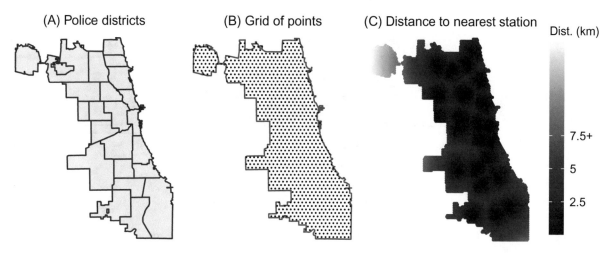

Figure 12.14: An assortment of views of Chicago police stations: (A) Outline of Chicago police districts, (B) Hexagonal grid of equally spaced records, (C) Raster plot of distance to nearest station.

Comparing distance distributions. To answer the original question, we plot kernel densities of the distance distributions as seen in Figure 12.15. The vast majority of crimes (red line) in this dataset occur within 2.5km of a police station, while our sampling of points from all of Chicago has a lower density in this distance range. This means that crimes tend to occur closer to police stations. *Is this causal?* It is hard to draw a firm conclusion.

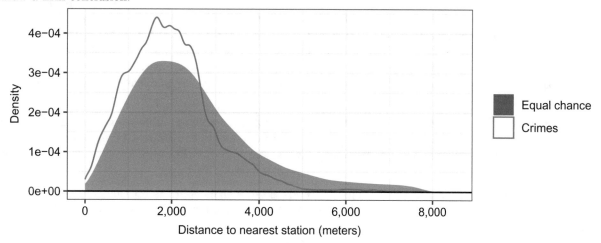

Figure 12.15: Comparison of distance distributions between equal chance and crime incidents.

There are many factors that could contribute to this trend. Perhaps police stations are placed in high-crime neighborhoods or perhaps police place more effort on areas near the station, *etc.* Drawing a causal inference is quite challenging without an experimental design.

Nonetheless, it may be informative to focus on a few crime types and other attributes of the data. Perhaps the likelihood of generating an arrest differs based on distance from a police station offers a clue. In Figure 12.16, we compare the distance distribution between narcotics incidents that led to an arrest versus those that do not. Interestingly, we see some evidence that the chance of an arrest *could* depend upon the distance between the incident and the police station. That said, there are many other relationships which we would want to explore more deeply before making any decisions.

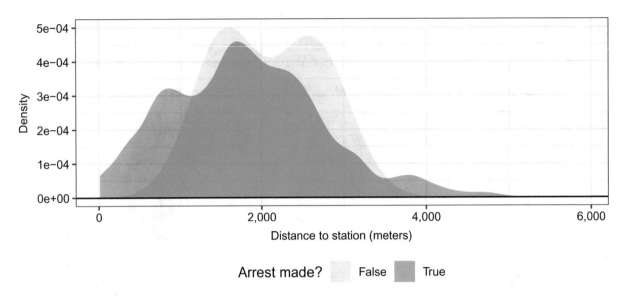

Figure 12.16: Distance distributions of narcotics incidents by whether an arrest was made.

12.5   Beyond this chapter

In this chapter, we presented approaches to process data for spatial analysis and create simple visualizations of spatial data. Indeed, being able to work with shapefiles and rasters opens a world of possibilities, extending data science beyond ordinary tabular data to the complexities of geography.

GIS Tools. While R provides many powerful tools for analyzing spatial data—especially the `sf` and `raster` packages—there are other standard GIS tools on which many geospatial analysts rely, such as `Quantum GIS` (open source) and `ArcGIS` (proprietary). Where these tools excel is their ease of use: analysts do not need to know how to program to manipulate vector and raster files. Both tools are designed with a graphical user interface, yet have the flexibility to be programmed to do more complex tasks.[13] For this reason, public agencies, especially in environmental and emergency management agencies, heavily rely on these tools to manage their GIS data on infrastructure assets and risks, but also to communicate emerging trends and operations plans.

For organizations that publish internet connected applications, `JavaScript` libraries such as `leaflet` are standard. Libraries like `leaflet` lower the bar for software developers to design, stylize, and render maps built on vector and raster data but also base map layers from providers like `Mapbox`, `OpenStreetMap`, and `ESRI`.

Spatial Models. The analysis of spatial data, however, is quite challenging as relationships can vary over time and space. For example, the effect that a new bar has on local drunk driving likely extends past the block on which the bar is located. *But how far?* Does the distance depend upon population density? What about proximity to other bars? Further, failing to account for the spatial structure of your setting can bias regression-based estimates for the effect of policies—essentially generating a spatial data form of omitted-variable bias. Much like autocorrelation over time, some questions require data scientists to consider *spatial autocorrelation*. This type of autocorrelation draws on nearby observations to help describe spatial variation. A crime hotspot would indicate some sort of spatial autocorrelation. Spillover effects from election wins, disease outbreaks among other phenomena are prime cases for analyzing spatial autocorrelation. However, spatial autocorrelation and other more advanced topics are beyond the scope of this text but are nonetheless important for advancing analytical goals in spatial data science and public policy.

These are important concepts that can vastly expand one's ability to deliver pragmatic insights and high quality research. For further reading on the topic, consider:

---

[13]To take advantage of the programming capabilities, one needs to have skill in programming in `Python`.

- *Applied Spatial Data Analysis with R* by Roger Bivand, Edzer Pebesma, and Virgilio Gómez-Rubio.
- *Spatial Data Science* by Edzer Pebesma and Roger Bivand.
- *Introduction to Spatial Econometrics* by James LeSage and R. Kelley Pace.

For a quick and gentle introduction to spatial statistical analysis, consider the open-source tool `GeoDa`. Designed for ease of use, the software analyzes spatial patterns (e.g., clustering, spatial autocorrelation, *etc.*) through an graphical user interface.[14] `R`, however, can accomplish most of the functionality of the software listed above, doing so scalability, flexibility, and reproducibility. Furthermore, its extensibility and integrations with interactive software (e.g., `HTMLWidgets`) makes `R` an appropriate general purpose spatial analysis tool. Nonetheless, applying effort to learn any of these tools will prove to be invaluable, especially as spatial data will become an integral part of public policy in the coming years.

---

[14]To download the GeoDa tool, visit https://geodacenter.github.io/.

Chapter 13

Natural Language

Although we are constantly exposed to written text and human speech, most of the data we encounter in policy settings is available in neat, tabular formats. In the public sector, both surveys and administrative data tend to rely on closed-form questions to quality control responses to a finite set of responses, such as yes/no and agree/neutral/disagree. Forcing respondents to a well-defined response makes for neat datasets, but ignores the richness of *natural language* that could be harvested from open-ended questions. Language carries a tremendous amount of information, encoded in nouns, verbs, adjectives, and other parts of speech. Combinations of these words allow language to carry multi-layered information, conveying sentiments, ideas, facts, etc.

Part of the reason for focusing on closed-form questions is the cost of interpreting and analyzing a large open-ended dataset. Historically, open-ended responses required an army of survey editors to quality control the responses, then another army of analysts to read, analyze, and annotate the responses so that structured data could be derived for quantitative analysis. As you can imagine, this can be a costly process and sustaining a consistent level of quality is challenging. It is relatively easy for one person to read and summarize every document—the encodings she applies are consistent with her interpretations and personal biases. But if the task is distributed to an analyst army, we quickly find challenges with *intercoder reliability* in which two or more reviewers can review the same document and reach the same conclusion. Language is complex, and when authors use different literary devices, vocabularies and nuances, two analysts may easily be led on divergent paths. In addition, if a certain quality of the data was not captured during the first pass, then the entire analyst army would need to backtrack and revisit past documents to reflect the updated thinking.

Despite these challenges with using text as data, let's be clear that *there is no better substitute to analyzing a corpus of text than human eyes*. The analysis of massive amounts of text, however, can benefit from advances in machine and statistical learning in the area of Natural Language Processing (NLP)—a discipline focused on automating the manipulation and analysis of speech and text. Recent advances have demonstrated possibilities that were previously believed to be science fiction. NLP has become so commonplace in modern web technologies that stories about their use are appearing in mainstream media.

Let's take a peculiar example from Facebook. Given growing challenges with privacy and security on social media platforms, Facebook built a chat bot called "Liam Bot" to help employees navigate challenging conversations about the company with family during the holidays (Frenkel and Isaac 2018). Chat bots are an NLP application that takes human-written text, parses it, then applies a trained prediction model to retrieve the best pre-written response. The input data does not need significant manual attention other than a set of answers and a number of different formulations of the associated questions. In the context of the holidays, a Facebook employee could send a question to Liam Bot when faced with a difficult conversation with a privacy-obsessed uncle—a far more economical option than offering a hotline.

While chat bots are not yet common in public sector, a working knowledge of NLP concepts can greatly expand one's range of motion. In this section, we introduce basic concepts for using natural language for

J. C. Chen et al., *Data Science for Public Policy*, Springer Series in the Data Sciences, https://doi.org/10.1007/978-3-030-71352-2_13

data science pursuits. Building upon data processing approaches used in Chapter 4 (Data Manipulation), we review how to process text into analysis-ready form. These basic data processing procedures are the gateway to a world of possibilities ranging from topic modeling to document search to sentiment analysis (Figure 13.1). We then explore two techniques that can easily be deployed for public sector applications: *sentiment analysis* and *topic modeling*. Sentiment analysis converts text into metrics of positive, negative, and other sentiment to provide deeper meaning, whereas topic modeling is an application of unsupervised learning to find mixtures of topics in documents. We close the chapter by contextualizing the application of these methods in the public realm and highlight notable developments that hold game-changing possibilities.

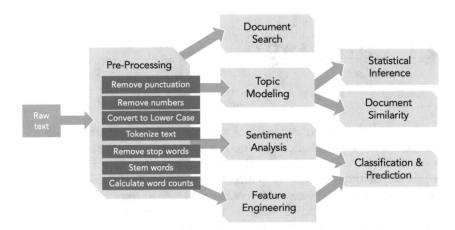

Figure 13.1: Basic text processing is the gateway to more advanced text analysis, such as document search, topic modeling, sentiment analysis to name a few.

13.1 Transforming text into data

13.1.1 Processing textual data

Speeches contain a wealth of information. As humans, we are taught to understand verbal and written communication—pick out the nouns, verbs, and adjectives, then combine the information to decipher meaning. Take the following excerpt from the 2010 State of the Union:

> Now, one place to start is serious financial reform. Look, I am not interested in punishing banks. I'm interested in protecting our economy. A strong, healthy financial market makes it possible for businesses to access credit and create new jobs. It channels the savings of families into investments that raise incomes. But that can only happen if we guard against the same recklessness that nearly brought down our entire economy. We need to make sure consumers and middle-class families have the information they need to make financial decisions. We can't allow financial institutions, including those that take your deposits, to take risks that threaten the whole economy.

We can infer that the excerpt was written in a time of economic hardship and the author recognizes that consumers and middle class need better information about the financial markets. Despite how clear this message may be humans, many analysts might not consider text to be data. Our minds analyze the text and infer what the author is signaling, sifting for key terms and phrases. The majority of words convey little to no meaning—they help make language more fluid. We can cut to the core meaning by dropping the "filler"

> ~~Now, one place to start is serious~~ financial reform. ~~Look, I am not interested in~~ punishing banks. ~~I'm interested in~~ protecting our economy. ~~A~~ strong, healthy financial market ~~makes it possible for~~ businesses ~~to access~~ credit ~~and~~ create new jobs. ~~It channels the~~ savings of families ~~into~~ investments ~~that~~ raise incomes. ~~But that can only happen if we guard against the same~~ recklessness ~~that nearly brought down our entire~~ economy. ~~We need to make sure~~ consumers ~~and~~ middle-class families

~~have the information they need to make~~ financial decisions. ~~We can't allow~~ financial institutions, ~~including those that take your~~ deposits, ~~to take risks that threaten the whole~~ economy.

The frequency of each unique word in each document can be tabulated, then structured into a matrix where each row represents a document and each column contains the frequency of each word. Of course, this process could be done manually, but imagine sorting through all 7,304 words in the 2010 address and scaling the process to the roughly *1.9 million words* in State of the Union addresses between 1790 and 2016. Every little detail about the data needs to be meticulously converted into a usable format. Let's take one line from above and dissect the processing from natural language into tabular data:

> "We need to make sure consumers and middle-class families have the information they need to make financial decisions. We can't allow financial institutions, including those that take your deposits, to take risks that threaten the whole economy."

Many NLP use cases are built on the assumption that the insights locked in the text can be learned from a *bag of words*—the frequency of words matter but their order might not. Thus, removing punctuation and numbers will reduce the size of the textual data. In addition, text can be further standardized by converting values to lower case—a computer may otherwise interpret "economy" and "Economy" as different strings. These basic transformations are applied to the text:

> "we need to make sure consumers and middleclass families have the information they need to make financial decisions we cant allow financial institutions including those that take your deposits to take risks that threaten the whole economy"

But perhaps most importantly is the necessity to break text into more manageable pieces known as *tokens*, or individual words or groups of words. It is only once words are tokenized that we can unleash the power of NLP techniques. Each space between each word can be used as a *delimiter* that can be used to parse a string of words (Table 13.1).

Table 13.1: Tokens from example sentence.

we	families	financial	those	that
need	have	decisions	that	threaten
to	the	we	take	the
make	information	cant	your	whole
sure	they	allow	deposits	economy
consumers	need	financial	to	
and	to	institutions	take	
middleclass	make	including	risks	

Some tokens carry little meaning, such as "the", "in" among others. These *stop words* are linguistic padding that makes language more easily spoken or communicated and can be removed without significant impact on the meaning of the text. By removing these words, we can also reduce the size of the data, computational burden and eliminate potential sources of noise. While there is a standard stop word list available in most data science software, some domains require custom stop word lists.

To maximize the signal from individual words, many tokens need to be standardized as they are variations of a root word. For example, the words "argues", "argued", and "arguing" are variations on the word "argue". If these variations are not distilled to their root, the value of a word may be understated. To address this problem, each word can be *stemmed* in order to expose common roots (Table 13.2).

Lastly, the importance of words can, in part, be summarized in terms of word frequencies. Not only can word counts reduce the size of the data, but they convert tokens into quantitative variables that enable analysis. When all these stops come together, how text is processed will directly influence the success of an analysis. If stop words are adequately removed and word stemming is accurate, then the word counts will be quite

Table 13.2: Terms after removing stop words and stemming.

need	middleclass	make	allow
make	famili	financi	financi
sure	inform	decis	institut
consum	need	cant	includ

useful. In this case, each "financial" and "make" appear twice in the text, perhaps indicating that there is an orientation toward action (make) for financial considerations (Table 13.3).

Table 13.3: Term Frequencies after removal of stop words and application of stemming.

Term	Freq	Term	Freq
financi	2	decis	1
make	2	deposit	1
need	2	economi	1
take	2	famili	1
allow	1	includ	1
cant	1	inform	1
consum	1	institut	1

Term frequencies will often be further transformed into a *Document-Term Matrix* (*DTM*), which lays out each document as a row and each token as a column with frequencies occupying the intersections. This format allows NLP models to treat each token as a variable and make comparisons between rows and between columns.

13.1.2   TF-IDF

Term frequencies from one document are not likely to carry profound insights; however, when term frequencies are compared across multiple documents, we can find identify which terms distinguish one document from all others. Conversely, we can also identify which words are likely stop words. *Term Frequency-Inverse Document Frequency* (TF-IDF) is one measure that can help in document comparison. TF-IDF is comprised of two measures. While we have already introduced term frequencies, Inverse Document Frequency is a measure of how rare a term is relative to the entire corpus:

$$IDF = ln(\frac{N}{n_{term}})$$

where N is the number of documents and n_{term} is the number of documents that contain a given term. We can also see that if the ratio $\frac{N}{n_{term}}$ is too large, the IDF value could overpower the term frequency, thus taking the natural log ln keeps the influence of IDF in check. A term appears in all documents will result in $IDF = 0$—indicating the term does not carry unique information.

The combination of TF and IDF allow their product—TF-IDF—to serve as one of the most commonly applied word weighting metrics:

$$\text{TF-IDF} = TF \times IDF$$

In practice, a value of TF-IDF = 0 can be treated like a stop word. When applied to *document search*, TF-IDF allows search applications to focus on the terms that matter most and efficiently rank and retrieve relevant files.

13.1.3 Document similarities

NLP has allowed organizations to quickly review documents at scale. Organizations often need to quickly find *similar documents* from a large catalog of documents. The process can at times be tedious and arduous, requiring much manual work. Lawyers, for example, often search for pertinent legal records to make their arguments more compelling. Meanwhile, academic researchers often conduct extensive literature reviews so they can articulate how their work contributes to their field of study. In both cases, they start with a set of documents and look for other documents that are also similar, often times using simplistic keyword searches to sort documents.

Suppose a lawyer has found an important document X and would like to find all other documents in a database that are related. A data scientist could conduct pairwise comparisons of term frequencies (X and Y) from each document. A Pearson's Correlation Coefficient is a logical first choice to facilitate the comparison:

$$\rho_{X,Y} = \frac{cov(X,Y)}{\sigma_X \sigma_Y} = \frac{\sum_{i=1}^{n}(X_i - \mu_X)(Y_i - \mu_Y)}{\sum_{i=1}^{n}(X_i - \mu_X)^2 \sum_{i=1}^{n}(Y_i - \mu_Y)^2}$$

Although a simple correlation is a convenient calculation, there are shortcomings. How similar a pair of documents depends on how many words overlap. In general, the vocabularies of two documents will rarely overlap, yet a simple correlation treats non-matching values as information. Notice that the vector of each X and Y ae each mean-centered (μ), which retains zero match values in the correlation calculation. In practice, *Pearson's correlation may artificially inflate the relationship between two vectors by treating zero values as relevant information.*

To hone-in on the relevant overlapping terms, we can instead rely on *cosine similarity*, which associates two vectors without mean-centering its values:

$$cos(\theta) = \frac{X \cdot Y}{\|X\|\|Y\|} = \frac{\sum_{i=1}^{n} X_i Y_i}{\sqrt{\sum_{i=1}^{n} X_i^2}\sqrt{\sum_{i=1}^{n} Y_i^2}}$$

Cosine similarity originates from vector calculus, calculating the angle between two vectors to measure their similarity. If vector X is parallel to vector Y, then they are similar. Otherwise, if X is orthogonal to Y, then they are dissimilar. Values of cosine range between 0 (not similar) and 1 (perfectly similar). Note that the measure is the cosine of the angle between two vectors, but not the strength of the relationship.

DTMs and any $m \times n$ matrix can be transformed into a matrix of cosine similarities that can map document similarities. From streaming services to document search, cosine similarity is a transformative measure that enjoys widespread use.

13.1.4 DIY: Basic text processing

DIYs are designed to be hands-on. To follow along, download the DIYs repository from Github (https://github.com/DataScienceForPublicPolicy/diys). The R Markdown file for this example is labeled `diy-ch13-text.Rmd`.

Textual processing, TF-IDF and document similarities are foundational for natural language applications. In this DIY, we illustrate how to put these concepts to use. Using a set of news articles, we illustrate the process of constructing a DTM, then use TF-IDF to find potential stop words. Lastly, we relate two or more documents to one another using cosine similarity. For a large scale database, these steps can help identify relationships between documents and easily facilitate qualitative analyses.

The news articles used for this DIY focus on the US' federal budget deficit as reported on in October 2019 (Rappeport 2019, Crutsinger 2019, Schroeder 2019, Elis 2019, Franck 2019) and were scraped from various news websites.[1]

---

[1]In the wild, these news articles would have been embedded in HTML files on websites, requiring some attention to the structure of the page.

To start, we load a combination of packages, namely the `tidytext` package to work with text, `dplyr` for general data manipulation, and `stringr` for manipulating string values.

```
pacman::p_load(tidytext, dplyr, stringr)
```

Processing. The five articles are stored in a CSV named `deficit-articles.csv` and should be scrubbed of all numeric values and punctuation using simple regex statements. While we recognize that numbers can carry useful information, we are more concerned with standardizing words in order to approximate their importance.

```
# Load file
deficit <- read.csv("data/deficit-articles.csv",
            stringsAsFactors = FALSE)


# Remove punctuation and numeric values
deficit$text <- str_remove_all(deficit$text, "[[:digit:][:punct:]]")
```

In one neat block of code, we unravel the text into neat word counts for each article by piping multiple commands together (`%>%`).

Tokenization. The code first tokenizes the `text` column into unigrams (single word tokens) using the `unnest_tokens` function (`tidytext`). A new column `word` is added to the data frame, expanding the number of rows from $n = 5$ to $n = 2989$. Note that each token is automatically converted to lower case.

Stop words. With the tokens exposed, stop words are removed. We retrieve a data frame of standard English stop words using the `get_stopwords`, then apply an `anti_join` (`dplyr` package) that retains rows that did not match terms in the stop word list. This step significantly reduces the size of the dataset to $n = 1792$.

Stemming. Using the `wordStem` function (`SnowballC` package), all words are screened and adjusted for stemming if appropriate. The result is assigned to a new variable `word.stem` using the `mutate` function (`dplyr`). As a sanity check, only stemmed tokens with more than one character are retained.

Tabulation. Lastly, the remaining unigrams are summarized as *term frequencies* (TF)— a count of how often each `word.stem` token appears in each article. The resulting dataset contains $n = 1005$ records, meaning that some terms appear more than once and contain relatively more of an article's meaning than other terms.

```
# Unigram in one go
unigram <- deficit  %>%
            unnest_tokens(word, text) %>%
            anti_join(get_stopwords(language = "en")) %>%
            mutate(word.stem = SnowballC::wordStem(word)) %>%
            filter(nchar(word.stem) > 1) %>%
            count(outlet, word.stem, sort = TRUE)
```

Document-Term Matrix. The tabulations can be further processed into a DTM using the `cast_dtm` function (`tidytext`)—useful for applications like topic modeling. We should also highlight that some NLP use cases call for a Document-Feature Matrix (DFM), which allows metadata about each document among other variables to be included in the matrix. DFMs are quite similar to DTM in structure, but are stored as separate object classes to facilitate other applications.

```
# Cast into DTM - requires tm package
pacman::p_load(tm)
deficit_dtm <- unigram  %>%
            cast_dtm(document = outlet,
                     term = word.stem,
                     value = n)


# Cast into DFM -- requires quanteda package
```

```
pacman::p_load(quanteda)
deficit_dfm <- unigram %>%
            cast_dfm(document = outlet,
                   term = word.stem,
                   value = n)
```

Distinguishing between documents. Term frequencies imply that higher frequency terms carry more importance. In our example, all five articles focus on the 2019 federal budget deficit and the words "trillion", "dollar", "budget", and "deficit" all have high TF values. These high-frequency terms prove particularly useful to distinguish between articles about deficits and any other non-deficit topic such as education and defense. In the case of a narrowly defined corpus exclusively focused on deficits, these terms can be viewed as stop words. TF-IDF can be applied to identify overly common words.

To use TF-IDF in R, we apply the `bind_tf_idf` function to our word counts, which calculates and joins the TF, IDF, and combined TF-IDF metrics for each token.

```
unigram <- unigram %>%
           bind_tf_idf(term = word.stem,
                   document = outlet,
                   n = n)
```

The effect of TF-IDF on word importance can be quite dramatic and visualized in a parallel coordinate plot as seen in Figure 13.2. The plot illustrates how a single token's relative value changes between simple term frequencies and TF-IDF. About three-quarters of unigrams increase in their importance once controlling for how common terms are across the corpus of articles (see blue coordinate pairs). In contrast, one-quarter of the terms reduced in rank, indicating that they hold less distinguishing information. These terms include obvious deficit-related terms such as *deficit, spend, year, trump, budget, tax*, among others.

TF-IDF can also be used to remove hard-to-identify keywords. In the deficit articles, approximately 10% of terms ($n = 100$) have TF-IDF values equal to zero. It is not to say these terms are unimportant, but removing these terms in this context could be beneficial.

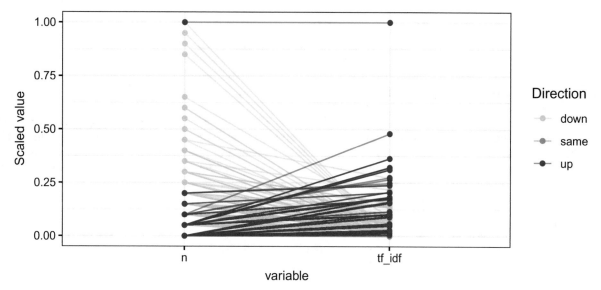

Figure 13.2: Comparison of word frequency and TF-IDF distributions.

Finding similar documents. In order to associate the articles with one another, we construct cosine similarities from the DTM. There is, however, an outstanding issue should the similarities be based on term frequencies or TF-IDF. Using the `cosine` function (`coop` package), we calculate the cosine similarities for each case.

```
# Load COOP
pacman::p_load(coop)

# Similarity based on termfrequencies
grams_n <- unigram  %>%
            cast_dtm(word.stem, outlet, n)
cosine(grams_n)

# Similarity based on TF-IDF
grams_tfidf <- unigram  %>%
            cast_dtm(word.stem, outlet, tf_idf)
cosine(grams_tfidf)
```

As is apparent in Figure 13.4, our choice of input metric emphasizes different qualities of the articles. The raw term frequencies help identify articles that are related in overarching topics, fixating on common words that describe the federal deficit. However, TF-IDF treats these common words as stop words, leaving only terms that are reflective of the author's style and attitudes rather than the big picture. Both are valid approaches. The most appropriate metric, however, is dependent on your analytical objective.

Cosine similarity is simply a relational metric that helps rank documents with respect to each document. By calculating the cosine similarity between all documents in a corpus, one can recommend other articles with similar content. For example, the scores in Figure 13.4 can be interpreted as a list of *similar articles conditional on the MarketWatch article.*

Table 13.4: Comparison of cosine similarity using TF-IDF and Term Frequencies. All values are compared against the deficit article written by MarketWatch.

Outlet	n	TF-IDF
NYTimes	0.5693	0.08263
The Hill	0.4759	0.03794
AP	0.6289	0.06353
CNBC	0.6377	0.1015

13.2   Sentiment Analysis

Although its roots can be traced to public opinion analysis in the early 20th century and computational linguistics research of the 1990s (Mantyla, Graziotin, and Kuutila 2018), *sentiment analysis* has only seen significant adoption in the 2000s with the advent of micro-blogging platforms and interest in quantifying sentiment at a societal level. *Sentiment analysis* extracts subjective signals from text, such as moods, feelings, and attitudes. When applied to Twitter data, for example, sentiment analysis can help track trends in public opinion (Pak and Paroubek 2010) and predict stock market movement (Bollen, Mao, and Zeng 2011). In the investment space, companies are keenly aware of the value of sentiment analysis in extracting information about company performance and trajectory. This new reality has led companies to be far more careful and place positive or less negative spin on their reporting (Trentmann 2019). In this section, we examine the basics of sentiment analysis, starting with sentiment lexicons that convert words into sentiments, then illustrate how to construct a sentiment score.

Flavors of sentiment analysis. Sentiment analysis can either be built on statistical models or human-made rules. In the case of supervised learning, the objective is to predict the sentiment label for each document. The sentiment label is determined by human reviewers in advance by reading a document and the input variables are a bag-of-words. Rules-based approaches, in contrast, rely on *sentiment lexicons*—or a reference table that connects keywords to sentiments—to annotate the sentiment for specific keywords. The former approach places sentiment prediction in a formal machine learning framework (often times a neural network), using a loss function to measure how a model can replicate the label. Its success, however, depends on the quality of the label—*is it generalizable and reliably labeled by the human reviewer?* In contrast, the rules-based approach is more convenient and economical, making it a reasonable starting point. For this reason, we dedicate the remainder of this section to rules-based approaches.

13.2.1 Sentiment lexicons

A sentiment lexicon is the bridge between keywords and sentiments, such as positive and negative, happiness, anger, etc. When joined with a tokenized document, each matching word is converted into a sentiment that then can be more easily analyzed on equal footing. For example, words like "abandoned", "attack", and "adverse" can indicate *anger*. A sentence with all three words can indicate more intense anger.

While time-consuming to construct, it is well-worth the effort to create your own lexicon for domain-specific use cases. There are, however, general purpose English language lexicons in extensive use in the field. We discuss three of which below

- The *Bing Lexicon* provides a rudimentary binary classification, by assigning words as negative and positive (Liu, Hu, and Cheng 2005).

- The *National Research Council (NRC) Emotion Lexicon* associates words with eight emotions (anticipation, trust, surprise, anger, fear, sadness, joy, and disgust) as well as two sentiments (negative and positive). The NRC lexicon was crowd-sourced using Amazon's Mechanical Turk (Mech Turk) platform (Mohammad and Turney 2013).[2]

- The *AFINN* dictionary assigns a score to words, ranging between -5 and 5, where negative scores are represented by negative values. Unlike NRC, AFINN was constructed by researcher Finn Arup Nielsen (Nielsen 2011).

As each dictionary is differently motivated, we would also expect their outputs to behave differently. Word choice, in particular, only minimally overlaps between dictionaries. The NRC lexicon has broad coverage of sentiments for $n = 6468$ words and the Bing lexicon covers $n = 6786$, but only $n = 2484$ words overlap between the two lexicons. One might assume that two lexicons can be combined to increase the word coverage, but each lexicon approaches sentiment encoding differently. The NRC lexicon covers a range of emotions and sentiments while the AFINN lexicon encodes the intensity of a sentiment on a numeric scale.

We illustrate the differences in focus between these three lexicons by scoring two exemplar sentences in Figure 13.3. The first sentence is clearly one that reflects a positive mood— all three dictionaries agree, but annotate different words with the exception of the word "delighted". Whereas NRC only flags one word as positive, Bing flags three and AFINN has a total of +4.

The second sentence taken from a Forbes magazine article about the life of the late Fed Chairman Paul Volcker (Carlson 2019) is more complicated. The sentence is positive as Volcker is credited for "killing" an era of heightened inflation in the economy yet all dictionaries score keywords as negative. This result illustrates that rules-based approaches will struggle with the innate complexities of natural language such as nuance and negation. Computer scientists have developed strategies for *negation handling* in which an algorithm checks for negation keywords (e.g., not, non-, no) in the proximity of a term. However, more layered language such as sarcasm is harder to detect (Gonzalez-Ibanez, Muresan, and Wacholder 2011). Nonetheless, lexicons extract meaning that would otherwise be locked in text.

13.2.2 Calculating sentiment scores

Sentiments can be summarized as a set of scores. In many ways, developing scores is a feature engineering exercise that captures sentiment from different perspective. We start with basic notation to distinguish between different types of tokens:

$$N_d = n_d^+ - n_d^- + n_d^0$$

---

[2]Amazon's Mech Turk asks internet participants to conduct many simple tasks for small payments. It is a favorite of social scientist experimentalists to create training samples (Samuel 2018).

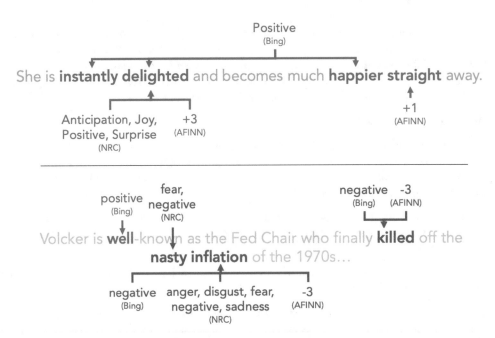

Figure 13.3: Two sentences and sentiment tags. The first sentence (top) is would yield a net positive score, although each lexicon fixates on different words. The second sentence (bottom) illustrates the challenges with rules-based sentiment analysis.

where N_d is the total number of tokens in document d, n_d^+ are positive tokens, n_d^- are negative tokens, and n_d^0 are neutral. In addition, each positive and negative token is assigned a sentiment value x_d. While AFINN dictionary directly assigns these values (between -5 and 5), NRC and Bing sentiments should be converted:

$$x_d = \begin{cases} -1, & \text{if negative} \\ 1, & \text{if positive} \end{cases}$$

These are the basic building blocks that can be used to quantify the qualities of the text.

A document's positivity and negativity is expressed as the proportion of tokens that are tagged for each sentiment:

$$Positivity_d = \frac{n_d^+}{N_d}, Negativity = \frac{n_d^-}{N_d}$$

The *net sentiment* is expressed as the normalized difference between positive and negative tokens, bounding the values between -1 and 1. However, if the number of sentiment tokens is small relative to the total number of tokens, the *net* metric can be small and hard to interpret.

$$Net_d = \frac{n_d^+ - n_d^-}{N_d}$$

Alternatively, the mood of a document can be estimated from the sentiment tokens, yielding estimates of *polarity*. The polarity can be computed from both frequency counts n and the sentiment values x—capturing word-level intensity. In practice, polarity and net sentiment are highly correlated

$$Polarity_d = \frac{n_d^+ - n_d^-}{n_d^+ + n_d^-}$$

$$Polarity_d = \frac{|x_d^+| - |x_d^-|}{|x_d^+| + |x_d^-|}$$

The amount of sentiment in the text can be informative but should also consider if many or only a few words drive the insights that we gleen. The *expressiveness* of a document is a useful metric that is estimated as the proportion of tokens that carry sentiment. For some political communication strategies, the goal is to deliver information devoid of any sentiment. The expressiveness metric is a useful way to keep track of how successful is one in keeping to an emotionless information campaign.

$$Expressiveness_d = \frac{n_d^+ + n_d^-}{N_d}$$

A deep understanding of the sentiment of text should be informed by *all* metrics. Focusing on the *net* or *polarity* alone endows only a narrow understanding of the text. Incorporating *expressiveness* to understand how much signal is present informs whether the sentiment is more than random noise. Decomposing the scores into positivity and negativity can give a clue about different arguments that exist within the same text. Language is indeed complex. When conducting a sentiment analysis, one should be cognizant of those complexities.

13.2.3 DIY: Scoring text for sentiment

DIYs are designed to be hands-on. To follow along, download the DIYs repository from Github (https://github.com/DataScienceForPublicPolicy/diys). The R Markdown file for this example is labeled `diy-ch13-sentiment.Rmd`.

In this DIY, we turn to a Wikipedia entry about the 1979 Oil Crisis that had a substantial effect on the Western World's economy.[3] The article on the Oil Crisis presents both positive and negative effects—a perfect example of how sentiment analysis can summarize a body of text. The first paragraph describes the magnitude of the crisis:

> The 1970s energy crisis occurred when the Western world, particularly the United States, Canada, Western Europe, Australia, and New Zealand, faced substantial petroleum shortages, real and perceived, as well as elevated prices. The two worst crises of this period were the 1973 oil crisis and the 1979 energy crisis, when the Yom Kippur War and the Iranian Revolution triggered interruptions in Middle Eastern oil exports.

Whereas the sixth paragraph softens the implications, turning to instances where the crisis had a far less adverse impact:

> The period was not uniformly negative for all economies. Petroleum-rich countries in the Middle East benefited from increased prices and the slowing production in other areas of the world. Some other countries, such as Norway, Mexico, and Venezuela, benefited as well. In the United States, Texas and Alaska, as well as some other oil-producing areas, experienced major economic booms due to soaring oil prices even as most of the rest of the nation struggled with the stagnant economy. Many of these economic gains, however, came to a halt as prices stabilized and dropped in the 1980s.

Our objective with this DIY is to show how sentiment evolves over the first six paragraphs as scored by the Bing Lexicon implemented in the `tidytext` package. To start, we read the raw text from the `wiki-article.txt` file.

---

[3]The Wikipedia article on the Oil Crisis can be found at https://en.wikipedia.org/wiki/1979_oil_crisis.

```
# Load package
pacman::p_load(tidytext, dplyr)

# Read text article
wiki_article <- readLines("data/wiki-article.txt")
```

To make use of the text, we tokenize the six paragraphs of text, remove stop words, then apply a left join with the Bing lexicon. The resulting table contains both matching and non-matching terms and aggregates term frequencies by paragraph `para`, `word`, and `sentiment`. In total, $n = 18$ words are labeled negative and $n = 11$ words are positive.

```
# Set up in data frame    wiki_df <-data.frame(para =
1:length(wiki_article),
                          text = wiki_article,
                          stringsAsFactors = FALSE)

# Join tokens to lexicon
sent_df <- wiki_df %>%
  unnest_tokens(word, text) %>%
  anti_join(get_stopwords(language = "en")) %>%
  left_join(get_sentiments("bing")) %>%
  count(para, word, sentiment, sort = TRUE)

# Label terms without sentiment
sent_df$sentiment[is.na(sent_df$sentiment)] <- "none"
```

Next, we write a basic function to calculate various sentiment metrics such as polarity and expressiveness. Packages such as `sentimentr` implement scoring, but as rules-based sentiment analysis is fairly simple, we directly implement metrics in the formula to illustrate their accessibility.

```
sentMetrics <- function(n, s){

  # Set input metrics
  N <- sum(n, na.rm = T)
  x_p <- sum(n[s =="positive"], na.rm = T)
  x_n <- sum(n[s =="negative"], na.rm = T)

  # Calculate scores
  return(data.frame(count.positive = x_p,
              count.negative = x_n,
              net.sentiment = (x_p - x_n) / N,
              expressiveness = (x_p + x_n) / N,
              positivity = x_p / N,
              negativity = x_n / N,
              polarity = (x_p - x_n) / (x_p + x_n))
  )

}
```

The function is designed to work on one document at a time, requiring a loop to score each paragraph in the article. We iterate over each paragraph using `lapply`, returning the results in a data frame `rated`.

```
# Apply sentiment scores to each paragraph
rated <-lapply(sort(unique(sent_df$para)),function(p){
            para_df <- sent_df %>% filter(para == p)
            output <- sentMetrics(n = para_df$n,
```

```
                                    s = para_df$sentiment)
             return(data.frame(para = p,
                               output))
         })
```

 # Bind into a data frame   `rated <-do.call(rbind, rated)`

Let's examine the results by plotting sentiment by paragraph in Figure 13.4. In the first line graph (Graph (A)), we see that the first four paragraphs are net negative, then turn net positive in the last two paragraphs. Notice that both polarity and net sentiment tell the same story, but the magnitudes of their values are quite different. In fact, the Pearson's correlation is $\rho = 0.964$. When we dive deeper into the positivity and negativity, we see that the switch in tone is widespread—earlier paragraphs have mostly negative terms while the tone softens in later paragraphs that describe less dire conditions in the world economy.

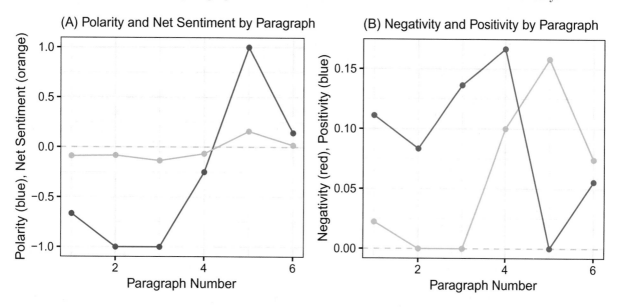

Figure 13.4: Sentiment scores for the first six paragraphs of the Wikipedia article on 1979 Oil Crisis. Graph (A) illustrates net sentiment and polarity. Graph (B) plots positivity and negativity.

13.3   Topic modeling

13.3.1   A conceptual base

Topics bring focus and cohesion to natural language, yet they are not explicitly defined or labeled. A State of the Union address, for example, is amalgam of many topics. But the precise number of topics is debatable—two or more experts might not agree on the precise number. To extract this information, topic models examine the co-occurrence of words and sometimes metadata about the text to identify likely topics. If were to use common unsupervised learning techniques (e.g., k-means, HAC), we would over-simplify the topic distribution as an entire document would be assigned to a single topic despite the likely presence of two or more topics.[4] To overcome this hurdle, topic models approach the problem from the lens of a *mixture model* in which one or more topics *could* be simultaneously present—the chance they are present is represented as a probability.

Mixture models are quite different from other techniques covered in this textbook. For the most part, we have

---

[4]Other clustering techniques can accommodate mixtures of non-exclusive clusters; however, these are beyond the scope of this text.

focused on *discriminative models*—techniques that find boundaries that separate classes or clusters from one another conditioned on a set of input variables X. For example, we could predict the labels associated with each document of a corpus by estimating their conditional probabilities— $Pr(label = 1|tokens)$. While the conditional model simplifies the world into a convenient finite set of discrete classes, it also tends to overlook the richness of the data's empirical distributions.

Topic models, in contrast, are *generative models* that are motivated by an interest in distributions. A document can be viewed as distributions of words and distributions of topics. A generative model learns by attempting to reconstruct the distributions from which the observed data was drawn—in a sense, it is forensic scientist piecing together the story from which a piece of evidence originated. By replicating the data-generating process, these models learn the interactions and nuances in text that are otherwise ignored in discriminative models. Interestingly, these models can also generate data. More formally:

> Topic models assume that each topic is defined as a probabilistic mixture over words.
> Each document contains a distribution of topics, represented as a topic proportion (θ).
> Each word (w) has a probability (β_{wk}) of belonging to topic k.

In policy contexts, θ and β are the minimum that is required to tell a compelling NLP story or enable a precision use case. The former (θ) serves as an indicator of how much of a topic is present in a document, while the latter (β) allows analysts to articulate a topic based on the words that most likely define it.

There is not just one topic modeling technique in use. In fact, there are a variety that have emerged since the early 2000s, the best known of which is Latent Dirichlet Allocation (LDA) as developed in Blei, Ng, and Jordan (2003). However, more recent developments such as Structural Topic Models (STM), as developed in Roberts, Stewart, and Airoldi (2016), extend the sophistication of topic models to capture nuances in text. Both techniques make use of common strategies that relate words, topics, and documents. Rather than covering its inner workings in extensive detail, we provide a cursory overview of common elements across topic models.

13.3.2 How do topics models work?

Topics models are computationally expensive and require patience to execute, especially when the number of tokens in a corpus can be insurmountably large (thousands if not hundreds of thousands). As illustrated in Figure 13.5, topic models iteratively search for K user-defined topics. Below, we walk through the common elements of a topic model as laid out in terms of inputs, initialization, and estimation. Note that most of the computational action occurs at the document level.

Inputs (*Top of diagram*). The empirical journey of a topic model starts with the data. Each document d contains N_d number of tokens that are used as the input data to determine which topics are present. This information is stored in a vector W_d—each document d has its own of a varying length N_d. Next, we define the number of topics K. When we refer to an individual topic k, we assume it is indexed between 1 and K ($k \in \{1, ..., K\}$). At this point, we do not know what is contained in each topic—only that we believe there to be K topics, which we may have prior ideas from reviewing the text. K is also a hyperparameter that can be tuned by minimizing statistical metrics (see Arun et al. (2010) and Cao et al. (2009)) although, once again, there is no better substitute than visual inspection of each topic's words. The choice of K will determine the success of the topic model and guides all steps that follow.

As we will see below, topic models learn by iteratively updating θ and β. Some topic models require a maximum number of iterations I. More iterations equate to more time spent waiting for results, but more stable and better quality results.

Initialization (*Middle of diagram*). To kick-start estimation, the topic model is seeded with some initial data that will be updated as we iterate. We start by creating a vector Z_d that indicates the topic assignment k for each word n in document d. This means that every token in vector W_d has a topic assignment reflected in Z_d. Since we do not know the true topic for each word, we randomly assign topics $k \in \{1, ..., K\}$—this is known as *random initialization*.[5]

---

[5]Other initialization methods can also be employed with prior knowledge.

Next, we populate the β and θ matrices with initial values. The β matrix is a $V \times K$ matrix of word probabilities. Each row v is a token (or word) from the corpus' vocabulary (all words) and each column is a topic k. The intersections are filled with a count of how often a token has been assigned to a specific topic. The counts for each value of β_{vk} can be calculated from the each document's topic assignment vector Z_d.

The second matrix is the *topic proportions* θ of dimensions $D \times K$. Each row d is a document, whereas each column is a topic k. Each cell θ_{dk} is the count that a topic k exists in document d, which can be estimated using Z_d. During estimation, θ is maintained as a matrix of token counts.

Estimation (*Bottom of diagram*). With the basic data in place, estimation proceeds through a series of loops, centering on each word w_{dn} and its topic assignment z_{dn}. For each word and topic assignment, we estimate the probability that the selected word belongs to topic k ($pr(w_n|z_n = k)$), which yields a vector of probabilities of length K. This vector of probabilities dictates how much of each topic is present in a multinomial topic distribution from which we draw one topic at random.[6]

For example, if topic $k = 3$ is estimated to be $pr(w_n|z_n = 1) = 0.75$, then there is a 75% chance that the topic that is drawn is topic $k = 3$. The new topic that is drawn z^* replaces z_dn, then in turn updates β and θ. Each iteration of the nested loop updates the multinomial topic distribution, which in turn forces the parameters to evolve.

Analysis . Once estimation is completed, β and θ are row standardized (each row sums to one), which yields two sets of probability distributions that can facilitate interpretation. A topic k can be interpreted based on the tokens in β with the highest probability of belonging to that topic. These keywords can be reasoned through to produce a topic label, though it depends on analyst interpretation. How often a topic appears can be inferred from θ_k, allowing one to sort a corpus based on the highest chance that a topic appears.

Indeed, topic models involve a fair amount of statistical machinery, but when applied to textual data, the results unlock deep and scalable insights.

13.3.3 DIY: Finding topics in presidential speeches

> *DIYs are designed to be hands-on. To follow along, download the DIYs repository from Github* (https://github.com/DataScienceForPublicPolicy/diys). *The R Markdown file for this example is labeled* `diy-ch13-topics.Rmd`.

Sorting through tens of thousands of paragraphs of text for the right nugget of information is time-consuming and expensive. Placing attention in the wrong part of a corpus can mean the difference between finding useful information and lost time. Topic modeling can summarize the gist of documents and facilitate more effective search, doing so on large corpuses of information. In this DIY, we illustrate how topic modeling can bring structure to an expansive set of text using the STM method using the `stm` package.[7] Our objective is to train a topic model, identify topics, surface quotes that provide a flavor for the topic, then recommend similar paragraphs for review.

Load. Our corpus is a collection of State of the Union (SOTU) speeches and letters delivered by 42 presidents.[8] The text was scraped from `WikiSource.org` and has been structured to include

- Basic identifiers such as an `id` for each SOTU, the title of the SOTU (e.g., Sixth State of the Union Address), the `year` of the address, a `paragraph` number, the `president`, and `link` to the source;

- The `text` storing each paragraph as a separate line;[9]

- An average sentiment score for each paragraph calculated using the Bing Lexicon

Topics and word choice have evolved over the last two centuries. The problems faced by the Founding Fathers of the US are different than contemporary presidents. We limit our analysis to speeches delivered between

---

[6]There are other inputs into the sampling procedure such hyperparameters that change the shape of the underlying distributions; however, these parameters vary from one technique to another, thus we have chosen to omit these in this overview.

[7]While LDA is the most common topic modeling in use, STM is arguably flexible, intuitive, and more extensible.

[8]As of 2019, the only presidents who did not deliver SOTU addresses or letters were James A. Garfield and William Henry Harrison.

[9]Minor adjustments were made to minimize errors. These adjustments can be seen in the data processing scripts on Github.

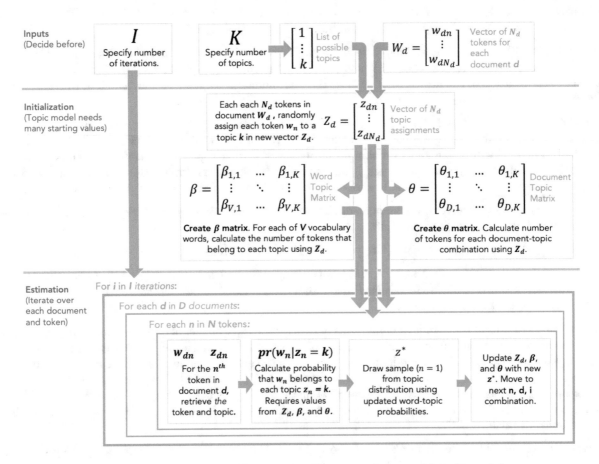

Figure 13.5: A simplified illustration of common elements of topic models.

1960 and 2019, or $n = 6219$ paragraphs.

```
# Load
pacman::p_load(stm, dplyr)

# Read 1960 to 2019
sotu_text <- read.csv("data/sotu.csv", stringsAsFactors = FALSE)
sotu_text <- filter(sotu_text, year >= 1960)
```

Process data. The `stm` package simplifies the process with a pre-built processing function.[10] The only user inputs required for processing are custom stop words and metadata. While a TF-IDF calculation can expose common stop words, a pure statistical calculation may be too aggressive. We instead choose to define a set that are known favorites of US presidents, defined in the `stop_custom` vector.

```
stop_custom <- c("applause", "thank", "you", "goodnight",
                 "youre", "im", "weve", "ive", "us", "that",
                 "know", "back", "one", "much", "can", "shall",
                 "fellow", "people", "government", "get", "make", "agency", "department",
                 "america", "united", "states", "american", "americans",
                 "tonight", "ago", "now", "year", "years", "just", "new",
                 "must", "without", "said", "will")
```

The `textProcessor` function in `stm` constructs a DFM from a vector of `documents` (text field in `sotu_text`). We make use of a STM's special ability to allow word choice to evolve *within* a topic depending on contextual factors as encoded in metadata about the document. We provide the function with a data frame of metadata on `sentiment` and the `year`. With the inputs provided to the function, `stm` removes common stop words, numbers, and punctuation; stems words; and converts text to lower case (`lowercase`). In addition, custom stop words are provided (`customstopwords = stop_custom`) and non-alphanumeric characters that can cause unnecessary complications in the modeling process are also removed (`onlycharacter = TRUE`).

```
ingest <- textProcessor(documents = sotu_text$text,
                        metadata = sotu_text[, c("year","sentiment")],
                        onlycharacter = TRUE,
                        customstopwords = stop_custom)
```

The SOTU corpus is further processed by sweeping rarely occurring words. By reducing the unnecessary sparsity of a DFM, we not only can cut the time required for the topic model to converge but also mitigate noise that would otherwise be introduced into the results.

```
sotu_prep <- prepDocuments(documents = ingest$documents,
                           vocab = ingest$vocab,
                           meta = ingest$meta,
                           lower.thresh = 2)
```

Model. With the data ready, we apply an `stm` model to the processed documents, vocabulary, and metadata as stored in the `sotu_prep` object. With $n = 6219$ paragraphs, there are likely a large number of topics. We arbitrarily choose $K = 25$ as a starting point and a maximum of $I = 100$ iterations. Furthermore, an initial seed is set so the analysis can be replicated.

```
sotu_stm <- stm(documents = sotu_prep$documents,
                vocab = sotu_prep$vocab,
                data = sotu_prep$meta,
                K = 25,  seed = 314,
                max.em.its = 100)
```

Analyze. The resulting `stm` object stores the learned topics and their associated words. To examine the

---

[10]For more advanced DFM processing, consider using the processes illustrated in this chapter using `tidytext` package or the `quanteda` package.

expected topic proportions, simply `plot` the `stm` object. The most common topic (#16), for example, appears to contain global security issues as characterized by "world", "peace", and "war". The second most common topic (#22) is budget and spending focused. Granted, the word combinations are only approximations of the underlying topic. If the analysis were re-run with a different *seed number*, we would likely obtain similar but not precisely identical topics—this is one of the challenges with unsupervised learning applied to a subjective area of study (Figure 13.6).

```
plot(sotu_stm, font = 1, text.cex = 0.8, main = NA)
```

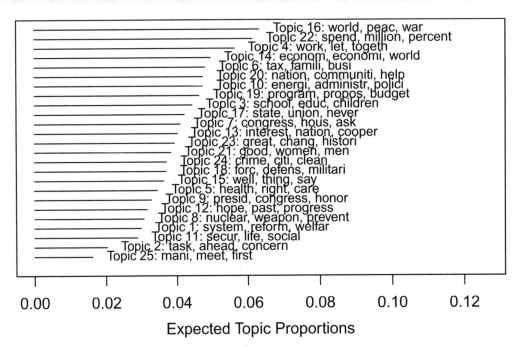

Figure 13.6: Preview of topics sorted by expected topic proportions in SOTU corpus.

Nonetheless, we can improve our interpretation of each topic by identifying paragraphs that have the highest probability of containing the topic of interest. This requires sorting the θ vector the topic k of interest in descending order, then retrieving the corresponding raw text for the highest probability. The `findThoughts` function simplifies the retrieval of example text that illustrates the concepts captured in a topic. When we focus on education (Topic #3), the clearest arguments come from George W. Bush and Barack Obama. When we turn our attention to military (Topic #18), we find details on military plans presented by John F. Kennedy and Jimmy Carter (Table 13.5).

```
thoughts <- findThoughts(model = sotu_stm,
          texts = sotu_text$text,
          n = 2,
          topics = c(3,18))
```

Table 13.5: First 150 characters of representative thoughts.

Topic	President	Text
3	George W. Bush	Third: We need to encourage children to take more math and science, and make sure those courses are rigorous enough to compete with other nations. We...
3	Barack Obama	Let's also make sure that a high school diploma puts our kids on a path to a good job. Right now, countries like Germany focus on graduating their hig...

Topic	President	Text
18	Jimmy Carter	the development of a new fleet of large cargo aircraft with intercontinental range; the design and procurement of a force of Maritime Prepositioning S...
18	John F. Kennedy	Thus we have doubled the number of ready combat divisions in the Army's strategic reserve–increased our troops in Europe–built up the Marines–added...

The probabilistic nature of a topic model means that topics will overlap with one another—some are more distinguishable than others. Using the topic probabilities Θ, it is a fairly straight forward procedure to calculate a topic correlation matrix using a Pearson's correlation `cor`. Visualizing these relationships can prove challenging, however.

One visualization that embraces complexity is the network graph, which uses space and negative space to show the interconnectedness and closeness between entities (referred to as nodes) based on linkages (edges). Whereas each topic can be treated as a node, the cells of a correlation matrix can indicate the strength of relationship between two topics. All topics are likely connected to one another, but even a weak linkage can introduce unwanted noise. Thus, the topic correlation matrix can be simplified by treating correlations below a threshold as if no relationship exists. Below, we rely on the `topicCorr` function (`stm` package) to processes the topic correlations, simplify the matrix by setting a correlation `cutoff = 0.02`, then format the data to be rendered as a network graph using `plot`.

In Figure 13.7, we find that one of the 25 topics is disconnected from all other topics. Meanwhile, the remaining 23 topics are correlated to varying degrees—this is the magic of a mixture model that allows topics to be related but with different foci. To illustrate nuances between topics, we render five topics (#3, #6, #11, #16, and #18) using the word `cloud` function. Topics 3 and 6 are network neighbors that have a common focus on families and children; However, the former focuses more on education issues, whereas the latter on finances. Topics 16 and 18 both relate to national security, but the former stresses peace and liberties while the latter focuses on military and defense.

```
# Topic correlations
sotu_cor <-topicCorr(sotu_stm,
                  cutoff = 0.02)


# Set a 2 x 3 grid
par(mfrow = c(2,3), mar = rep(2,4))


# Plot network diagram of topics
plot(sotu_cor,  vlabels = 1:25,
      vertex.color = "lightblue",  vertex.size = 25)
text(0.5, 1.5, "Topic Correlation",  cex = 1.2)


# Plot word clouds for five topics (requires wordcloud package)
pacman::p_load(wordcloud)


  for(i in c(3, 6, 11, 16, 18)){
     cloud(sotu_stm, topic = i,  max.words = 15, col = "slategrey")
     text(0.5, 1, paste0("Topic ", i), cex = 1.2)
  }
```

Searching for text. With topic proportions θ, we can more accurately correlate documents with one another. A document's topic proportions are a set of coordinates that summarize its complexities and allow for it to be related to any other document. Suppose a specific paragraph from President Bill Clinton's 1997 speech captures a concept that you would like to further explore in presidential speeches:

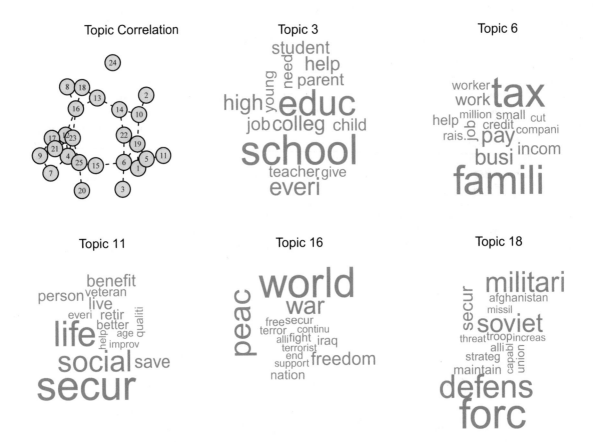

Figure 13.7: Network graphs show how two or more topics are correlated. When plot a few topics as word clouds, it becomes clear that some topics are more similar than others and these similarities are clearly represented by the proximity of topic nodes.

> *We must continue, step by step, to give more families access to affordable, quality health care. Forty million Americans still lack health insurance. Ten million children still lack health insurance; 80 percent of them have working parents who pay taxes. That is wrong. My balanced budget will extend health coverage to up to 5 million of those children. Since nearly half of all children who lose their insurance do so because their parents lose or change a job, my budget will also ensure that people who temporarily lose their jobs can still afford to keep their health insurance. No child should be without a doctor just because a parent is without a job.*

This is paragraph #4121 of $n = 6219$ paragraphs in the corpus, focusing on affordable health care and the risk of losing coverage. Using the topic probabilities (sotu_stm$theta), we calculate the cosine similarity (coop package) to estimate how every paragraph is related to all other paragraphs. This similarity matrix can feed a function to recommend relevant paragraphs conditional on a specific document ($sim(\theta|d)$), requiring a vector of text, a cosine similarity matrix (sim), the paragraph number (item_index), and the number of records to be returned (n).

```
# Calculate Cosine Similarity
pacman::p_load(coop)
sim_mat <- cosine(t(sotu_stm$theta))

# Function to return top N most similar records
simRecs <- function(text, sim, item_index, n){

# Retrieve similar text, removing the target document from consideration
top_rec_index <- which(rank(-sim[,item_index]) <= n + 1)
top_rec_index <- top_rec_index[top_rec_index != item_index]
top_recs <- text[top_rec_index]

# Retrieve top scores
top_scores <- sim[top_rec_index, item_index ]

# Return results
results <- data.frame(top_scores, top_recs)
return(results)
  }

# Return similar documents conditional on paragraph 4121
simRecs(text = sotu_text$text,
     sim = sim_mat,
     item_index = 4121,
     n = 5)
```

Table 13.6 presents the three most related paragraphs to #4121. The results are striking: they capture a broader topic than just health care— financial well-being and stability. The topic also seems to be one that is always a priority from one president to the next. In fact, the example text surfaces quotes from Presidents Lyndon Johnson and John F. Kennedy, helping to draw links over time. This approach does not always return perfect results, but it is far more efficient than manual annotation and expands one's reach to beyond what is possible by analyst.

Table 13.6: Top 3 most similar paragraphs relative to Paragraph 5024.

Score	Year	President	Text
0.9696	1998	Bill Clinton	We have to make it possible for all hard-working families to meet their most important responsibilities. Two years ago, we helped guarantee that Americans can keep their health insurance when they change jobs. Last year, we extended health care to up to five million children.
0.9573	1964	Lyndon Baines Johnson	We must extend the coverage of our minimum wage laws to more than 2 million workers now lacking this basic protection of purchasing power.
0.9522	1962	John F. Kennedy	For our older citizens have longer and more frequent illnesses, higher hospital and medical bills and too little income to pay them. Private health insurance helps very few—for its cost is high and its coverage limited. Public welfare cannot help those too proud to seek relief but hard-pressed to pay their own bills. Nor can their children or grandchildren always sacrifice their own health budgets to meet this constant drain.

13.4   Beyond this chapter

13.4.1   Best practices

The NLP techniques presented in this chapter are a zero-to-one improvement—a first step towards using text as data. Rules-based sentiment analysis is a brute force approach that is only as good as how clearly humans can articulate the relationship between words and sentiments. Topic modeling incorporates sophisticated statistical techniques to find structure and latent concepts in text, but their outputs are too abstract for direct consumption by policy teams.

As these techniques are newer and more complex than tried and true methods like regression, data scientists need to demystify NLP and articulate their value to policy audiences. Below, we have outlined a few example uses of NLP:

- *Feature engineering.* Open-ended responses are often left untouched. Sentiment analysis and topic analysis can transform open-ended responses and augment a dataset with new variables that capture signal that would be overlooked.

- *Sentiment tracking for elections.* Sentiment analysis can be applied to news and social media feeds to develop sentiment indexes. These indexes can in turn be used as inputs into a nowcasting model that correlates contemporaneous polling results with the sentiments online. If the relationships are stable, the nowcasting model could provide an approximation of candidate favorability in near real time.

- *Summarizing public comment.* Governments often request for public feedback on proposed plans and legislation. While policy teams will opt for slow, expensive manual review of thousands of open responses, topic modeling can be an economical alternative to find topical structure among the responses and relate documents to one another.

- *Indexing documents.* Imagine a policy analyst finds an important document during the course of her research and needs to find all other related documents. It would be infeasible to read tens of thousands of documents. Instead, topic modeling can index and organize a large corpus of documents for review. By applying cosine similarity to the topic probabilities, the analyst could rank documents by relevance, then more effectively prioritize her document search.

In each of the examples, notice how NLP serves as a middle layer that processes text into a more useful form. The outputs do not speak for themselves. Thus, the success of a NLP project is very much dependent on how one plans the use case.

13.4.2 Further study

While language is so commonplace in society, humans often underestimate its complexity. For a computer to achieve the same understanding has yet to be seen, but there have been remarkable advancements that are made possible through deep learning. We end this section by introducing a game-changing advancement known as *language embeddings*.

In 2013, a group of researchers at Google developed *Word2vec*—a framework that constructs word embeddings by training a two-layer neural network on a corpus of text (Mikolov et al. 2013). Some words carry a meaning that is more like some than others—there is a spatial relationship between each discrete word. Language embeddings make this spatial relationship possible. Each word is encoded as a k dimensional vector. For example, words defense and pizza can be encoded as five-dimensional embeddings:

$$defense = \begin{bmatrix} 1.3 \\ -0.2 \\ 4.2 \\ 2.3 \\ -3.8 \end{bmatrix} \quad pizza = \begin{bmatrix} 0.1 \\ 2 \\ -1.2 \\ 0.2 \\ 5.4 \end{bmatrix}$$

Unlike topic probabilities, embeddings are not interpretable like topic probabilities produced by topic models. However, embeddings can be treated as a spatial coordinate that allows comparison between a pair of words. When used in conjunction with a language model, word embeddings enable use cases such as auto-completion used in smartphones and word processing applications. In 2014, Google researchers extended the framework to allow documents to be mapped into a k dimensional embedding (Le and Mikolov 2014). The *Doc2vec* algorithm allows any text document to be concisely summarized and compared using a few numeric values rather than relying on a large document-term matrix. Indeed, these advancements have improved tech company data products, such as recommendation engines and search.

The ecosystem of NLP algorithms has rapidly expanded and software capabilities are regularly open sourced for public use, such as Stanford University's Global Vectors (GloVe), Google's Bidirectional Encoder Representations from Transformers (BERT), and Facebook's fastText. While the technology has evolved and is more accessible, these advancements are not yet widely used in public sector due to the lack of natural use cases and clear paths to adoption. Nonetheless, knowledge of these developments can be useful for when the right opportunity presents itself to innovate through NLP.

To build a base in NLP techniques, *Machine Learning for Text* (Aggarwal 2018a) provides a survey of most of the key developments and their inner workings. We also recommend reading the following seminal articles for commonly used NLP methods:

- Mikolov, Tomas, Kai Chen, Greg Corrado, and Jeffrey Dean. 2013. "Efficient Estimation of Word Representations in Vector Space". *arXiv.* https://arxiv.org/abs/1301.3781.

- Devlin, Jacob, Ming-Wei Chang, Kenton Lee, Kristina Toutanova. 2018. "BERT: Pre-training of Deep Bidirectional Transformers for Language Understanding". *arXiv.* https://arxiv.org/abs/1810.04805.

To stay up to date on the latest developments, visit the *Computation and Language* section of *arXiv.*

Chapter 14

The Ethics of Data Science

14.1 An emerging debate

In 2017, Ali Rahimi and Benjamin Recht received the "Test of Time Award" at the Neural Information Processing Systems (NIPS) conference—one of the top research conferences for machine learning and artificial intelligence. The award recognized their paper entitled "Random Features for Large-Scale Kernel Machines", which found a way to speed up the process of training a non-linear model by projecting the problem into a lower dimensional space that could be solved with linear methods. Since the time of the original paper the machine learning field had dramatically evolved. Rahimi reflected upon the current state of the field in acceptance speech, taking aim at an issue that polarizes the field:

> We've made incredible progress. We are now reproducible. We shared code freely and used common tasks benchmarks. We've also made incredible technological progress – self-driving cars seem to be around the corner, artificial intelligence tags photos, transcribes voicemails, translates documents, serves us ads. Billion-dollar companies are built on machine learning. In many ways we're way better off than we were 10 years ago. And in some ways we're worse off.
>
> There's a self-congratulatory feeling in the air. We say things like "machine learning is the new electricity". I'd like to offer another analogy. Machine learning has become alchemy. (Rahimi 2017)

Rahimi's statement was indeed provocative. It likened modern day magic (machine learning) with an ancient pseudoscience, but his argument is not without merit. Alchemists believed it was possible to turn base metals into gold, attempting to do so without the tools of the scientific method. In short, their approach was arbitrary. Rahimi argued that modern day black box algorithms are being developed in a similar tradition— we see inputs and outputs but have little understanding of the magic in between. Rahimi's criticism was aimed at some practitioners' disconcerting lack of interest in *why* certain phenomena happened in algorithms. Instead, the tuning of neural network architectures has a tendency to be iterative trial and error until some optimal hyperparameter set is identified. Thus, the intuition behind what makes one deep learning algorithm better than another is often lacking.

Without clear intuition and interpretability, we place faith on a black box's outputs alone, which in turn means we forego careful review and understanding of the many equations, assumptions, and the data. It could be the case that the training data have biases that impose adverse policy positions on individuals. The underlying model could unfairly favor one group over another without even including variables that describe protected groups. Or perhaps the mathematics of the algorithm itself are not proven or understood, giving way to unpredictable errors in the system. Rather than being the drivers of our scientific destinies, we become passengers of happenstance when adopting the use of black boxes in public policy. Rahimi argued for understanding the behavior of algorithms, which can be achieved through simple experiments and formulating theories.

The following day, Yann LeCun—another top mind in ML—offered a different perspective in a public Face-

© Springer Nature Switzerland AG 2021
J. C. Chen et al., *Data Science for Public Policy*, Springer Series in the Data Sciences,
https://doi.org/10.1007/978-3-030-71352-2_14

book post:

> *Understanding (theoretical or otherwise) is a good thing. It's the very purpose of many of us in the NIPS community. But another important goal is inventing new methods, new techniques, and yes, new tricks.*
>
> *In the history of science and technology, the engineering artifacts have almost always preceded the theoretical understanding: the lens and the telescope preceded optics theory, the steam engine preceded thermodynamics, the airplane preceded flight aerodynamics, radio and data communication preceded information theory, the computer preceded computer science. Why? Because theorists will spontaneously study "simple" phenomena, and will not be enticed to study a complex one until there [is] a practical importance to it.[1]*

The debate between Rahimi and LeCun resurfaced an uncomfortable tension between empiricism and theory—should engineering new possibilities come first or should we instead focus on the theory that explains its inner workings? The juxtaposed positions illustrate a clear divide in how much (or little) attention is paid to transparency and interpretability. Much of the machine learning community follows the mantra, *if it does better than a benchmark, it is newsworthy.* Indeed, the rapid pace of the peer-reviewed publication cycle reflects the mantra exactly.[2] But does this pose an obstacle for transparency? LeCun goes on to argue that rigor and transparency may be required depending on the context:

> *Believe me, the vast majority of decisions made by ML systems do not require explanation. If explanation were produced, no one would have the time or motivation to look at them. Explanation are clearly required for decisions that affects people lives in major ways, and which they can influence if given a reason. They are also required for judicial decisions and decisions that are ultimately made by a person. But for most things, they are not useful.*

How does this relate to public policy? While LeCun's pragmatic position is well-taken, it overlooks the tense interface between the Public's perception of technology and the speed of innovation. The technology wielded by governments should be for the public good and treat everyone equally. When the creator of the technology is unable to offer an explanation of how or why it works, it is not possible for governments to ensure that the black box is an equitable black box. This, in turn, creates risk for decision makers and ethical dilemmas. As we will cover in this chapter, there are many situations in the utility of the black box suffers due to bias and opaqueness.

The goal of data science is to better society by giving people access to better information. We can achieve more rapid, precise decisions by placing data science products in as support tools to influence decisions, but it leads us down a brave new path where technology and policy meld together. We need to ask hard questions about how data science is applied. Does data science have the ability to serve all members of society rather than a select few? How can data scientists ensure that they are producing fair and ethical products? In this chapter, we review four topics that are shaping the ethics of data science, namely *bias, fairness, transparency* and *privacy*. As the scope of the debate is evolving, we focus on foundational concepts and provide examples where the state of practice is mature.

14.2   Bias

In 2014, Amazon started development on a supervised learning algorithm to rank resumes on a scale of one to five stars based on previous candidate outcomes—in effect, a resume screening algorithm. While the path forward seemed clear, there were lurking biases in the data. At the time, 77% of Amazon's technical workforce was male. And even though gender is not an explicit input into the algorithm, its outputs suggest

---

[1]See https://www.facebook.com/yann.lecun/posts/10154938130592143.

[2]At the time of writing this textbook, when using the term ML in the tech sector, practitioners usually refer to neural networks and deep learning. The same term in less cutting edge sectors includes traditional ML (as covered in this book) and new developments. However, the need for transparent models is an emerging necessity regardless of the form of ML.

that biases were implicitly learned. For example, the word "women's" and names of some all-women's colleges were more heavily penalized. While some obvious biases could be mitigated, data scientists found that word choice patterns were also indicative of gender. For example, some male candidates chose strong active verbs such as "executed", which would perpetuate the biases that Amazon sought to limit. Ultimately, the project was unable to deliver reasonable results, erroneously recommending unqualified candidates. Amazon then opted to close down this experimental initiative (Dastin 2018).

Bias is synonymous with a data science project that has gone off the rails. However, bias is a nuanced concept. It can arise from many different sources, usually originating from within the data on which a model or algorithm was developed. At the heart of it:

> *Bias is a condition in which a dataset does not reflect the target universe to which it will be applied.*

The broadness of this definition captures the essence of bias: *it is complex*. It could arise from a variety of sources, and identifying and addressing bias requires practitioners to draw from a nuanced vocabulary. In this section, we describe three types of bias, the tell-tale signs of their presence, and the effect they can have on data projects.

14.2.1 Sampling bias

Current landscape. Sampling bias occurs when the data do not reflect the share of groups within a target population. When a group is over- or under-represented, sampling bias can lead to misleading inferences and erroneous predictions. While this might seem to be a mere statistical problem, it has had very real consequences.

The 2016 U.S. presidential election is one such case in which sampling bias led to misleading inference. Prior to Election Day, national polls indicated that the likelihood of Hillary Clinton winning the presidency was between 71% and 99%. However, real estate mogul Donald Trump was the victor. *What happened?* A study conducted by the American Association for Public Opinion Research (AAPOR) found that among the main sources of inaccuracy was over-representation of college-educated voters.[3] As college-educated voters tended to favor Clinton, the estimates biased in her favor. Furthermore, the voter turnout patterns changed between 2012 and 2016 in a way that favored Trump, suggesting the sampling frame did not capture the behavior of "likely voters" (Kennedy et al. 2018).

Sampling bias has been well-known since the advent of survey research and has been the focus of decades of sampling design research. Nonetheless, biases will always appear when humans collect data. Detecting sampling bias starts with defining and constructing a *sampling frame*—a list of all elements (e.g., people, places, etc.) that comprise the population of interest. Well-designed samples are collected adhering to a stratification scheme that subdivides the population into observable qualities (e.g., income bracket, gender, race, education). This allows researchers to assess how closely the sample matches the known observable characteristics of the target population and quality control the data collection process. Typically, hypothesis tests (e.g., chi-square test, Kolmogorov-Smirnov test, etc.) are employed to assess the similarity between sample and population. When proportions are not aligned, the data may have sampling bias that can lead to inaccurate insights.

MIT Media Lab's Gender Shades project offers a striking reminder of the value of a sampling frame. The project focused on quantifying the bias in facial analysis datasets. By examining two phenotypical reference datasets for gender and skin type (IJB-A and Adience), Buolamwini and Gebru (2018) estimated that the benchmarks were skewed toward light-skinned subjects— 79.6% of IJB-A and 86.2% of Adience. These skewed distributions are not necessarily biased—it depends on whether they match the population of interest. To illustrate the effect of class imbalance, the researchers systematically tested commercially available phenotypical classification algorithms on a more balanced benchmark dataset. Their results suggest that

---

[3]The study found a number of contributing factors; However, for the sake of brevity, we focus on sampling biases in this example.

misclassification rates can be as high as 34.7% for darker-skinned females, but close to 0% for light-skinned males. This important research highlights the perils of pre-trained algorithms and reference datasets: while they are intended to be generalizable, they might not reflect the populations on which they are applied.

Adjusting the bias. While the best strategy is to collect data without sampling bias, post-collection adjustments can be applied under certain conditions. Let's consider the tamest case in which each stratum is represented in the data (e.g., each level of education attainment in a population) and the sample size within each stratum is large (e.g., $n \geq 120$). The intuition behind adjustments are simple: each record in the sample represents some number of subjects in the population. For example, a full population census with a 100% response rate implies that each record represents precisely one person. A 5% stratified random sample would ensure that 5% of each observed stratum in a target population is represented, thus each record represents $n = 20$ but holds equal weight. If the sample proportions do not precisely reflect the population, then sampling weights can be employed to adjust the sample.

A sample weight (w) is constructed from the sample size (n) and population size (N) for each stratum of interest (s):

$$w_s = \frac{N_s}{n_s}$$

Proportional representation in the sample allows for self-weighting, when all strata carry equal weight, $w_1 = w_2 = ... = w_s$. Otherwise, the sample weight can be directly applied to adjust estimates of key parameters for inferential purposes and also mitigate bias in algorithms.[4] When stratum $s = 1$ is under-represented relative to stratum $s = 2$, then we should assume each record in $s = 1$ will carry greater weight than $s = 2$ $(w_1 > w_2)$.

To illustrate how to identify and adjust sampling bias, we simulate a set of cross-tabulations intended to estimate income by education attainment. When examining Table 14.1, we can see two sources of bias. Whereas two-thirds of the population does not have a college education, the same group captured in the sample is substantially under-represented accounting for only 50%—a 17-percentage point difference. A far lesser source of bias lies in the slight deviations in the mean income. If these biases are not addressed, the mean income would be over-estimated by 12%. In contrast, an average weighted to reflect the population would reduce the difference to only 1%:

$$
\begin{aligned}
\hat{y}_{all} &= \frac{n_{\text{college}} \times w_{\text{college}} \times \hat{y}_{\text{college}}}{N} + \frac{n_{\text{no college}} \times w_{\text{no college}} \times \hat{y}_{\text{no college}}}{N} \\
&= \frac{40 \times 250 \times 84500}{10000} + \frac{40 \times 500 \times 41000}{20000} \\
&= 55500
\end{aligned}
$$

Table 14.1: Example of sampling bias in hypothetical education attainment strata.

Stratum	Pop. (N)	Actual Mean Income	Sample (n)	Initial Estimate	Weight (N/n)	Re-weighted Components
College	10000	84000	40	84500	250	28167
No college	20000	42000	40	41000	500	27333
Total	30000	56000	80	62750	375	55500

---

[4]Most statistical methods and some machine learning algorithms provide facility to incorporate weights. In cases when this is not possible, each record can be duplicated proportional (e.g., `rep` function) to their sampling weight to artificially impose the weights structure.

This example is a simple case, however. It is common for data to exhibit *coverage bias* in which entire groups are omitted from the data due to poor data collection design or respondent non-response. Furthermore, if the patterns in the data differ between respondents and non-respondents, the data can also suffer from *non-response bias*. For an in-depth treatment of sampling, read *Sampling: Design and Analysis* (Lohr 2019).

14.2.2 Measurement bias

Measurement bias occurs when the data collection process skews values in a consistent direction. The skew depends on defining what is "true" and depends on the subject matter and context. We describe two varieties of measurement bias that are discipline-specific: *instrumentation bias* and *survey-related bias*.

Instrumentation bias. Let's motivate instrumentation bias using sensors that collect data from physical phenomena. Sensors come in many forms such as radiometers, radar sensors, gyroscope sensors, among others. Instrumentation bias occurs in sensor data due to differences in implementation and imperfections in the sensor itself. While some sensors are tested and calibrated by manufacturers before they are distributed, this is not always possible and can require the data user to calibrate sensor data themselves. For example, Nighttime Light (NTL) imagery has been collected for decades by the US' Defense Meteorological Satellite Program (DMSP). Even though the satellites capture imagery of the same locations on Earth for overlapping periods of time, the brightness will differ from one instrument to the next. In order to construct a consistent NTL time series, the imagery from each successive sensor needs to be analyzed for biases then adjusted to a common benchmark (Zheng, Weng, and Wang 2019, Toohey and Strong 2007, Li, Zhou, and Zhao 2020).

In Figure 14.1, we simulate instrumentation bias by comparing two hypothetical sensors (blue and gray lines) that measure the same phenomenon. The vertical difference between the two lines are a clear indication of bias. In order to produce a consistent time series, simply substituting one series for the other will cause to a level shift, leading to poorer data quality and false inferences. Instead, the bias can be removed by training a simple linear regression:

$$y_{\text{grey}} = \beta_0 + \beta_1 y_{\text{blue}} + \varepsilon$$

Notice the regression model includes a β_0 to quantify the vertical shift and β_1 to correct for growth differences. Applying the resulting model will mitigate observed biases in the blue line, yielding in a bias-adjusted orange line:

$$\hat{y}_{\text{orange}} = \beta_0 + \beta_1 y_{\text{blue}}$$

While this calibration technique is simple, it is an effective step toward mitigating instrumentation biases when two or more coincident data series need to be combined.

Survey-related measurement bias. Unlike instrumentation bias, measurement bias is more challenging to detect. In fact, tracing its source is easily obscured by many factors. For brevity, we sub-divide survey biases into two broad problem areas: *cognitive challenges* and *rater reliability*.

Cognitive challenges. The clarity and neutrality of a question can influence the quality of responses. Leading questions, for example, can influence respondents to provide a particular answer —often dropping hints of what might be expected. For example, a political campaign asked the following question "Should President Trump and his campaign do more to hold the Fake News media accountable?" (Trump-Pence Campaign 2020). The term "Fake News" is a charged term that will elicit different responses depending on one's political leaning. Meanwhile, questions such as "Have you ever attended a Trump Rally?" are neutral and clear.

In addition to the phrasing of the question, what it asks of a respondent can bias the response. In particular, when respondents are asked to recall a past event or memory, the response might be affected by recall bias. For example, asking respondents how much time they spent in traffic in the last week will likely elicit upward biased responses. In contrast, patients who are asked by their doctor about their alcoholic beverage consumption are likely to provide downward biased responses.

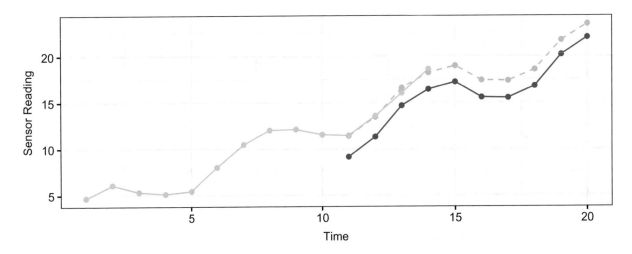

Figure 14.1: Simulated data from two sensors (blue and gray) exhibit a consistent bias.

In general, questionnaires should be put through cognitive testing before they are released into the field. Cognitive testing focuses on understanding if questions can produce accurate answers and identify inefficiencies that obscure a question's meaning (Alaimo, Olson, and Frongillo 1999). In many cases survey designers will experiment with different formulations of the same question to understand respondent cognition. But there are some simple rules of thumb that can vastly improve response quality.

Mathiowetz, Brown, and Bound (2002) argue that simple, single-focus questions are easier to follow than compound-focus questions. Simplifying questions reduces the mental overhead to produce an answer. As it is more challenging to achieve reasonable accuracy when asking respondents to recall complex life experiences (e.g., recalling income from multiple jobs), the questionnaire could be designed to break down complex concepts into a series of questions, helping the respondent to systematically recall specific experiences. Furthermore, the respondent intake process can involve baseline measurements to obtain current or near-past information to avoid long-term memory retrieval. Indeed, the art of asking questions is very much one that is inextricably tied to human psychology.

Interrater reliability. Individual interviewers might deviate from survey protocols (e.g., read question as written, avoid leading question, neutral body language) and add their own spin to the data collection process. For example, an enthusiastic interviewer may positively bias responses while an interviewer with a disapproving attitude may negatively bias responses. In other cases, raters (e.g., respondents, interviewers) may be asked to make judgment calls. For example, Street Conditions Observation Unit (SCOUT) conducts visual surveys of New York City streets to rate street cleanliness conditions—a sort of sanitation audit. One could imagine that two raters could give vastly different ratings to the same street segment. For the ratings to be useful performance management, inspector judgment must be consistent and reliable (SCOUT 2020).

Interrater Reliability (IRR) is a common method for assessing the consistency of two or more raters. It examines how often raters agree in their ratings using measures of correlation. For continuous responses on a numeric scale, IRR can be measured in terms of a correlation coefficient:

- Continuous responses can be measured using a Pearson's correlation coefficient. For example, two restaurant inspectors are asked to give 30 restaurants a numeric score for their cleanliness. A high positive correlation would indicate that their rating approach is aligned.
- Ordered discrete responses can be assessed using a Intraclass Correlation Coefficient. For example, the same two restaurant inspectors are then asked to provide letter grades (A through F).
- Non-ordered discrete responses can be compared using Cohen's Kappa statistic (Cohen 1960). For example, the same two restaurant inspectors are asked to issue grades in terms of pass and fail.

In all cases, IRR is merely a check. Perfecting the art of asking questions requires extensive planning and design around every aspect of the survey experience, from formulating questions to training raters.

14.2.3   Prejudicial bias

While post-hoc re-weighting and calibration can reduce the effect of sampling and measurement bias, these remedies might not be sufficient to detect and correct social biases. *Prejudicial bias*, in particular, occurs when cultural influences skew the portrayal of certain groups, sometimes as stereotypes and caricatures—social constructs that require substantial context and knowledge of the domain. These biases are hard to detect, but when people are faced with them—often times when interacting with a data product—they are easy to spot.

As data products become more ubiquitous, the effect of prejudicial biases becomes more real. Berlin-based Algorithm Watch, for instance, found biases in Google's computer vision services. While an image of a dark-skinned hand holding a thermometer was labeled as a "gun", the computer vision services labeled a similar image with a light-skinned hand as "electronic device" (Kayser-Bril 2020). This finding indicates that the training sample may portray dark-skinned subjects wielding weapons more often than light-skinned subject or perhaps the labels were inappropriate, thereby perpetuating unfavorable stereotypes.

Datta, Tschantz, and Datta (2018) found biases in Google's Ad Settings—a feature intended to allow users to set preferences in the sort of ads that are displayed. The researchers found that when the gender setting was set to female, the number of advertisements for high-paying jobs was less than when setting it to male. Meanwhile, Bolukbasi et al. (2016) trained NLP algorithms on Google News articles and found that the resulting word embeddings reflected deep gender stereotypes.[5] Whereas the embeddings indicated that the terms "maestro", "skipper", and "protege" are the most similar to the word *he*, occupations most associated with *she* include "homemaker", "nurse", and "receptionist".

Addressing biases already ingrained in data is quite challenging and there is no clear cut fix. Some researchers have attempted to employ crowdsourcing to identify and remove biased instances (Knight 2019) while others have developed mathematical fixes (Bolukbasi et al. 2016). Governance researchers suggest that environmental changes in the data collection process and use case design can help mitigate biases (Nicol Turner Lee and Barton 2019). Until a consensus is reached, there is no better measure to counter bias than designing use cases with well-defined target populations, inviting people with different views to vet and engage in vigorous dialogue about the project's intents, and conducting rigorous and continuous spot checks to assess data quality.

14.3   Fairness

Fairness is an idea that many can support, but the spirit of the idea alone is not sufficient to operationalize it. One of the best known fairness controversies involved the COMPAS algorithm. Produced by a company called Northpointe, COMPAS was designed to predict whether a criminal defendant will recidivate in the future. These risk scores are then used to inform decisions taken by judges in court. A study conducted by *ProPublica* of 10,000 criminal cases found disparities in the model's accuracy. Despite efforts to minimize the impact of race variables in the COMPAS model, white re-offenders were twice as likely misclassified as low risk than black re-offenders. Black defendants were also twice as likely to be misclassified as high risk of committing violent crime compared to white defendants (Larson et al. 2016). In response, Northpointe argued that recidivism risk scores are equal regardless of race, which allows judges to interpret scores without concern about racial bias (Corbett-Davies et al. 2016). The case of COMPAS raises a valid concern for responsible data products: there are divergent standards of fairness in use and a consensus on a gold standard has yet to be reached.

In data science, we can think of fairness as follows:

> *An algorithm or model is considered to be fair if its outputs are not influenced by protected variables.*

Central to this idea are protected variables such as race, ethnicity, gender, sexual orientation, or other traits. Algorithms should treat each group within a protected variable the same, but the question is what

---

[5]See Chapter 12 for a refresher on language models.

specifically should be equal? We describe two standards in this section: *fairness in predicted scores* and *fairness in predictive accuracy*.

14.3.1 Score-based fairness

Score-based fairness revolves around *statistical parity*, which evaluates whether the chance of being assigned a given predicted score is uninfluenced by a protected variable G (see Table 14.2). If a model can satisfy this condition, the same proportion of each protected group will be classified as positive (Corbett-Davies et al. 2017, Zemel et al. 2013). Training data will almost certainly contain observable differences between groups— some of which are deeply rooted in societal inequities that cannot be ignored. However, some differences are due to *legitimate factors* (L) that can help adjust scores and enable a like-for-like comparison. When adjusted scores are equal or similar, we can deem the algorithm to satisfy *conditional statistical parity*.

For example, an Urban Institute study of 60 American cities found that 38 cities have differences in median credit scores of one hundred points or more when comparing largely white and largely nonwhite areas (Ratcliffe and Brown 2017). The lives led by these different subpopulations follow divergent paths, marked by differences in economic realities and access to opportunities. But scores could be adjusted for the number of credit inquiries (l_1), income bracket (l_2), and the number of open credit card accounts (l_3).

Table 14.2: Scored-based fairness conditions expressed for two groups (A and B).

Concept	Condition
Statistical parity	$Pr(\hat{y}|G = g_A) = Pr(\hat{y}|G = g_B)$
Conditional statistical parity	$Pr(\hat{y}|G = g_A, L = l_1) = Pr(\hat{y}|G = g_B, L = l_1)$

Both statistical parity and its conditional formulation can be evaluated as an inequality. To satisfy a parity condition, the difference between the probabilities of two groups (Pr_1 and Pr_2) should be less than a defined threshold $p \in [0, 1]$:

$$Pr_1 - Pr_2 \leq p$$

Note that the threshold p is arbitrary. We can take advantage of precedent, however. Some government agencies have enacted affirmative action guidelines that can inform the value of p. The U.S. Equal Employment Opportunity Commission (EEOC), for example, recommends the *80% rule* in which a protected group should achieve at least 80% the rate of the highest-scoring group (U.S. Equal Employment Opportunity Commission 1979). In this case, we can assume that the threshold can be set to $p = 0.2$.

While statistical parity ensures fairness in predicted scores, it ignores whether a model can guarantee equal access to accuracy for all protected groups. We need to expand the inputs considered to also include observed outcomes.

14.3.2 Accuracy-based fairness

Accuracy-based fairness seeks equal access to accuracy across groups. As seen in Table 14.3, virtually any accuracy metric drawn from a confusion matrix or error function can serve as a quality of interest. *Predictive parity*, for example, seeks a consistent Positive Predictive Value (PPV) for all protected groups (Verma and Rubin 2018). The PPV is defined as the proportion of predicted positive cases that are found to be true positives:

$$PPV = \frac{\text{True Positives}}{\text{Predicted Positives}}$$

As one can imagine, a higher PPV indicates that an algorithm is more effective in anticipating positive cases and useful for decision making, especially in high stakes use cases in criminal justice, medicine, and targeting. Poorly performing models with $PPV < 50\%$ are particularly egregious as a coin flip can deliver better results. By identifying disparities in PPV, we can minimize disproportionate unfair treatment.

Table 14.3: Accuracy-based fairness conditions expressed for two groups (A and B).

Concept	Condition
Predictive parity	$Pr(Y = 1 \| \hat{y}, G = g_A) = Pr(Y = 1 \| \hat{y}, G = g_B)$
Predictive equality (FPR Balance)	$FPR_A = FPR_B, \quad TNR_A = TNR_B$
Equal opportunity (FNR Balance)	$FNR_A = FNR_B, \quad TPR_A = TPR_B$

PPV is just one type of accuracy. In fact, any accuracy metric—such as False Positive Rate (FPR) and False Negative Rate (FNR)—can allow a data scientist to hone-in on different qualities of a model, which in turn requires one to make clear value judgments on what constitutes an acceptable cost. Let's take the example of an algorithm that predicts re-offense of inmates who are candidates for parole. The algorithm will invariably make mistakes. When the parole algorithm yields a large amount of false positives (i.e., someone is predicted to re-offend but ultimately does not), it can unfairly deprive an inmate time outside of prison and opportunities to contribute to society. Ensuring equal FPRs across groups is referred to as *predictive equality* (or *FPR Balance*). A false negative (i.e., someone is predicted to be safe for release but ultimately re-offends) re-introduces a risky individual into society. Achieving balance in terms of FNR is known as *equal opportunity* (or *FNR Balance*).

14.3.3 Other considerations

A clever data scientist could attempt to achieve fairness across all three main conditions, namely statistical parity, equal opportunity and predictive equality. However, Kleinberg, Mullainathan, and Raghavan (2017) found that all three conditions cannot be achieved simultaneously. In their study of the fairness of the COMPAS algorithm, Corbett-Davies et al. (2017) indicated that:

> . . . *Optimizing for public safety yields stark racial disparities; conversely, satisfying past fairness definitions means releasing more high-risk defendants, adversely affecting public safety.*

Both studies suggest that there will always be some degree of quantitative unfairness. There is no substitute for careful deliberation. Selecting the appropriate metrics to balance is a discussion that policy makers and data scientists need to have to ensure progress through algorithmic intelligence while minimizing societal harm.

Fairness rarely is discussed when applied to policy-focused statistical analysis, yet models often form the basis of policy options presented in memoranda and reports. Thus, it could be argued that fairness tests should be applied in policy contexts. After all, policy recommendations are a simpler form of prediction. As they are optimized for clarity, inference and narrative, models may be overly parsimonious and may not achieve meaningful levels of predictive accuracy—the fairness of these models may be suspect and biased. While these shortcomings will need to be addressed in the future, achieving meaningful fairness in policy and strategy settings will likely require a ground change in the field.

14.4 Transparency and Interpretability

Before implementing a new algorithm or model in government, public officials may need to have a transparent view into its inner workings, especially given the societal risks posed by bias and unfairness. Acknowledging these growing concerns, some governments have made concerted pushes for transparency. The French government, for example, passed the Digital Republic Act (law no. 2016-1321) that requires transparency around decisions taken based on "algorithmic treatment" (Edwards and Veale 2018). Meanwhile, the City of New York passed Local Law 49 of 2017 that created a task force to provide recommendations on transparency of automated decision systems used in municipal agencies (Vacca et al. 2017).[6]

While the specific definition of transparency remains elusive, we can generally assume that it involves documentation about how the model functions and sharing of that information with a third party. Simply providing information is not sufficient to build trust, however. We can imagine that a mathematical white paper would not likely be useful for less technically inclined audiences, while a high-level description that discloses only one model accuracy metric would not provide enough detail for technical audiences. Instead, we need to consider the steps beyond ordinary transparency—achieving a level of *interpretability* and *explainability* to convey understanding.[7]

14.4.1   Interpretability

Interpretability is prized in any field that makes high stakes decisions, such as public policy, criminal justice, medicine, and epidemiology. Practitioners need to be able to articulate how exactly a model functions. Not only does it lend confidence in a model's abilities, but it provides end users a chance to limit liability. Most data science endeavors in government will start with an interpretable model. For example, FDNY's first generation FireCast model was a logistic regression. Firefighters and inspectors gave data scientists some input on what variables should be included in addition to ones that data scientists found to be correlated. After training the model, the impact of each variable could be clearly audited by data scientists by simply reviewing the model coefficients.

While the FireCast example illustrates one case of an interpretable model, Lipton (2018) notes that:

> *The term interpretability holds no agreed upon meaning, and yet machine learning conferences frequently publish papers which wield the term in a quasi-mathematical way.*

In other words, it is a quality that we can recognize when we see it. Nonetheless, a definition is important even if each research article offers a similar yet different explanation. Kim, Khanna, and Koyejo (2016) suggests that interpretability is:

> *. . . the degree to which a human can consistently predict the model's result. . .*

While Miller (2019) proposes that interpretability is:

> *. . . how well a human could understand the decisions [of an autonomous system] in the given context. . .*

These first two definitions capture a human-centered perspective; however, they also leave ample room for interpretation. In contrast, Rudin (2019) provides a definition that grounds interpretability in the qualities of a model:

> *. . . An interpretable machine learning model is constrained in model form so that it is either useful to someone, or obeys structural knowledge of the domain, such as monotonicity, causality, structural (generative) constraints, additivity or physical constraints that come from domain knowledge. Interpretable models could use case-based reasoning for complex domains.*

This more expansive definition highlights qualities that can be used to classify modeling techniques. In particular, we focus on monotonicity and additivity as they are principal determinants of classic interpretable models.[8]

Monotonicity allows for the relationship between input and target variables to always move in a consistent direction over the range of an input. For example, a regression model could indicate that, on average, income

---

[6]With a bench of academic experts leading the charge, the bureaucracy and politics of local government prevented even the most basic information to enable the task force's work—ironically, leading to a lack of transparency for the transparency investigation (Kaye 2019).

[7]The terms "Interpretability" and "Explainability" are both used in the debates centered on transparent machine learning. While they have proximal meanings and are often used interchangeably, some researchers argue that there are differences.

[8]Recent research has begun to blend non-interpretable models for causal inference, such as Athey and Wager (2019). Thus, we remove causality from this discussion.

rises monotonically from age 18 through 45. Monotonic relationships are useful for public policy problems: *as a policy treatment is applied more, the effect on a target population will generally change in one direction.*

Additivity allows each variable to make a distinct contribution to a model. By adding together the partial contributions of each variable for a specific subject (e.g., person, company), a model can both describe and predict a target variable. For example, a model finds that smoking reduces average birth weights while birth weight increases with long gestation periods. Adding together the two effects given an individual's circumstances produces an additive prediction.

When monotonicity and additivity are combined, we find that only one technique satisfies these constraints: *linear regression*—the workhorse of the social sciences. Regression coefficients isolate the partial effect of an input variable on a target variable, lending themselves to interpretation of the relationship between an input variable and an output. The coefficients of two nested regression specifications can help triangulate whether a measured effect is stable or true. We can follow each record's values as it enters the trained regression and replicate precisely how it arrived at its output at each step of the way. This quality, in turn, allows for transparent what-if analyses.

Other modeling techniques have one of these qualities but not both. Logistic regression, for example, satisfies monotonicity but the logistic transformation makes the formulation non-linear. Meanwhile, tree-based models, neural networks, and other models could preserve monotonicity if such a truly monotonic relationship is present in the data, but otherwise fail to preserve other qualities.

For the intents of public policy, regression is the *de facto* choice for an interpretable model.

14.4.2 Explainability

While an interpretable model is a public policy favorite, it may not necessarily be the best solution for all problems. These three prediction use cases illustrate the challenges:

- Imagery data are merely collections of pixels with different colors and brightness. To identify discrete objects in satellite imagery, a regression cannot identify these latent qualities. Instead, pixels should be convolved using deep learning to unlock deeply complex representations.
- A national model for predicting housing prices is trained on dozens if not hundreds of variables. The precise price for an individual house depends on its qualities and the context of its local market. Given the large number of markets in a country, building a regression that molds to each local market's conditions will be time-consuming and less effective than applying a random forest or gradient boosting algorithm at scale.
- After a disaster, an emergency authority may have limited intelligence on where damage has and has not occurred—the data might only cover small pockets in the service region. A visual survey will take time to complete, so anticipating the extent of damage requires interpolation. If regression is used, its monotonic properties will erroneously assume that damage grows with latitude and longitude. Instead, case-based techniques like kNN can be more effective for interpolating damage patterns.

While these techniques are not interpretable, they are well-suited to extract predictive gains when micro-decisions need to be made. Nonetheless, a data scientist needs to *explain* how the model arrived at its conclusions. For these *explainable models*, the rationale is made clear by building *another model to describe the predictive model.*

At the time of writing this text, the data science field has yet to converge on a gold standard for explainable models. In fact, the techniques that have been developed offer explanations that focus on different aspects of a model rather than only on one. Below, we describe a selection of explainable models.[9]

Permutation-based variable importance quantifies the importance of each input variable in terms of its contribution to a model's accuracy. Each variable's importance is cleverly estimated by comparing performance on a test sample with a modified version of the test sample in which a single variable's values are randomly shuffled. The greater the decrease in accuracy relative to the unshuffled predictions, the more important the variable. While variable importance shows contribution to accuracy, it does not indicate the direction of

---

[9]For more on these three explainable ML diagnostics, revisit Chapter 10.

influence (e.g., X increases Y, X has no effect on Y). For practitioners in policy environments, the inability to describe the direction of a relationship leaves much to be desired.

Partial dependence plots expose the shape of an input variable by calculating the marginal relationship of the input variable at discrete increments on the target. The output is a plot that exposes the curvature of the relationship, whether it is non-linear, monotonic, or other. While the plot illustrates the shape of the relationship, a non-linear pattern can be hard for policy practitioners to grasp.

Local Interpretable Model-Agnostic Explanations (LIME) provide a record-level breakdown of how each variable contributes to an additive prediction (Ribeiro, Singh, and Guestrin 2016). These model agnostic explanations are built on a regularized regression that is weighted to the area around a selected point of interest. The learned coefficients can then be interpreted as additive components that approximate the prediction. For example, a machine learning model trained on a dataset with $k = 12$ input variables is applied to score record $i = 102$ in a test sample, yielding a prediction of $\hat{y}_{i=102} = 0.9$. The LIME approximation evaluates each variable, finding that

- x_4 contributes 0.7;
- $x_1 2$ contributes 0.2;
- $x_1 3$ reduces the prediction by 0.2; and
- the remaining deviation of 0.2 is captured in a bias ($\beta = 0.2$).

This additive break down of partial effects is quite intuitive. There are shortcomings, however. The approximation can be an oversimplification, especially when the regularized regression yields a large bias (i.e., the LIME model achieves low accuracy for a point of interest). Furthermore, LIME produces explanations for individual points, thus computing the explanation for every point in a dataset may prove to be impractical.[10] Instead, consider identifying exemplars that illustrate how the underlying model functions under different conditions.

These explanations are approximations. As all models make assumptions, each explanation offers only one perspective of what happens inside a model, but certainly is not the definitive explanation.

Guiding thoughts. To succeed as a data scientist in public sector requires engaging with stakeholders and ensuring they understand the inner workings of a data science project. Without building mutual understanding, sophisticated data projects might not receive the necessary social and political support that enables success.

When embarking on new projects, we recommend starting with an interpretable model. Not only do interpretable models enable clear narratives, but that narrative serves as a platform on which trust can be built. For non-technical stakeholders, interpretable models are a wonderful introduction to the field of data science. As stakeholders become more comfortable with the subject matter, more complex techniques can be considered.

Explainable models have a role when the qualities of the data require a more complex model to extract predictive gains—when interpretable models are unable to fulfill the aims of a policy objective. The explanations offered by these models will not be as clear or as accurate as interpretable models. Nonetheless, the explanations give stakeholders the closure they need before a machine learning project can proceed.

As data scientists, we must remember that the vast majority of people are not familiar with the technical aspects of our craft. When people are not familiar or knowledgeable about a new discipline, they may come to dismiss it—it is a human reaction. Ultimately, it is the responsibility of the data scientist to build bridges. Interpretable and explainable models are merely useful tools in that pursuit.

---

[10]See Staniak and Biecek (2019) for the implementation in **R**.

14.5 Privacy

14.5.1 An evolving landscape

Data science can only enhance public policy if the Public entrusts their data to government agencies and believes that the resulting data applications derive value for society. This trust is rooted in a ubiquitous concern for *data privacy*:

> *Data privacy focuses on an individual's control over their personal information, including the collection, storage and use of that data. Successful data privacy enables data scientists to use datasets without compromising the privacy of the individuals whose information is contained in the data.*

When using data for anything, we should assume that *something* is disclosed. The question is whether that information is enough to re-identify and compromise an individual. For this reason, the privacy field has not only developed best practices and regulations to guide the use of data in a safe way, but also technical strategies that minimize the chance of disclosure.

Different privacy standards. Around the world, data protection regulations attempt to safeguard data by limiting how it can be used. In the United States, there are data privacy laws that are domain-specific, such as the Health Insurance Portability and Accountability Act (HIPAA). The purpose of HIPAA is to protect patient confidentiality, but there are gaps in the law that can be exploited. At the time of writing this text, HIPAA does not cover Direct To Consumer (DTC) genetic testing companies. Due to this gap in the law, a 2018 study conducted by Vanderbilt University researchers found that 71% of DTC genetic testing companies in the United States used patient data for projects other than testing. 78% of surveyed companies shared genetic data with third parties without consumer consent (Hazel and Slobogin 2018).

While some loopholes can be exploited, certain governments have taken steps toward cohesive privacy policies, especially to afford broad protections to citizens while providing corporations with well-defined uses of data. For example, the European Union's General Data Protection Regulation (GDPR) addresses transfer of personal information for professional and commercial use. GDPR outlines key aspects surrounding privacy such as lawful uses of data (e.g., perform tasks for public interest, "legitimate interest"), the rights of data subjects, and the responsibilities of "controllers" and "processors" (Council of European Union 2014). While broad regulations ensure coverage, they do require data controllers and processors to adopt new practices and incur financial costs (DeNisco Rayome 2019). When companies are deemed to be in breach of data privacy laws, the laws provide consumers the right to legal action. In 2019, French data protection watchdog CNIL issued a $57 million fine to Google as it failed to comply with GDPR regulations (Dillet 2019). In 2020, Oracle and Salesforce were hit with $10 billion lawsuits for GDPR breaches (Page 2020).

Interestingly, government agencies entrusted with personally identifiable data are also governed by specialized privacy laws. In particular, statistical agencies that are tasked with conducting research on sensitive data and publish tabulations and micro data must institute and uphold disclosure avoidance laws such as the US' ("Public Law 107-347: Title V - Confidential Information Protection and Statistical Efficiency (116 Stat. 2962)" 2002) and the EU's Commission of European Communities (2002). As we will describe later in this section, the science of disclosure avoidance has evolved as more data and computational technologies have become available, leading some agencies to take arguably radical steps to achieve data privacy.

14.5.2 Privacy strategies

Whereas data scientists tend to be concerned when increased data privacy reduces the signal in the data, data privacy professionals' chief concern is re-identification—when freely disseminated data is used to identify individuals. As more data is made available publicly, this risk becomes more real. For example, Rocher, Hendrickx, and de Montjoye (2019) demonstrate a technique that can re-identify 99.98% of Americans using only 15 demographic attributes. For a government agency, the only worse scenario is if re-identification occurs due to the failure of an agency's dissemination activities—unwittingly allowing confidential information to be published.

Agencies go to great lengths to avoid the disclosure of identifiable information, especially if they are required to publicly release insights and data. However, as more data is suppressed or randomized, the quality of the signal in the data will degrade. Below, we describe three strategies that help reduce the chance of disclosure; however, the risk will always remain as long as data exists.

Aggregation reduces the risk of disclosure by reporting only summary statistics. These high-level records are useful for publicly communicating trends without transferring personal identifiers. For example, Google's Community Mobility Reports aggregate data collected from location settings to infer how people's behavior has changed during the COVID-19 pandemic. Aggregated by country, state/province and county, the mobility indexes show benchmark travel behavior to offices, parks, retail stores, and other facilities relative to pre-pandemic levels (Fitzpatrick and DeSalvo 2020). In government, statistical agencies such as the US Bureau of Economic Analysis and the Bureau of Labor Statistics publish monthly aggregate statistics about the economy drawing from surveys and administrative records.

While aggregated data omits personal identifiers, re-identification still remains a risk, particularly among *small cells*—subsets of a population that have only a few observations. For example, because there are few large scale aircraft manufacturing companies, publishing a statistic about the total cost of aircraft-grade aluminum in every county of Washington State would likely disclose sensitive business information about The Boeing Company's operations in Snohomish County. Likewise, publishing a tabulation of cancer patients by demographic characteristics (e.g., gender, ethnicity, age) might identify a specific patient if a cell has a size of $n = 1$.

To mitigate the risk of disclosure, statisticians have developed *cell suppression* techniques to remove "sensitive cells" that might directly expose private information (e.g., cells with < 5 observations) as well as "non-sensitive" cells that can be combined to infer sensitive values (Cox 1980). The detection of non-sensitive cells is reliant on modeling techniques that examine the unsuppressed table, apply primary suppressions of sensitive cells, then identify which secondary suppressions are required to deduce the primary suppressions. For more on suppression techniques, consider reviewing *Statistical Disclosure Control* (Hundepool et al. 2012) and the `sdcTable` package.[11]

Pseudonymization. Data cannot be anonymized fully without stripping it of its utility. Recognizing the needs of organizations to be able to work with detailed data, pseudonymization helps to safeguard individuals by replacing sensitive identifiers with ones that are artificially generated. In practice, the identities of individuals and their mapping to pseudonymized identifiers are stored in a reference table to which only few have access, while all other data tables mask identities using pseudonymized identifiers. Thus, pseudonymization does not eliminate the risk of identification, but rather is a risk reduction measure.

Artificial identifiers take on a number of forms—each with benefits and costs:

- A *simple integer* can be assigned to each new record, but can expose when the record was generated (i.e., sequence) and encounters scalability issues as records become large (e.g., a billion records requires 10-digit identifiers).
- A *random number* generated identifier eliminates the risk of disclosing sequence information, but two random numbers could be drawn by chance (i.e., *collisions*) and accidentally lead to duplicates.
- Meanwhile, a *cryptographic hash* maps text fields into fixed length identifiers. While they preserve data quality, they can be cracked through "brute force attacks" in which an attacker cracks the code through large scale, repeated guessing (Jensen, Lauradoux, and Limniotis 2019).

In all cases, simply applying pseudonymization does not mean data are safe for public release. Variables that are not explicitly recognized as personal identifiers could also identify an individual. For example, AOL released 20 million web queries from 650,000 users. Since many users will search their own name to see what information has been published about themselves, it is then easy to tie the users' search habits to specific individuals (Arrington 2006). Pseudonymization should instead be viewed as a necessary aspect of data stewardship that safely enables data science within an agency. For a hands-on introduction to pseudonymization, review the DIY in Chapter 5.

---

[11]For more information on this R package, visit https://cran.r-project.org/web/packages/sdcTable/index.html.

Differential Privacy approaches data privacy in terms of a mathematical representation of the privacy loss as the direct result of disclosing information from a database. From examining results of queries, a database is considered to be differentially private if it is not possible to determine that a specific individual was included in the underlying data. Statistically, the distribution of each variable in the dataset would appear unchanged if an individual is removed from the data (Dwork 2011, Dwork et al. 2014).

As laid out in Dwork et al. (2006), ε-differential privacy adds well-placed random noise (ε) on top of the query results from a database. If more noise is added to a query, the results are less precise, thereby anonymizing the data and making it less useful. Conversely, less added noise increases the precision of results and sacrifices anonymity. These "noisy sums" cannot be viewed only as a one-time query to a database, but rather in terms of multiple queries. While a data scientist can receive different noisy results when repeatedly submitting the same query to a database, the results could be averaged to remove the effect of noise and retrieve the 'true' statistics. Thus, unlimited access necessarily means absolute disclosure prevention is not feasible. Instead, a *privacy budget* should be set to reflect a data owner's concern for disclosure—how much privacy loss is acceptable? This budget limits the total number of queries that can be submitted to a database before access is revoked. In turn, this budget forces explicit decisions on the trade-off between privacy for individuals versus the usefulness of the data.

In a relatively short period of time, ε-differential privacy has seen adoption by tech industry giants such as Apple (Greenberg 2017) and Google (Lardinois 2019). Perhaps the most notable push in government has been made by the U.S. Census Bureau, which collects and disseminates socioeconomic data. In 2018, the agency announced plans to apply differential privacy principles to the results of the 2020 Census. As Census data is the lifeblood of social science researchers, prominent academics have voiced concern that the protections will damage the quality of the data (Mervis 2019). Meanwhile, Santos-Lozada, Howard, and Verdery (2020) found that had differential privacy has been applied to the 2010 Census, the results would have been dramatically different for population counts of ethnic and racial minorities in smaller, less densely populated regions. As differential privacy is only in a nascent state in government, the debate about how it should be used will need to run its course before best practices emerge.

14.6 Beyond this chapter

Data ethics is indeed a sprawling field of inquiry that has yet to reach maturity. There is no doubt that challenges will be studied and reported on in the research and popular press. Until consensus is reached, we recommend following developments in the machine learning field as they are published. A modern data ethics classic, *Weapons of Math Destruction* (O'Neil 2016) is an award-winning book that asks hard questions about how algorithms and data should be used in an ethical manner.

The state of the science is becoming deeper and more specialized by the day. To dive into the technical solutions, we recommend reading articles that are presented at the Association for Computing Machinery (ACM) Fairness, Accountability, and Transparency (FAccT) conference that occurs each year.[12] In fact, much of the research cited in this chapter is sourced from ACM FAccT.

Lastly, ethical data science relies on thoughtful design—one needs to answer hard and tedious questions to truly understand the strategic intent and purpose of a project. In *Appendix A*, we have enumerated a comprehensive list of questions that can inform the design of a data science initiative. By working through each question, one can develop a Concept of Product Operations Plan that helps data science teams articulate the goals of the data project, how it will be used and sustained, outline ethical challenges, and serve as a charter.

---

[12]Previously known by the acronym FAT prior to 2020.

Chapter 15

Developing Data Products

15.1 Meeting people where they are

Over the last thirteen chapters, we have provided a technical blueprint for working and transforming data into useful information and tools. A data scientist can find patterns in data through exploratory data analysis, extract the empirical and causal narrative, and use prediction to drive prescriptive action. However, there are aspects of data science that are missing from this picture. It is rare that the data scientist sets public policy, nor is it typical that a data product directly implements the change. Whereas policy makers need insights to make set strategy and operatives in the field implement that strategy, data scientists provide the technical expertise, data intelligence, and tooling to enable users to achieve their goals. The output of a data science project is a *data product*—whether it is insights for reports, dashboards for monitoring, derivative data, prioritization lists, or some other work product. It is possible for data scientists to dream and implement a new data product, but there is one core rule on which success depends: *the data product must be aligned with a specific user's needs*.

Let's take the example of a data science initiative focused on tort cases at the end of the Bloomberg Administration in New York City. There are risks when running a government. Every so often, torts occur when accidents and incidents cause harm or loss to people. The claimants can file a tort lawsuit against the government to seek closure. In NYC, the average annual cost of tort payouts between 2008 and 2012 was over $500 million—a challenge that mayors have continued to inherit from one administration to the next. Although the NYC Law Department started 6,388 cases in FY 2012, the number of pending cases in the backlog was 16,850 (NYC Mayor's Office of Operations 2012). The City's attorneys tended to settle cases, which not only reduced the backlog but also minimized the risk of higher payouts when going to trial. In law, there are rarely magic bullets. Settlements equate to a large bill for the taxpayers, yet not all cases have the same payout potential. Although efficiency was prized when faced with a large backlog, as each settlement closed, the already sizable cost to the taxpayer incrementally ballooned.

To improve the backlog situation, the Mayor's Office turned to one of its data teams to work with attorneys to investigate the problem. The data team, acting as an internal consultancy, initially focused on predicting whether a case could be "won"—a convenient starting point from a research perspective. However, legal experts highlighted that it was the wrong question—settlements resulted in neither "wins" or "losses". The team shifted its attention to predict cases that could result in a payout and their expected cost. To do this, they assembled a treasure trove of metadata and textual data from case notes and administrative data, then processed it into a training sample. The tort sample contained rich detail, such as the type of matter, the city agency involved, the time elapsed since incident, the court, the law firm, the firm's prior litigation history, the number of plaintiffs and defendants, amount of legal research required, among other text-mined variables. Given the unprecedented view into the inner workings of the legal system, strikingly clear patterns emerged from the data. For one, the type of matter was enormously predictive of outcomes regardless of the quirks of the case. For example, officer action cases tended to have a higher chance of payout than civil rights

© Springer Nature Switzerland AG 2021
J. C. Chen et al., *Data Science for Public Policy*, Springer Series in the Data Sciences,
https://doi.org/10.1007/978-3-030-71352-2_15

cases. The behavior of claimants likewise provided clues. The amount of time between an incident and case filing was inversely correlated with payout potential—perhaps an indication of the quality of evidence. The data scientists simplified these insights by training a pair of prediction models that could predict a payout would occur with a AUC = 0.83 as well as predict the expected cost with a $R^2 > 0.5$.

Indeed, the data scientists saw a clear path to improving lawsuit outcomes. By training an algorithm, a screening program could seamlessly sort through thousands of cases, prioritize them, then provide a confidence bound on the expected settled amount to inform negotiations (Figure 15.1). An upper and lower bound cost estimate could then help lawyers navigate the negotiation process, understanding what is too high or low. Despite these promising results, the City's attorneys did not see the value in using prediction models for their day-to-day activities. There was a feeling that the models only showed the obvious—when presented with a case, an experienced lawyer can sense the chance of a payout and the amount it will be paid, thus a model was a redundant use of technology.

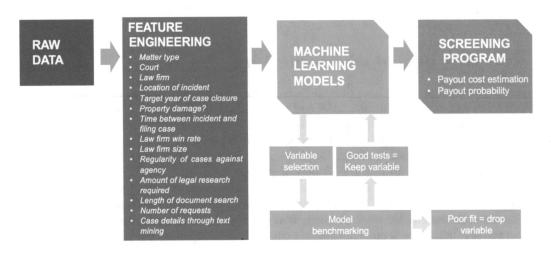

Figure 15.1: A simple flow of a data science project for predicting lawsuit outcomes.

Both perspectives are valid, however. On the one hand, prediction models can provide as much insight as expert opinion. When a data scientist is not careful to frame and communicate the value of the research, the project runs the risk of falling into the *So What Zone*—when a project is viewed as a purely academic exercise. On the other hand, policy audiences tend to overlook the fact that prediction algorithms offer other competitive advantages, namely precision and scale. Whereas human productivity is limited by time and number of available legal experts, a trained machine learning model could be effortlessly applied to a massive caseload in a matter of seconds.

In the end, the data scientists were unsuccessful in making the case for predictive tactics. The project invariably disappeared into the ether of city government.

The science rarely speaks for itself. Ambitious data science projects crumble when the focus is placed on the latest algorithmic advancements rather than first establishing trust, understanding, and buy-in with potential users. More often than not, the end users of data science have not been trained in the discipline and are unable to appreciate the wonders of these scientific advancements. Success thus hinges on a data scientist's willingness to *meet people where they are*. If the target audience operates at a strategic level, then data science should enable deeper understanding of the policy problem at hand. Even a simple memorandum can do the trick. Alternatively, if thousands of complex public safety decisions need to be made every day, then a data product be productionized algorithm that triages decisions to simplify coordination. The goal of data science in the public realm is to design a *data product* to fulfill the needs of a well-defined user.[1]

---

[1]In tech teams, data products refer to a feature of an application that make use of prediction algorithms and data. In government, data products can take on a broader range of forms as the pathways to impact are not only restricted to a piece of software.

In this final chapter, we reflect on practices that support successful data products. We start with thoughts on how to *design for impact*—a holistic view of data science as part of a policy and implementation. Whereas policies embody the values and direction of a society, the policy implementation involves identifying and establishing *decision points* at which authorities can triage resources and affect societal outcomes. Data science works best when built to support decision points—providing more information to make smarter decisions. The discipline works best when it is designed to influence actions. In the second half of the chapter, we turn our attention to the shapes that data products can take. Policy audiences tend to request dashboards and presentations, but the best data science initiatives should be appropriate for the context and the moment. We describe four data science work products and in what contexts are they best suited. The chapter closes with final thoughts on the future of data science and public policy.

15.2 Designing for impact

There are many ways to affect change in public sector, but using data has typically held a role that informs strategy rather than being used in daily operations. As we have illustrated, causal inference and prediction can be applied at scale and enable a new generation of public services. However, social and public organizations tend to be risk adverse, operating in accordance to well-defined processes. For sophisticated data science projects to bring to bear a markedly improved level of accuracy and precision, data scientists need to be masters of process, communication, and diplomacy. They need to find the *decision points* that dictate *what* should be done.

Some decision points sit within everyday problems. FireCast, for example, was designed to fit within a well-defined Standard Operating Protocol (SOP), specifically to serve as a recommendation system to guide firefighters to buildings with the highest risk of fire. Other decision points are more strategic in nature. The process of setting toll road prices, for example, may involve an econometric analysis on price elasticities and the effect of price increases on vehicle volume. When co-created with *decision makers* that set direction and *users* who will rely on the data product, data science can be a game-changing resource in innovative public service delivery.

During the Obama Administration, technologists who were recruited from Silicon Valley focused their efforts on identifying user needs. While many of these teams were engaged in web and software development than data science, the approach is also effective for data product development. Data science projects can evolve through the following four-step process as illustrated in Figure 15.2.

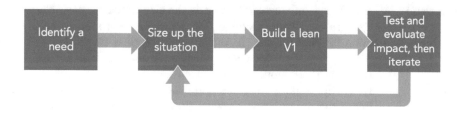

Figure 15.2: A four-step process for designing a data science project.

15.2.1 Identify a user need

All too often, policy teams describe users as "the public" or "decision makers". While convenient for policy narratives, these generic labels are not specific enough to articulate a clear user need for a data product. The mark of a true need is whether you can name a specific person and a specific pain point.

If the objective is to guide inspections in the field, data scientists must go into the field to see what inspectors see. If the objective is to streamline call center operations, then the data scientists must sit with call operators to see their workflow. Having a personal connection with the target user allows you to develop a functional understanding of the problem space: *what specifically is being decided, how would a prediction or a parameter estimate inform the decision, how often does the decision occur, what is the cost of being wrong, what are the attitudes toward using data and if there is even interest to try something new.*

While these questions can help lay the foundation of a data project, they only scratch the surface. For a more in-depth treatment, review ***Appendix A***.

15.2.2   Size up the situation

Governments technology projects have a reputation of going over budget and never ending. To maximize the chance of a success, *size up* the problem. In rank and file organizations, a senior officer will arrive on scene (e.g., a fire, a battle) and evaluate the problem and identify potential solutions, then determine how to engage. By answering a set of fundamental questions, you can rapidly assess if a proposed data science project will be worth the time investment. Some of the core considerations are laid out in Table 15.1.

Table 15.1: Questions to inform a size up exercise.

Area	Considerations
Use case	Is the intention to infer patterns, predict outcomes, or prescribe action?
Decision cadence	What is the decision time window? Can data be refreshed in time for each decision? If there is not an operational time window, is this a one off project?
Data acquisition	Where is the data coming from? Can it be reliably obtained without becoming a bottleneck?
Qualities of the data	Are there many variables? If so, is the number of variables large relative to the sample size? This informs the choice of the model. How many records are complete? Are entire records missing or are records missing in random cells? If the former, you might consider dropping records; otherwise, consider imputation.
Targets (continuous)	Are there many unique values? If not, maybe the target should be transformed into a discrete variable to simplify and clarify the signal? If there are many unique values, is there a central tendency? Is the coefficient of variation large or small? If far greater than one, you will likely require detailed data to capture extremes.
Targets (discrete)	Are classes imbalanced? What is the size of the smallest class? If less than 5%, can two or more similar cells be combined to improve the signal of the minority class? Does the output require class prediction or a probability ranking for prioritization? If the former, consider balancing the classes to facilitate prediction.
Infrastructure	How will the data be stored and processed? Is there reliable computational infrastructure to ingest, process, and store the data? Are compute resources available to efficiently analyze and make available data outputs?
Output	What should the output look like? How will the output of the product be used? Should the results be made available in a particular form, like an API or a report?
Maintenance	How many different datasets are required to create the training sample? Are there many dependencies? If a dataset is corrupted or no longer available, what is the viability of the project? What data quality measures can be constructed?

There are, of course, many other questions that should be considered. But if the results of the size up exercise are favorable and can be answered handily, then it may be worth the effort to build a "V1".

15.2.3 Build a lean "V1"

All too often, data scientists strive for perfection at the prototype stage. While technical excellence is prized, it has been known to prevent products from seeing use by the people who need it most. The "V1" (short for "Version 1") of a data product needs only to be a proof of concept (POC)—a tangible vision of what could be if resources were put behind it. Therefore, V1s should be lean products. The faster one can *show the "thing"*, the faster the project will garner support.

To keep yourself on track, consider adopting a practice from software development known as a *sprint*, which is a time-boxed period of time of two to four weeks to deliver discrete tasks toward a goal. The concept of a sprint is often used in connection with *Agile Software Development*, which is a methodology for delivering a product and refining it through iterations. In so doing, sprints prevent excessive perfectionism and allow products to evolve as more is learned about the problem space.

Start by listing all tasks and processes that must be completed in order to produce a V1, then work backwards to map out timelines. For example, building a forecasting algorithm for the economy might involve

- investigate how to ingest and process alternative data sources;
- identify which variables are most correlated with the target and develop a variable selection procedure;
- identify which algorithms are most appropriate and estimate the trade-offs of interpretability versus predictive accuracy;
- develop a procedure to train hundreds of thousands of different models in parallel;
- evaluate cost options of parallel computation;
- construct an outlier detector and data quality check;
- integrate all modeling steps with a scheduler; and
- write a research memo and presentation to communicate results.

The *backlog*, or master list of tasks, is prioritized so that dependencies logically flow from one to the other while other tasks can be scheduled in parallel. In a team setting, one team member should serve as the *scrum master*, one who can keep guide the team to prioritize the backlog, keep teammates on task, and remove obstacles. During sprint planning, each team member takes a task from the backlog, then is given an additional task once their first task is completed. Throughout the sprint, the scrum master calls for daily *standups* to have each team member report on progress and blockers so that obstacles can be mitigated as early as possible. At the end of each sprint, the team comes together for a *retrospective* (*retro* for short) to debrief on successes and discuss what can be improved.

Delivery is the principal policy of data science—making data products is the best possible way to show the superior outcomes that data can enable. Even if one is working as a team of one, being able to articulate and enumerate a backlog instills discipline. A speedy but accurately delivered product is sure to earn attention and respect.

15.2.4 Test and evaluate its impact, then iterate

Upon developing V1, present the results to close allies so that you can get open and honest feedback. These trusted individuals are the *sniff test* that help areas for improvement so that the product can be refined through additional sprints. Close allies can also help hone the quality of presentation.

Early feedback will tend to reflect flaws in communication and framing rather than the quality of the product itself. It is not uncommon for data scientists to provide a technical description of the work rather than a functional narrative. While the technical details are important, they need to be provided with some context. A data scientist working on a Random Forest model for predicting a medical problem might fixate on purely technical aspects of the project:

> "The Random Forest achieved an F1 on the OOB and Out-of-Sample of 0.8 and 0.81—which is a partial truth as the minority class comprised 1% of the sample. While the variable importance was x_1 was 100 compared with x_2 is 24. Compared with prior attempts, we were able to train the model in only one-third the time."

When the audience is unfamiliar with data science, the framing needs to be adjusted. Start with the context first, then progress through a short narrative that is structured like a scientific paper (i.e., motivation, methodology, results, then implications):

> "The incidence of medical condition y has been growing in recent months reaching z cases. Early warning diagnostics are believed to be inadequate, meanwhile the number of reported cases has grown from 1:1000 to 1:100. To improve early warning tools, we constructed a flexible algorithm on currently available blood panels to predict medical condition y. The results suggest that variable x_1 contributes the relative most to predicting this—roughly 4.2-times more than the next greatest contributor. Model performance is nuanced, however. If we use this model for prediction, we know that 90% of all positive cases can be predicted, but only 70% of predicted cases are in fact positive. If deployed for diagnosis, the algorithm will rely on data from a blood panel and provide probabilistic estimates of the presence of the condition."

As you prepare to present to each new audience, spend a few moments to consider how they may react. Think about what they are likely to think and feel and how do you speak so that you can get the point across without compromising the technical details. Whether the point landed will be clear from the reactions. Audience members who did not understand the data science project is will comment on cosmetic aspects of the project (e.g., color of graphs) rather than the inferences (e.g., X caused Y). Whereas, expert audience members will ask questions relating to project design, sampling, and math in general.

Not all feedback will be valid. Interpret it as you may, but be sure to identify the most compelling arguments to focus on and drive improvements. Iterate the product direction, then run another sprint to improve.

15.3 Communicating data science projects

During the early stages of implementing analytics at FDNY, long time analysts at the Department told the newly established data science team to "Meet people where they are". While the team of millennials were prepared to build futuristic data products, they heeded the advice of their peers. In practical terms, meeting people where they are means to provide data insights in a form that people are ready to consume as not everyone is a data scientist. In this section, we focus on how to structure and frame *presentations* and *written reports* for public policy and government audiences—an art form that is not typically part of a data scientist's training. These two formats are crucial for communicating ideas in the public sector and garnering support.

15.3.1 Presentations

In all phases of a data science project, presentations are a typical medium for communicating to audiences. Presentations are particularly well-suited to persuade audiences of an argument or thesis, weaving the personality of the speaker with the visuals and words. Whether you would like to launch a project, provide an update, or brief a target audience, presentations are an optimal format to pitch and sell an idea.

Structure. For most people, a perception exists that presentations are "easy to do"—quite the contrary. Regardless if the audience is familiar with your endeavor, the presentation should always begin with a few moments to contextualize why people have gathered and describe the central theme that should underlie the entire presentation. The audience must be able to easily place the information in a mental scaffolding and it is your job to create and place every talking point in that easily accessible scaffolding. One way to create that scaffolding is to present a *road map* or *outline* of what will be discussed, then return to that outline throughout the presentation to help the audience regroup.

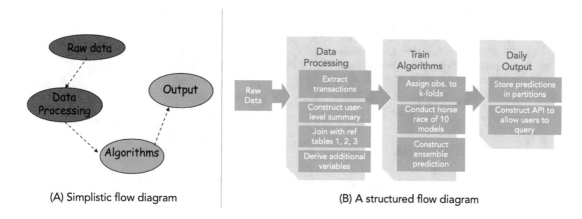

(A) Simplistic flow diagram   (B) A structured flow diagram

Figure 15.3: Two distinct visuals in a deck.

Design. Speakers often present slides packed with bullet points, which then cue the audience to read snippets of text rather than listen to the narrative. Many times presenters will simply read the slide, negating the opportunity to use one's voice to humanize the argument. Sloppy visuals can be designed without much thought about how to bring to bear the power of esthetics in communicating concepts.

The choice of visual elements convey meanings and signal certain soft qualities. For example, Figure 15.3 presents two flow diagrams of how data flows through real estate price prediction project. The first diagram (left) over-simplifies the data science project, failing to convey useful information. In addition, poor esthetic design choices signal the work of an amateur. The second flow diagram (right) is furnished with some technical depth to illustrate the mechanics of the project. Furthermore, the color palette makes clear distinctions between elements and steps of the process without distracting from the meaning. Together, these qualities signal a command over the subject area. The payout of spending time of the visuals can be quite substantial.

Even the way statistics are presented has an impact on how audiences treat the information. Regressions, for example, will no doubt be used for causal inference and correlative analyses. However, only certain users will be able to navigate and interpret a regression table. The table in Figure 15.4, for example, presents the results of a housing price model in a format that is appropriate for academic research audiences, providing point estimates and uncertainty around the estimates. By presenting the full table, the speaker gives the audience a chance to think for themselves and should likewise give more time in the presentation to reflect on the results.

Policy audiences, in contrast, are likely more interested in crisp insights, cropping in on specific elements of the regression table—keep it simple and tell the story. At the same time, it is important to show just enough technical firepower to signal that the work was well-designed and implemented. The bar chart in Figure 15.4 plots the price of housing relative to a reference level, color coding each neighborhood's estimate by its statistical significance to show nuance of interpretation. By making use of negative space, the choice of this particular graph is effective to emphasize the clear price spreads between neighborhoods, so much so that the most expensive of the neighborhoods in the analysis are over 100% more expensive, on average, than the lowest.

For a worked example, refer to an example presentation available at https://www.github.com/DataScienceForPublicPolicy/demo-presentation.

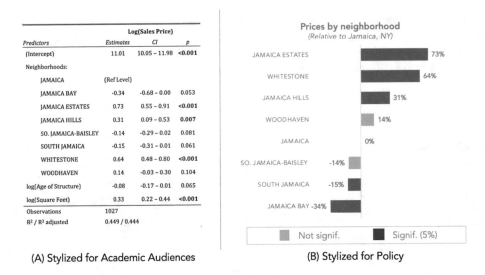

Figure 15.4: Different ways of presenting regression results.

15.3.2   Written reports

While slide presentations have become part of the culture of most organizations, the ultimate medium for facilitating deep understanding is a well-written report. A memorandum (or memo for short) is a natural format to describe the problem in a technical and accurate way while having enough space to elaborate for audiences that are less familiar with the subject matter. Writing a memo is a great exercise for formalizing and articulating an overarching strategy, forcing data scientists to structure an otherwise exploratory process in terms of a real-world application. It is, however, not an opportunity for one to bask in their intellectual glory but rather should be focused on the audience. When done well, memos can inform policy makers of a more technically valid way forward. And in strategy environments, memos are sometimes viewed as political currency, serving as a reference point for building arguments.

A well-written memo provides unambiguous information to guide discussions. The statistician and visualization expert Edward Tufte begins his workshops with a "study hall" involving reading material that ensures participants have time to absorb the same information. Similarly, Amazon founder and CEO Jeff Bezos has a preference for memorandum rather than presentations (Umoh 2018):

> The reason writing a 'good' four page memo is harder than 'writing' a 20-page PowerPoint is because the narrative structure of a good memo forces better thought and better understanding of what's more important than what.

In any case, written reports are a valuable tool that can reduce bureaucratic friction by virtue of demystifying data science.

Structure. The structure of a memo is informed by its purpose. However, in all cases, the introduction should clearly state the purpose and background to set readers' expectations. Even when provided to a routine cast of characters, memos should start with two to three lines to quickly jog their memories. Without the purpose and context, readers may quickly get lost in the details that follow. While it may seem redundant, an empathetic data scientist will recognize that there are competing priorities. Small recaps can go a long way to lubricate the policy process.

There is often a perception that writing for policy audiences means stripping the content of all heavily technical concepts. Quite the contrary. As data scientists are still fairly new in policy, explaining technical concepts in a *relatable* way can help build coalitions around your work. Use analogies, imagined scenarios, among other literary devices to concretely illustrate the point and why it matters to the reader. As more people truly understand how to make use of your efforts, the support for data science will invariably grow.

Thus, the remaining two to four pages of the memo should furnish the reader with a working knowledge of the problem or decision at hand.

Briefing memos are designed to inform stakeholders of a topic area or progress of a project—a sort of crash course in four pages or less. The body paragraphs explore different aspects of the problem space. Like a standard grade school essay format, each paragraph starts with a clear topic sentence and argument, supported by evidence woven into narrative to accentuate the argument's merits. For example, a data scientist may be interested in implementing a nowcasting project for public opinion polls and would like to draft a briefing memo to seed the idea. The memo could follow this structure:

- *Introduction*: What is nowcasting and why is it valuable? How do nowcasts provide preemptive intelligence on otherwise lagging metrics?
- *Body Paragraph 1*: What can a successful nowcast accomplish for our organization?
- *Body Paragraph 2*: What is prediction and what is entailed? What can we expect from an initial version of a nowcast?
- *Body Paragraph 3*: Why is there a trade-off between interpretability and accuracy?
- *Conclusion*: What are some ways we can move forward with nowcasting?

An *options memo* presents paths that decision makers can take to address a problem, weighing the pros and cons of each option. In a data science context, the options are different ways that a project can evolve. For each option, the author should provide an overview of the potential solution, how likely the option is to succeed, a high-level sketch of how to action upon the option, and an enumeration of notable dependencies and contingencies. If appropriate, the data scientist should recommend a course of action. Following the previous example, imagine that the data scientist is now ready to propose three options to decision makers:

- *Introduction*: Provide a recap of the nowcasting concept and highlight three distinct options.
- *Body Paragraph 1*: Describe Option #1—real-time nowcasts are conceptually attractive, but may be cost prohibitive to maintain.
- *Body Paragraph 2*: Describe Option #2—less frequent monthly nowcasts equate to higher accuracy.
- *Body Paragraph 3*: Describe Option #3—doing nothing equates to no additional cost but no improvement.
- *Conclusion*: Recap the options, and if appropriate, make a recommendation.

In contrast, a *research memo* has a more scientific focus, laying out the aims, approach, results, and implications of a data science project. For quantitative researchers who have worked in policy circles, a research memo is usually the end product of a data project, presenting a descriptive or causal analysis of a policy phenomenon. Sometimes, the memo can be modified to also serve as a peer-reviewed article if cleared by the chain of command. For data scientists, however, a research memo presents an opportunity to socialize new data science applications, especially to show the return on investment and potential impact of V1 applications. Building upon the nowcasting example, our industrious data scientist would like to scale her nowcasting project beyond V1. Her first step is to draft a research memo:

- *Introduction*: What was the aim of constructing a real-time nowcast? What made it different from other projects? What was the main finding?
- *Body Paragraph 1*: Data & Methodology—What approach was taken and why? What considerations factored into the approach? How accurate and stable are the data to support the use case?
- *Body Paragraph 2*: Results—How successful was the project? Did the performance meet expectations?
- *Body Paragraph 3*: Discussion—How can the project be applied? What are the weaknesses and strengths over alternatives?
- *Conclusion*: Recap what was learned, areas for improvement, and next steps.

Style and Design. Judging the style and design of a memo tends to be a subjective affair. However, consider this: *written documents that have authority tend to be well-formatted and well-worded*. For one, optimize for readability and avoid packing the document with too much information. The memo header should be clearly addressed to stakeholders while the subject line should be straight to the point, providing a clear idea of what the document is about. Use space generously (e.g., one-inch margins, 1.5 spacing) and choose appropriate font sizes (e.g., 11-point to 12-point font for body text). The font type carries an implicit meaning. Serif fonts carry more austerity and authority while sans-serif fonts carry a more modern connotation—choosing the appropriate font sets the tone.

Graphs and tables can illustrate and support the memo's principal arguments, but they can also distract from the main point or reduce the available space if their value is not clear. Thus, consider incorporating visuals when the point of a paragraph can be markedly bolstered by its inclusion. The color choices should be carefully selected so as not to become the focal point of the document. While colorful graphics can make your document more attractive, the main takeaway of your memo should not be your color choices but rather the content of the text. Use color wisely.

Word choice and tone also contribute to how the document will be perceived. This text, for example, frequently uses "we" and "you" as a means to make the content more approachable and conversational. Memos meant to serve as a reference point should be written with an impersonal tone (e.g., avoid "we", "you") and avoid contractions that are common in everyday vernacular (e.g., "don't", "we'll", etc.). The flow of the writing has an impact on the reader's ability to absorb knowledge. If possible, reduce the use of parenthetical and footnotes that detract from the flow.

Lastly, the perception of competence can easily erode with repeated grammatical and spelling errors. Proof-read, edit, and spell check the document. The best test of the polish of your work is if a reader who is unfamiliar with the subject matter can absorb the content while finding minimal stylistic and technical errors.

How can R help. When many statistics and graphics are needed to craft the argument, calculating, copying, and pasting each piece of information is time-consuming and tedious, especially when the dataset is updated or if assumptions are changed. `RMarkdown` is built to integrate code and text so that an entire analysis can be refreshed and update a smartly designed document. In fact, this entire textbook was written using an extension of `RMarkdown` known as `bookdown`. For a worked example, refer to an example written report available at https://www.github.com/DataScienceForPublicPolicy/written-report.

15.4   Reporting dashboards

Executives and managers often ask for a dashboard as the end product of a data science project. The visualization format has an esthetic allure that signals progress—*something* has been done. However, dashboards are seldom used more than once or twice making it an expensive investment considering its development and maintenance costs. This failure is rooted in a misunderstanding of the fundamental value of a dashboard: *they measure context around a specific decision point*. Every metric in the dashboard should be well-researched and communicated to the user so all are aware of its exact meaning, which in turn provides unambiguous input into a specific action. This requires thorough investigations as to how the information will be used.

Dashboards are not only a tactical tool, but are also a technical feat of data engineering. The data needs to be wrangled into an analysis ready form, requiring ingestion, cleaning, processing, and aggregation before being piped into the dashboard. If intended for daily use, the data engineering process must be well-tested with exceptional error-handling. The ROI of a dashboard is realized over time. As they are expensive to maintain, the cost can be amortized through habitual use, serving as an evolved tool for implementing policy.

Structure. To get the most from a dashboard, let's consider the case of a *real* dashboard in a car. A car's dashboard presents the speed, the fuel, engine status, among other data that help the driver make decisions on the road. Without the dashboard, the driver can only make gut decisions, although some may argue that drivers can be cavalier with their driving decisions even *with* a dashboard. Nonetheless, a metrics-driven dashboard can be used to monitor an organization's performance and inform what actions should be taken at specific decision points. Furthermore, a dashboard can be host to both predictive and causal inferences. The only rule is to ensure that the information is relevant to the decision point.

Dashboards should focus on a single decision point, providing just enough information to contextualize the state of affairs while minimizing information overload. Each visual element should add to the user's understanding. The monitoring dashboard in Figure 15.5 provides the real-time status of water levels along a stream in New York City. It is comprised of a times series plot that compares the actual level (blue) against an alert threshold (red), rudimentary summary statistics of how likely a flooding event will occur, and automatically generated commentary to help the user interpret the data and suggest courses of action.

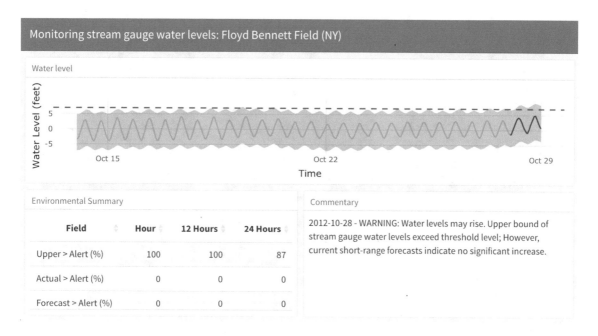

Figure 15.5: Stream gage monitoring when warning logic is triggered.

The dashboard, however, sits at the tail end of a data science project. Data scientists must write the underlying code to conduct and automate an entire analysis that can be executed, without error. In many cases, fresh streams of data can be downloaded through an Application Programming Interface (API) that can provide querying access in a database. As illustrated in the process flow in Figure 15.6, the machinery of this particular dashboard is written in RMarkdown, weaving together a number of processing steps:

- ingest data from a real-time API, then parse and clean it;
- lay out a dashboard;
- perform a time series decomposition analysis to extract probabilistic bounds;
- forecast water levels with a 24-hour horizon;
- compute implications relative to alert level with pre-determined recommended actions; then
- render two JavaScript-based packages to render interactive visualizations.

When executed in RStudio, the code runs through the analytical script, then renders a HTML file that visualizes the results. Note, however, that this dashboard does not dynamically update unless the underlying code is re-executed.

To update the dashboard on-the-fly requires the project to factor in *hosting costs*—expenses associated with continuously operating a server that allows the dashboard to be available at any time. While R is not a language designed for general purpose web applications, the R shiny framework allows R to serve as the engine behind a visual analytics dashboard. While shiny is built for data scientists, hosting a shiny application tends to be slow—interpreted statistical languages like R have their performance limitations. Commercial software such as Tableau and Qlik, in contrast, are built on top of servers that are optimized for speedy computation. However, commercial software also come with subscription and operation costs.

Design. It is easy to construct a dashboard with graphs. It is hard to design one that responds to new conditions in the data. Mastering the element of timing is key.

While the commentary box is the least visually interesting element, a data scientist can provide a *nudge* when the data changes. As the data approaches or reaches pre-defined alert thresholds, the dashboard will use if-else logic to construct a commentary and instructions that are appropriate for the moment. In Figure

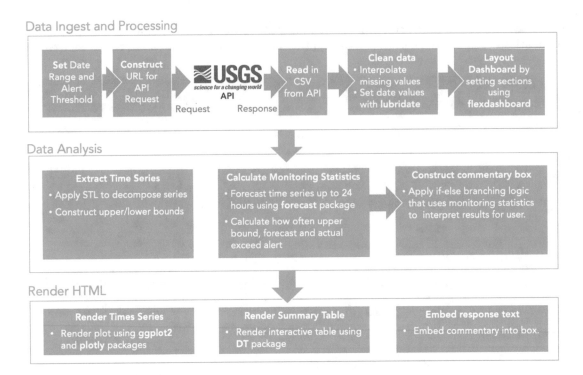

Figure 15.6: Underlying data processing flow that generates the RMarkdown dashboard.

15.7, the commentary has been extracted from three days with different water conditions. Before flooding, the dashboard uses the upper bound of water level confidence interval to warn of potential flooding. Just before or as the data and forecasts suggest that flood waters will rise, the dashboard moves more direct instructions. The commentary is short and to the point when water levels are normal.

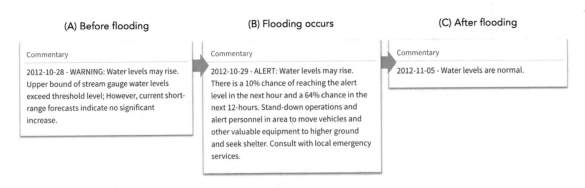

Figure 15.7: In addition to the plots and statistics, dynamic commentary can help contextualize what actions the user should take given the patterns being presented.

How R can help. With recent advancements in R, an assortment of packages can accommodate most dashboarding use cases. Using RMarkdown along with the flexdashboard package as the base, a dashboard can incorporate visual elements that can produce a dashboard, it only updates when the flooding dashboard is re-run in RStudio. For the worked example for the flooding dashboard, refer to the project repository at https://www.github.com/DataScienceForPublicPolicy/flooding-dashboard.

15.5 Prediction products

Most data science projects in government produce strategic insights intended for presentations and memos—their principal role is to inform strategy, direction, and decisions. But every so often, we see the possibility of a *tactical data product*, one in which a prediction model can directly influence the actions that people take in the field.

Because public sector agencies can operate reactively, it is only after an incident (e.g., fire) or event (e.g., election) that action is taken to address a problem. However, if a predictive model is accurate enough, it can predict the yet-to-be observed statuses for individual cases. The predictions can remove the cloud of uncertainty and enable others to take action, usually provided in the form of *lists* and *scoring engines*.

15.5.1 Prioritization and targeting lists

Much like in the case of FireCast and election microtargeting, a prediction model applied to a universe of interest yields ascore for each record. These scores can be sorted to create a prioritization or targeting list to drive a policy intervention—both powerful and confrontational. Through lists, agencies can affect change with a degree of precision (e.g., inspect a building given a high chance of a fire, provide counseling before recidivism, etc.).

The question then is how to effectively make use of these scores? It ultimately depends on the policies that govern the problem space. In some cases, all items on a list need to be addressed at some point, but some need more urgent attention than others. These cases are best addressed by *prioritization list* in which items are rank sorted by the outcome of interest. In the case of FireCast from Chapter 8, for example, individual buildings were sorted based on the highest chance of fire, then firefighters can focus their attention on the highest risk buildings.

Alternatively, a *hit list* limits the universe to only records that are most likely to result in the outcome of interest, which improves the effectiveness of targeted interventions. Political campaigns, for example, have limited budget but need to market candidates to the electorate. Each canvasser's effort costs time and money. Thus, focusing on "convertible" members of the electorate may yield higher hit rates than other population segments. While these use cases are powerful and provide actionable intelligence, there are ethical implications of targeting—particularly around fairness. Hit lists are an explicit statement of policy priorities and the decision to deploy these lists should be carefully considered.

Simply building a hit list does not guarantee its use. Data scientists need to build trust in these lists by providing the intuition behind the results. Lists also imply that someone is giving someone else an order to carry out a task, which can be perceived as a loss of autonomy. One approach to build trust is to help end users explore recommendations to understand why they are the obvious choice. Consider developing an exploratory visualization to facilitate exploration of hit list results. For example, Figure 15.8 shows a snapshot from an interactive dashboard using the prioritizations from the downed trees example from Chapter 10. The interactive visualization uses the `crosstalk` package to integrate a slider bar, which filters the list based on the prioritization score. The filtered results are then rendered on a map using the `leaflet` package while individual records are shown in a table built on the `DT` package. The dashboard's functionality could be expanded to incorporate more filters to help users drill down and learn more about how the prioritization works. To get started with exploratory dashboards, visit the following link for this worked example: https://www.github.com/DataScienceForPublicPolicy/interactive-list.

15.5.2 Scoring engines

Some use cases need predictions to be available on-demand, which requires models to be pre-trained and be available to score new data at a moment's notice. Scoring engines have common place for extracting new variables from unstructured data (e.g., keywords from photographs, sentiment from text) and have been known to apply computer vision for augmented reality applications. A pre-trained model alone is not enough to make a scoring engine. It needs to be integrated with software that make the algorithm usable for a well-defined application. Like Iron Man, the scoring engine may know what to do but needs the machinery

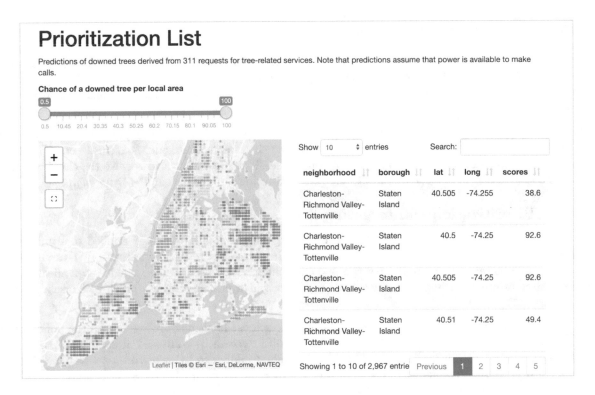

Figure 15.8: Prioritization visual tool.

around it to make it functional. We can illustrate the complexities of a software application in a pair of examples in Figure 15.9.

When a consumer submits a credit card application, for example, the scoring engine sits in the middle of the process. The information in the application is extracted and checked if it has valid information, then pushed through the same transformations and processing that was applied to the data for training. Sometimes additional variables were engineered from the data, which then need to be derived from the application. Only once we transform the applicant's data into the format that is expected by the scoring model do we push new data into the pre-trained model. For transparency, the pre-trained model is likely a logistic regression, which is quick to score results. However, some more complicated algorithms require more processing power to score new data. The pre-trained model returns a single credit score that feeds into the financial institution's decision logic, which in turn returns decision to the applicant.

Computer vision services simplify access to neural network algorithms that have been pre-trained to examine the patterns in photographs and return predicted tags and keywords as well as other "embedded" information. These have become commonly available data science services made available by technology companies such as Google, Microsoft, and Amazon. Computer vision services are typically a good tool for enriching a dataset, by using another algorithm to predict information that would otherwise require an army of human analysts to collect. A data scientist could loop through a large number of photographs or video, sending each to a computer vision API. The API pushes the photograph into a data pipeline to process the image (e.g., resizing, color adjustments, convert to matrix), then passes the processed data to a pre-trained neural network. The model returns a list of tags that are associated with the photograph's contents. The vector of tags is passed to processing once again to package the data into a standardized form, then sent back to the data scientist's software for further use.

In both cases, the first step in making available the scoring engine is to construct an Application Programming Interface (API). APIs simplify communication between two software applications by standardizing requests. Specifically, it standardizes the communication with a server-side application and the application likewise standardizes the output that results from the request. Computer vision, credit card applications, and even

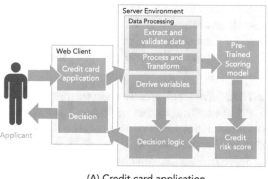 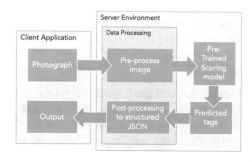

(A) Credit card application   (B) Computer vision

Figure 15.9: Scoring engines sit within more complex data applications that require software development expertise.

visualization services like `plotly` rely on APIs to simplify the data science process. In fact, companies like Amazon have mandated that all within company process be made possible through APIs to standardize and control information flows (Mason 2017).

`R` and software development are not commonly seen in the same sentence. However, there have been advancements to make APIs a possible output of a data science project. To get started with a simple worked example of an API at the Github repository: https://www.github.com/DataScienceForPublicPolicy/interactive-list.

15.6 Continuing to hone your craft

The types of data products in use are so wide-ranging that your skill set is not likely to cover the full gamut. For instance, a data scientist who focuses on communicating complex insights for strategic business problems might not have the expertise to develop large scale productionized prediction systems. Likewise, a data scientist who has more engineering skills may miss the nuances in policy. There invariably will come a time when a stakeholder asks for a data product that you are not yet ready to deliver. This is why many technical leaders treat data science as a team sport—everyone can pitch in. While teams comprised of complementary skill sets can build better data products, team members need to also have overlapping skills. There is strength in redundancy—more people who are expert in a subject area can lead to more robust vetting of ideas for even stronger products. While data scientists can be specialists, they should be familiar with all aspects of data science and software engineering pipelines.

Indeed, the most employable data scientists can switch between two or more coding languages and frame problems from policy, statistical, and engineering perspectives. In this text, we have emphasized pertinent aspects of data science for public policy—but to work across domains requires more study. Below, we highlight a few skill areas that should factor into data science.

Dataset Design. While the age of big data has been filled with great potential, the representativeness of data is something that cannot be handled by software, but rather a command of basic probability and statistical theory. In fact, the key issue that is often overlooked in data science projects is whether the *sample design*—or how the data is collected—is appropriate. Policy makers may at times suggest that more frequent, larger samples from self-selected sources are better than smaller, randomly sampled data sources. Whether one data source is better than another is dependent on the application. When using any sample, one assumes that it is a suitable approximation of the target population.

For example, a model designed to help Medicare recipients should not be built on a sample of college students. Understanding the basics of *set theory*—or the study of collections of objects—can come in handy when planning your data. The generalizability of the data, on the other hand, depends on how the sample is collected. Data collected through *random sampling*, or sourced through random selection from a universe of all Medicare recipients, would arguably be more generalizable than a *convenience sample* collected from a regional health care provider. For further reading, consider *Sampling: Design and Analysis* (Lohr 2019) as well as *Experimental and Quasi-Experimental Designs for Generalized Causal Inference* (Shadish, Cook, and Campbell 2002).

Statistical Analysis and Modeling. `Stata`, `SPSS`, and `SAS` have traditionally been the empiricist's weapon of choice in public policy. While these software enable estimation and causal inference, they are best suited for academic research and traditional social science workflows. Knowledge of an analytical language such as `R` and `Python` can not only cover analytics, but the rest of the data science pipeline. `R` is the language of choice of statistically minded data scientists focused on inference and smaller scale prediction—appropriate for strategy problems and moderately paced prediction projects (e.g., decision time frame is a day rather than real time). `Python`, in contrast, makes large scale data product possible—appropriate for any problem. The choice of language tends to be rooted in one's technical upbringing, but having some command of both languages is valuable in the modern workplace. For further reading, consider *Python for Data Analysis* (McKinney 2017) to expand to one additional data science language.

Data Processing. While much of data processing could be handled with `Python` or `R`, there are clear challenges as data becomes larger and more complex. An 100 terabyte dataset is not easily extracted and transformed on a laptop with 500 gigabytes of storage and eight gigabytes of memory. To handle large datasets requires heavier infrastructure, such as *cloud computing and storage* on Amazon Web Services or Microsoft Azure. With larger infrastructure comes larger costs that tend to be billed based on the type of cloud service (e.g., server build, storage size) and how long it is used. To use the infrastructure, we need computational software that coordinates multiple servers to simultaneously work on processing and analytical tasks. `Apache Hadoop` is a set of software utilities that enable distributing processing. Rather than having one server to store and process a large dataset, Hadoop distributes the data storage over multiple servers (or nodes), then executes a processing command for a batch of data *simultaneously* across servers. Distributed computing reduces processing times by many fold, approximately proportional to the number of nodes. `Apache Spark` is another framework that enables *real-time* data processing over clusters of servers, but works with data in-memory. The two frameworks are popular and address different problems, some requiring immediate attention while others can wait a little longer for processing. The languages typically used to work with both `Hadoop` and `Spark` include `Python`, `Java`, `R`, and `Scala`. For further reading, consider *Spark: The Definitive Guide* (Chambers and Zaharia 2018) to scale your work in production environments.

Querying. Data are often stored in databases that keep information in a standardized, manageable form. To extract subsets of relevant data, knowledge of a querying language is a must. `SQL` (Structured Query Language) is the go-to language for tabular databases, allowing data scientists to subset, join, aggregate, and process data. In addition, SQL can be used to manage databases. However, the data may be too large to efficiently query in a typical relational database. For large amounts of data, a data warehouse is a more efficient solution. When processing data with `Apache Hadoop`, for example, `Apache Hive` is a software that handles large-scale data warehousing, which also relies on a form of SQL to query data.

Visual Communication. Visualizations should help people develop an intuitive understanding of the problem space and data science project. However, some visualization forms of more appropriate from some audiences than others. Likewise, the technologies are different as well.

- *Notebooks.* Code notebooks are an interactive format to weave code, visuals, and text. In `R`, `RMarkdown` makes it possible to create documents, presentations, and dashboards. For more technical audiences, `Jupyter` Data scientists can share Jupyter notebooks with one another in `R` or `Python`, illustrating their research and coding processes in an interactive, replicable environment.
- *Dashboards.* Similar to the `flexdashboard` example, dashboards are a good choice for supporting operational decisions. For more scalable solutions, dashboarding software like `Tableau` and `Qlik` are more easily maintained.
- *Custom visualizations* are useful for telling stories in new and exciting ways that are not otherwise possible through standard dashboards. Newspapers such as The New York Times and The Washington Post rely on interactive visuals to tell their stories. Each interactives visualization is built with a trio

of web development languages, namely Hyper Text Markup Language (HTML) to construct a web page, Cascading Style Sheets (CSS) to apply styles to the page, and JavaScript (JS) to enable interactivity. While HTML and CSS are fairly easy to learn, JavaScript along with visualization libraries like D3.js are arguably more challenging to master, but are well-worth the effort.

Verbal Communication is critical for the success of any data science pursuit. Faster you become comfortable speaking and pitching ideas to large audiences, the more support you will foster for your pursuits. There is no better way to learn to communicate complex data science concepts than to force yourself to speak at meetings and events. But more importantly, learn to modulate your tone and word choice depending on the audience. To get into the habit of speaking as a technical leader, aim to speak to one technical audience and general audience each year. As one can imagine, the range of technical audiences can be quite wide. Academic audience will likely fixate on the intellectual contributions and if the work fits in the traditions of the field. In contrast, a forum of practitioners might be interested in the impact of your work and the architecture of your approach. In any case, great speakers know their audiences and their material like a plane on auto-pilot—rehearse and practice with trusted colleagues who represent those distinct audiences.

15.7 Where to next?

Data science is increasingly absorbed by organizations working in various parts of society. But as an organization becomes more mature with its use of data science and as teams grow larger, it is also likely that one's latitude to work on all aspects of data will be far diminished. As a result, a data scientist in an already data-driven company tends to be a specialist. In contrast, public sector agencies tend to have smaller budgets and smaller data science teams. The lean budgets force data scientists to be more creative and clever with their craft. In addition, the societal problem space is far larger than most private sector companies. This combination makes public sector a target-rich environment, affording an ambitious data scientist the opportunity to be involved in virtually all aspects of the data science pipeline. In short the public sector is a great training ground to become a versatile data scientist.

The additional freedom comes with the need for a keen awareness of society's attitudes toward data science. Most people, regardless of industry, are not familiar with what goes into data science. Public discourse around data science and machine learning has taken a more defensive stance, focusing on how artificial intelligence will lead to job loss and a fourth industrial revolution. Artificial *general* intelligence, or algorithms that can think generally like humans, is still science fiction for the foreseeable future. Instead, what is possible at present is the use of data science to help bring increased accuracy and precision to policy decisions—jobs are not lost, but rather the effectiveness of people is improved. Until people become accustomed to this new paradigm, data science is viewed as a threat.

In public sector, working the data science way is still quite novel and relatively abstract. The extent to which data science plays a role in government and social sector is dependent on communicating an inclusive vision of how data science fits into public service. Data science is not the solution to policy problems, but can be a game changer in how governments deliver services. But public servants and constituents need to see its value, and driving adoption is a matter of *meeting people where they are*. Every data scientist has the responsibility of convincing others that their proposed course of action is the right, most equitable, most accurate and most natural. It is only through building trust with your stakeholders will you be able to affect change.

Chapter 16

Building Data Teams

Like private sector, data science has swept government by storm. Teams have been recruited under a simple assumption: *good things can happen if you give smart, eager people some data.* Indeed, the earliest data scientists in public sector had the opportunity of a lifetime to explore untapped datasets and affect change. But the technical competence alone has not enough to affect change—much of government is not currently set up for data science. The technical inputs for data science (e.g., data, software, and infrastructure) were often not mature enough to accommodate data science projects. For example, data can be trapped in hard-to-use PDF files. In other cases, a dataset can be scattered across a disparate assortment of Microsoft Access databases. And in the worst case, data simply does not exist. Without these fundamental ingredients, agencies are forced to start from square one and engineer their own data and systems. Even when analyses and products are successfully produced, stakeholders might not be able to appreciate quantitative insights or support targeted action.

While data scientists can quickly build prototypes, the goal of innovation is to deliver a *sustained change.* Thus, to succeed in data science in government requires more than just technical skill. One needs to have empathy for those who have successfully operated public service for decades without the use of machine learning. It also requires the patience to first understand an agency's culture and processes, then adapt data science strategies to that environment. Data scientists need to hone their communication skills and build trusted relationships with internal partners. By mastering soft skills, data scientists can help agencies evolve and appreciate what is possible through data science.

This chapter is aimed at two audiences within public sector agencies: *managers* and *data professionals.* We outline structural issues that managers should consider when building their first data science team. We also provide aspiring data scientists a glimpse of what lies ahead. To start, we describe how to establish an operating baseline—a mapping of key factors that determine the viability of a data science team. We then describe a selection of operating models that are suited for different policy objectives and set up the relationship that data scientists will have with the rest of the organization. We end the chapter with a review of roles in data science teams and how to structure the hiring process.

16.1 Establishing a baseline

While Davenport and Patil (2012) suggest that the data scientist role is the sexiest job of the 21st century, not all organizations will be ready to reap the benefits. Not only do data scientist roles tend to be expensive to employ, but there is a steep learning curve that new managers will need to climb before they can effectively harness and deploy these specialized roles in the field. Without proper preparation, one runs the risk of having expensive human capital on staff without a clear path to affect change.

Before hiring any data scientists, prospective hiring managers should map out the opportunities within an agency by establishing a baseline. The baseline maps out all of the key stakeholders stakeholders and whether

J. C. Chen et al., *Data Science for Public Policy*, Springer Series in the Data Sciences, https://doi.org/10.1007/978-3-030-71352-2_16

their needs can be addressed through data science. These needs motivate a team mission, inform the first project ideas, and guide prioritization of a project roadmap (i.e., the progression of projects). Perhaps most importantly, the baseline contains the pertinent details that underlie a funding pitch to agency executives.

Scouting the problem space. Data science teams need the right conditions to flourish. To make sure your agency is ready to invest in data, let's start with scouting the problem space by considering these five foundational issues:

- Which *problems* do internal *stakeholders* need help to solve? The real objective of this question is to identify stakeholders who will be the allies who open doors, shape the problem space, and voice support for the data science team. There will also be unwilling stakeholders—these can be won over after initial successes.

- Who are the *people* who should be hired to solve problems? While it can be tempting to hire a PhD in astrophysics and an expert econometrician, smart people can get bored easily and expect the autonomy to explore problem spaces. But hiring inexperienced staff might require a significant amount of guidance and looking after.[1]

- What *data* are available? Ideally, data is available in a machine-readable format (e.g., CSV, XLSX, JSON) that requires less preparation to use. However, if data are stored in non-machine-readable formats (e.g., PDFs, images, and hardcopy), the cost of launching a data science initiative will be substantially higher and the time to first success will be delayed. Data availability is thus important as it will inform how you will manage expectations.

- What *space* will the data science team occupy in the agency? The team will need to report to an office. How the data science team is positioned relative to senior executives will determine their autonomy, but also how much stakeholders will trust them.[2]

- How will *funds* be raised to hire and sustain the team? There are "different types of money" in government. Some funding lasts for a single appropriated budget year and can be renewed each year, while other funding has no time limit. These budgetary nuances matter as they can influence whether staff are permanent employees or hired on a fixed term.

The answers to these five variables can inform how to structure the team, its mission, and its operations. But *where do we find the answers?*

Searching for answers. Data science projects should be technically exciting, but it is easy to create ideas that have little practical value. One could construct mental models of which agency issues can be solved using data science, but the fact of the matter is that gaining support on new greenfield issues is resource intensive and political. Instead, consider building the team's baseline by reviewing the agency's strategic plan. Not only does the strategic plan describe what the leadership team deems to be important, but it also indicates which officials are in charge of priority issues. Data science projects will be far easier to pitch and fund if efforts are put towards addressing the strategic issues. Then, it becomes a matter of figuring out how data science fits with the key stakeholders.

When engaging in stakeholder conversations, take the opportunity to learn about strategic issues and build relationships. Taking a learning posture forces one to listen and gather information about a problem space. This relaxed posture also allows prospective stakeholders to engage within their comfort zone. Otherwise, hard selling a data science initiative from the get-go can be seen as political or aggressive, which in turn can cause prospective stakeholders to take a defensive stance in order to side-step the risk of being dragged into a science experiment.

Through these discussions, it will become clear that data science is not be relevant to all priority issues. To narrow the field, we recommend ordering the strategic issues list based on pragmatic considerations:

- Focus on issues that can show momentum and progress early in the team's tenure. Early and frequent signs of progress will inspire confidence in the newly formed team, thus initial bets should avoid scenarios where data is sparse or unavailable—without data, collecting *new* data will be time-consuming and will only slow the delivery of wins.

---

[1] We will describe different data science roles later in this chapter.
[2] In general, higher up the ladder one is, more cautious stakeholders will be.

- Prospective projects should be coupled with clear decision points. By placing data products as a tool for reducing uncertainty and increase confidence in decisions, we also ensure that the data team plays a concrete role in a recognized process in the agency's operations.
- New teams are fragile. Any amount of resistance from stakeholders can shake the team's confidence. To protect the team, prioritize prospects that are sponsored by stakeholders who are willing to try something new.

The intersection of these three criteria marks the *sweet spot*.

Drawing from your newly gained knowledge of the strategic issues, focus on sweet spot issues and develop a few high-level project ideas.[3] These project ideas should be conceived only at a high-level, whereas "in the weeds" details should be left to the future data science staff to explore and apply their craft. In other words, *set the vision, but not the solution*. With these project concepts in hand, schedule follow up meetings with stakeholders to confirm what was heard in the exploratory discussions, then describe what is possible. Notice that technical details only enter the picture after foundational relationships have been established.

Articulating the mission. From day one, a new data science team will need a mission to guide how it selects its first initiatives. Without a mission, teams could easily drift from one task to the next without leaving lasting change. Some missions are lightning focused on one issue, while others are quite broad. For example, the mission of the Alan Turing Institute—the United Kingdom's national institute for data science—is as follows:

> *Our mission is to make great leaps in data science and artificial intelligence research in order to change the world for the better. Research excellence is the foundation of the Institute: the sharpest minds from the data science community investigating the hardest questions. We work with integrity and dedication. Our researchers collaborate across disciplines to generate impact, both through theoretical development and application to real-world problems. We are fueled by the desire to innovate and add value.* (Alan Turing Institute 2020)

The Alan Turing Institute has a national-level mandate to support almost any field and is allocated the funding to enable its far-reaching mission, thus a broad mission is appropriate. But not all teams will have the resources to sustain such a broad mission. In fact, successful new teams will typically be resourced to focus on a specific area within an agency's operations and will only expand to new areas once it has proven itself.

To identify the right-sized mission, we can draw from the project ideas developed from stakeholder insights. Start by mapping each project idea to one of three types of use cases as described in Chapter 14. As a refresher, Table 16.1 summarizes these use case types along with their objective and outputs. In addition, determine if project ideas will support one or multiple stakeholders. The diversity of use cases and number of stakeholders will point toward a broad or narrow mandate. If project ideas are concentrated within one area, then it may make sense to craft a narrow mission. For example, if projects require machine learning to identify unfair practices among financial lenders and help target regulatory activities, then the mission could be:

> *Our mission is to ensure that consumers have access to fair financial services without fear of unfair lending practices. We protect consumer welfare by monitoring and targeting bad actors who reduce the quality of the lending market.*

In contrast, project ideas could cover all three use case types, then a broader mission statement would be more fitting. In either case, articulate *a* mission, then revise it as needed.

---

[3]See Chapter 15 for guidance on designing data products.

Table 16.1: Three categories of data projects.

Category	Description	Outputs
Reporting	Provide the metrics and tooling that help stakeholders monitor and understand operational issues and make inform decisions.	Dashboards, tables
Insights	Parse and analyze complex, layered questions to identify key patterns that re-risk strategic decisions.	Tables, written reports, visualizations, slide decks
Prediction products	Engineer a product that draws on data and prediction algorithms to address a user need.	APIs, software, visualizations

In addition, project ideas should be organized into a project roadmap—or a plan that outlines *when* each project will be started and delivered. The order of the projects is important. Some projects will have dependencies while other projects have political importance. Begin by classifying each project idea by their time sensitivity—i.e., which projects will need to be addressed first and which are longer term investments. While there is no absolute right way of constructing a roadmap, prioritizing projects by time sensitivity is a reasonable approach. For example, a housing agency has a growing backlog of maintenance requests and its resident population is growing steeply. An algorithm is needed to optimize how maintenance crews are assigned to projects to quickly reduce the backlog. This means that before the algorithm can be researched, the data team will first need to build data pipelines that collect data on how maintenance is currently being handled.

Roadmaps are an evolving management document. They are the operational plan that reflects how the team will fulfill its vision. And while it is tempting to change the roadmap to shift the team's attention, keep in mind that it should be updated at most once a month in order to give teams enough time to make meaningful progress.

16.2   Operating models

The data science team's operating model not only sets ways of working but also helps stakeholders understand how to interact with the team. Some operating models give a senior executive more control over the roadmap while other models afford data experts the latitude to prioritize projects. In this section, we describe four operating models.

16.2.1   Center of excellence

From the perspective of an agency executive, perhaps the most desirable team model is a Center of Excellence (CoE). Staffed with technical experts, a CoE model concentrates deep expertise in one office and is afforded the autonomy to identify and research key problems. As CoEs focus primarily on research, the payoff will take time to be realized. And since it is challenging to schedule deadlines for research projects, deliveries are infrequent and should instead be viewed as an intellectual investment for the good of the agency's knowledge base.

CoEs are often staffed with a cross-section of data professionals (e.g., data scientists, data engineers, etc.) and attract candidates from rigorous academic backgrounds who crave academic freedom. While agency stakeholders can benefit from osmosis of cutting edge practices, that freedom can be easily spent on unrelated foci if not channeled. To bring structure to that freedom, successful CoEs have a well-defined mission. *Will the CoE focus on researching the economy? Or will it focus on engineering software for reducing climate risks?* Whatever it may be, the mission must also be accompanied by governance that helps agency stakeholders get the most from the CoE.

Figure 16.1 illustrates how a CoE can be operated. CoE staff can propose projects as inspired by the issues that the agency faces—they determine which issues are most worth their time. The head of the CoE approves each project's direction and briefs relevant stakeholders on progress. When pressing matters arise,

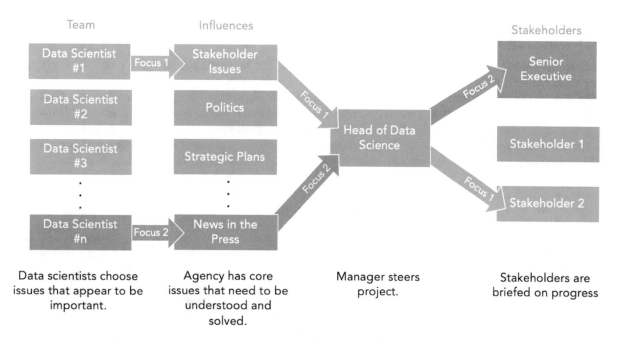

Figure 16.1: Center of Excellence. Data scientists identify problems worth solving, which are then scoped and steered by the CoE's manager. The manager then briefs relevant stakeholders on progress.

the manager can assign projects to team members; however, if done too often, staff may feel the loss of freedom to conduct the research and seek opportunities elsewhere.

Pros. Assuming funding is available, obtaining approval to build a CoE is an attractive idea as it injects fresh blood into an agency. For prospective staff, the intellectual freedom is appealing, and with the right caliber staff, research outputs can be technically impressive and make for suitable content for peer-reviewed research articles—a mark of technical excellence. The center serves as a source of innovation for other units in the agency, actively educating staff on new possibilities.

Cons. Finding and attracting top talent is expensive, thus intellectual freedom and other perks will no doubt be selling points. However, with too much research freedom means that projects can stray away from relevant policy issues, especially as CoEs tend to operate as standalone offices that are not beholden to operational priorities. This can lead to an unfavorable perception that a CoE is not responsive to stakeholders. Furthermore, as project deliveries in research environments are infrequent, these offices are not typically well-suited for operational tasks.

In short, the CoE model is an excellent option for generating data science thought leadership, but not appropriate for operational excellence.

16.2.2 Hack teams

While the CoE model can attract impressive talent, some organizations might not be able to appreciate advanced data science. If senior technical experts are hired into such an environment, they may never fully be able to exercise their deep expertise, become frustrated, then seek opportunities elsewhere. Furthermore, the CoE model lacks the nimbleness to respond to operational needs. A *hack team* is a less expensive variant that is staffed with junior talent who are eager to learn. While the term "hack" is associated with the darker side of technology, in this context, the term refers to exploratory programming—or experimenting to see the possibilities with code and data.

Usually sponsored by an agency executive, the team excels at driving technical awareness: *promote data science and show what's possible*. For example, during the third term of the Bloomberg Administration in New York, most data science teams were hack teams that sat in divisions of the executive branch (the Mayor's Office), supporting portfolio agencies through data analysis and predictive modeling. As the first wave of data teams in NYC government, much of their efforts was spent on introducing agencies to working and thinking with data.

Hack teams are led by a manager who has strong soft skills. The manager can navigate politics and articulate how data fits into policy issues, sell the promise of data science in a wide variety of problem spaces, and maneuver the team to work on strategic projects. Figure 16.2 illustrates the operational flow in a hack team. New project ideas originate from executives, who recommend issues for investigation by the team's manager. The manager triages the top priorities and assigns to hack team staff. In effect, the hack team is treated as the executive's personal team of "fixers". Since junior-level data scientists are hungry to learn, they are willing to take on any requests but will often bend the scope to suit their intellectual curiosities.

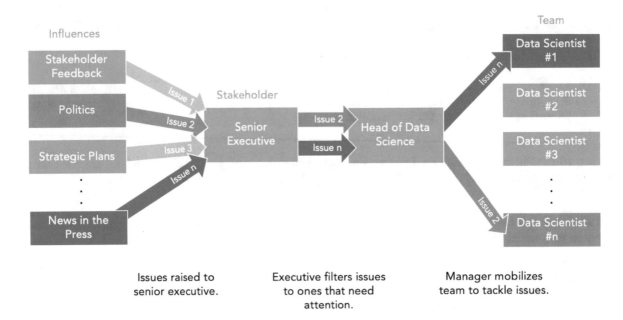

Figure 16.2: Hack Teams. Project ideas originate from the issues that a senior executive must address. The executive filters issues and only passes top priority issues to the team manager, who triages and assigns tasks to staff.

Pros. By hiring junior talent, staff will naturally need time to grow and mature as data professionals. Typically, junior staff tend to gravitate to the most technically advanced techniques without fully understanding how they work. Meanwhile, stakeholders will demand intuitive explanations rather than technical wizardry. This push and pull relationship forces both stakeholders and junior staff to grow together. Hack organizations are great career launching pads for junior staff and can be a source of emerging leadership talent.

Cons. Government executives may treat a hack team as "parachute squad"—they focus on short-term problems, then move onto the next problem. Data team managers will be pressured by executives to cover a sprawling portfolio without acknowledging that quality solutions need time to execute. Eager junior staff are then dropped into large scale problem spaces and are given significant latitude to develop a solution, but then must move onto the next priority in short order. In many cases, these teams are in constant "firefighting mode" rather than proactively working toward a sustainable policy solution.

While more economical, the focus on junior talent also comes with growing pains. As junior data scientists undertake their first real-world projects, they can make avoidable technical mistakes that lead to poor decisions. Furthermore, these teams tend to be more analysis focused and face challenges when the task requires the development of maintainable, scalable software. To guide the team's development, we recommend hiring an expert, more senior data scientist to provide technical mentorship.

In short, Hack Teams are a great structure to introduce agencies to data science, but can fall short on technical sophistication and lasting change.

16.2.3 Consultancy

An internal consultancy consolidates technical talent into a central office that triages data-related requests from stakeholder offices. Perhaps the simplest consultancy is request-based: Each request is triaged by the team manager who then assigns the best-qualified team member to liaise with the internal client, propose a path forward, then execute. Under this model, data scientists and engineers will float from one project to the next. (see Figure 16.3).

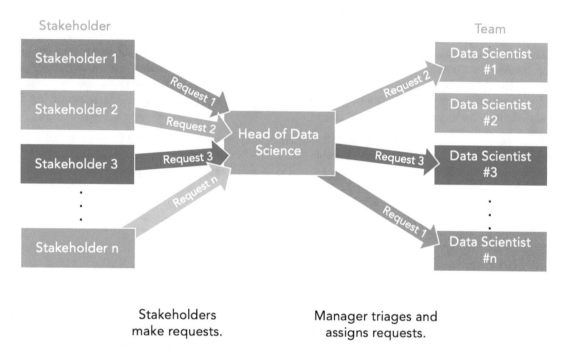

Figure 16.3: Consultancy model. Individual stakeholders submit requests to the data science manager, who triages and assigns the task to the most qualified data scientist.

Pros. Data scientists who like to have an ever-changing variety of projects tend to enjoy working for consultancies. Not only can the subject matter change, but working with different types of stakeholders teaches data scientists to be diplomatic and adapt to their surroundings. Consultancies are attractive for their economies of scale as multiple stakeholders can pool resources and have access to a centralized service. Furthermore, consultancy models do not burden data scientists with the politics within the stakeholder's organization—the objective is to deliver upon request.

Cons. There are organizational challenges that arise when operating a consultancy. If established in a demanding environment, the consulting team is beholden to multiple stakeholders rather than just one, raising questions about if all offices have equitable access. Because stakeholders can submit requests at will, the data science team can be inundated with a large number of requests. And as requests mount, data scientists will be pushed toward quick ad hoc deliveries and disincentivized from long-term research and development. For this reason, projects will tend to be short-form analyses rather than larger scale software development. Managers will need to triage and load balance requests so that data scientists can maintain their work tempo while fairly addressing the needs of stakeholders.

In high demand environments, managers may need to "break the thread"—a situation where different data scientists are assigned to a project at different phases. Context and institutional knowledge can be lost along the way. Alternatively, a consultancy established in a low demand environment will likely spend significant time with lead generation and business development in order to onboard new projects.

In short, consulting models give power to offices that can afford research services, but teams will prioritize short and quick turnaround projects.

Pro Tip. The funding mechanism for a consultancy has a direct impact on time spent on administration and accounting requirements. If the team is fortunate, a single office will provide "base funding" to operate the office. Where funding is not fully available, agency executives can levy an internal tax of stakeholder offices to fund the consultancy; however, not all offices will be supportive of such a funding arrangement. Alternatively, a consultancy can raise money through a "fee for service" scheme that gives interested stakeholders access to data services. For this to be financial viable, the demand for services needs to cover costs, requiring managers to account for time spent on every request and increasing management overhead. While this can be burdensome, it sometimes is the only politically and financially viable option.

In order to balance the budget, managers are forced to balance the demand for services and the operating costs. The demand equates to revenue that covers the teams' costs. If demand follows an even-keeled pace, then the team's costs can be managed with ease. In contrast, irregular demand (e.g., many small projects, large breaks between projects) has implications for the team's financial solvency. Thus, managers should opt to build a constant stream of billables by advocateing for retainer funding—a regularly occurring fee that allows stakeholders to use services when required.

When consultancies rely on billables to fund their operations, then managers need to be aware of the *billable rate* and the *utilization target*. The hourly billable rate is the "fully loaded" cost of employing a staff member, including variable and fixed costs (e.g., total annual cost of salary, overhead, management, infrastructure, etc.).[4] The billable rate is expressed as

$$r = \frac{\text{Salary} + \text{Fringe} + \text{Overhead}}{\text{Work Hours per Year} \times \text{Target Utilization Rate}}$$

The financial benefit of operating an internal consultancy is the lower financial cost relative to private consultants. If r is too high relative to private alternatives, then agency stakeholders could seek external help. Thus, the manager should control r by adjusting the utilization rate and the employment costs. If the demand for services is low, the utilization rate will also be low, which translates to a higher hourly billable rate to account for costs. To ensure that the team delivers economies to the agency, the staff salary cost would need to be lower and, in turn, less competitive with the data science market. While consultancies sound good, they are quite a challenge to operate in government settings.

16.2.4   Matrix organizations

While an individual data scientist can build a data product on her own, bringing her work to life requires social and technical infrastructure. If the product is a new algorithm intended to be used on-demand, then the data scientist will have to rely on a data engineering team to build data pipelines to feed the algorithm and a software engineering team to productionize the service. In addition, stakeholders will need to be convinced that the data product is worth their time. Indeed, launching a product is a feat of coordination. Matrix organizations can solve some of these coordination and inter-office challenges.

As shown in Figure 16.4, matrix organizations lower barriers, doing so by allowing each team member to occupy a dual role:

- Each data scientist is embedded in a project team along with members of other offices. The project team has a defined mission and focus area with a project lead who guides the team's activities. These *cross-functional teams* are a cohesive unit that have their own mission independent from the data organization.

---

[4]In private sector, r is multiplied by a target profit margin, which is inappropriate for an internal consultancy.

- Since cross-functional teams lack a deep bench of similar talent, data scientists may feel alone and unsupported. To balance, data scientists are managed by a central data science organization. This allows data scientists to have a "home base" of like-minded technical experts who can support their growth.

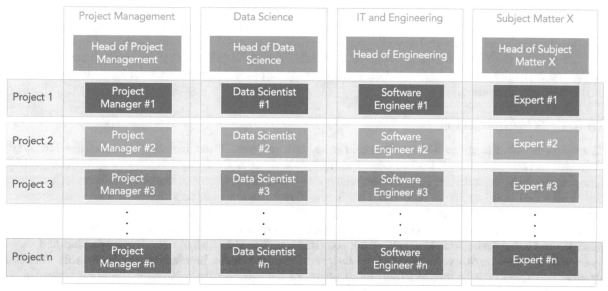

	Project Management	Data Science	IT and Engineering	Subject Matter X
	Head of Project Management	Head of Data Science	Head of Engineering	Head of Subject Matter X
Project 1	Project Manager #1	Data Scientist #1	Software Engineer #1	Expert #1
Project 2	Project Manager #2	Data Scientist #2	Software Engineer #2	Expert #2
Project 3	Project Manager #3	Data Scientist #3	Software Engineer #3	Expert #3
⋮	⋮	⋮	⋮	⋮
Project n	Project Manager #n	Data Scientist #n	Software Engineer #n	Expert #n

All staff are assigned to a project,
managers check in.

Figure 16.4: Matrix organizations: All staff are assigned to a project team. The direction of projects is set by the project lead, but data scientists are managed by the data science department.

Pros. A matrix model removes the "us versus them" barrier: data scientists can integrate with other organizations and stakeholders can feel they are working with their "own" people. This arrangement also works well for other data professionals such as data engineers and data analysts. The cross-functional structure removes bureaucracy and gives individual teams the authority to make decisions specific to their mission and have dedicated resources to move the project forward.

Cons. The centralized data science organization will take on a technical support role rather than serve as the driver of change. Instead, the project lead and the lead's home office will drive the direction of project endeavors. Since data science managers do not have control over project work, they have less visibility into the day-to-day due to the working arrangement. Thus, managers will spend far more time to sync with staff and keep track of developments.

On the staff level, there are occasionally communication challenges in cross-functional settings. Each role in a cross-functional team is trained in a different way of working. Thus, data scientists will operate differently than software engineers and policy advisors. Each data scientist must become their own communication strategist to convey what their discipline can offer and how they fit in. Some will excel while others will shy away from this responsibility. In the worst case, project teams might not understand how to utilize a data scientist and the skills are not adequately employed in the team's work.

Pro Tip. Matrix organizations give much less authority to the data science manager. The manager will find the role especially challenging if her interests lie in setting policy. Instead, consider hiring someone who will be satisfied with empowering a team through technical mentorship. Matrix organizations also give hiring managers the chance to use the project team's mission as a unique selling point—some missions are more attractive and serve as intellectual compensation for ambitious talent.

In short, matrix organizations reduce friction between departments and allocate adequate resources to mobilize projects. Data scientists fit in this model only when project teams understand how to make use of their craft.

16.3   Identifying roles

If we search the internet for data science roles, we will likely find that many postings have the same job title but different responsibilities. These differences reflect the specific needs of each organization. Some organizations will focus more on delivering insights while others build intelligent systems. Thus, there is not just one single archetype of a data professional, but rather a range. In this section, we provide one perspective of the types of roles that government data science teams employ and provide a brief introduction to hiring in the civil service system.

16.3.1   The manager

The manager sets the vision for the team and prioritizes projects. Titles include "Data Science Manager", "Director of Data Science", "Head of Data Science", "Director of Data Analytics", among others. Perhaps more importantly, the manager is an enabler who opens doors so the team can explore possibilities throughout the agency and build technical capacity. This is not to say the manager sets the team loose—sometimes it will be necessary to nudge the team toward organizational priorities. Often times, the manager has strong communication skills, in both meetings and in front of audiences. In early stage data science teams, the manager tends not to be technical, but rather someone who has the trust of executive leadership and serves as a guide into new territory. As the team evolves and the organization becomes more comfortable with data science, managers may be sourced from more technical backgrounds to help steer the team toward more sophisticated use cases.

Pro tip: We advocate for hiring managers who come from technical backgrounds to set a data-driven culture from the very start. However, if hiring a first-time manager, be prepared for a transition. In other words, be patient. Manager roles are focused on empowering staff, but the new manager may still want to spend significant time on research and code. While this is a common pattern, it needs to be carefully managed. A manager who focuses too much time on the technical aspects will inadvertently foster an environment of "competition" with their direct reports—this has adverse consequences for team morale. The transition needs constant coaching. The conversion is worth the effort, not only for the agency, but for the broader data science field.

16.3.2   Analytics roles

Within agencies that are making their first foray into data, newly established data teams are often tasked with conducting analyses. These teams tend to be staffed with *data analysts* and *statisticians*:

Data Analysts conduct *descriptive analyses*. Their charge is to craft a narrative from what they can see from the data. Data analysts are skilled with transforming data to find answers. For example, data analyst could be asked questions such as

- What is the fire department's response time trend over the last few weeks?
- Are service requests resolved within 30 days of receipt?
- How many people in county X stayed at a shelter in the last 48 hours?
- How often has economic growth exceeded Y% in a quarter?
- Which are the greatest correlates of variable Z?

While these questions are simple, they are valuable inputs that inform stakeholders' initiatives. Some analysts will gravitate towards `Excel` and `SQL`; however, large scale problems will require programming languages like `R` or `Python`. In any case, data analysts form the first line of support for analytical requests.

Statisticians focus on making *generalizable inferences* from data.[5] What sets statisticians apart from data

---

[5]In some organizations, statisticians also have data scientist titles. Computational statisticians are a special case that rely on machine learning to draw inferences and have many overlapping skill with "full stack data scientists".

analysts is their deep understanding of probability and statistics. As full population datasets are expensive and rare, statisticians can determine whether inferences from a sample can be generalized to a target population. In addition, statisticians are skilled with using statistical models (e.g., linear regression) to estimate relationships between variables and decompose influences (e.g., coefficient β_k). By estimating the contributory effects on an outcome, agencies given a view into the inner workings of phenomena that would otherwise not be possible from descriptive analyses.

For example, statisticians could answer questions such as

- Is candidate A ahead of candidate B in the polls?
- What is the effect of policy A on population X?
- How many more infections should we expect in the next 48 hours?
- Can these five surveys be re-weighted to infer societal attitudes toward issue Z?
- What are the topics covered in this corpus of speeches?

To accomplish these tasks, statisticians rely on programming languages like R, `Python`, and `Matlab`. Some statisticians are trained in the quantitative social sciences using higher level languages such as `Stata`, `SPSS`, and `SAS`.

Together, statisticians and data analysts form the backbone of inference-focused organizations. They can conduct descriptive analyses and draw causal inferences.

16.3.3 Data product roles

In contrast to insights-oriented roles, product-oriented roles require a different skill set that not only rely on statistics and probability, but also blends knowledge of software development, computational infrastructure, and architecture. Because of the broad yet deep knowledge required to build functioning products, this type of data science team tends to be expensive to operate. Here, we narrow our focus to two skill groups: *data scientists* and *data engineers*.

Data Scientists have knowledge of data analysis, statistics, and engineering—generally with a stronger command in some areas than others. Data scientists are expected to conduct analyses to understand the data, then build models that feed intelligent services. They serve as a bridge between insight and use case. In short, the role is a catch all.

Example tasks include:

- Develop an API (or even a list) that returns a list of the highest risk buildings in a local area.
- Prototype a prediction engine that nowcasts key KPIs.
- Flag all aircraft that are present in aerial imagery.
- Predict which businesses will need financial loans in the next six months.

The product-oriented toolkit is more expansive that insight-focused teams, including analysis languages (e.g., R, `Python`, `Julia`), software development languages (e.g., `Python`, `Java`), a large-scale data processing framework (e.g., `Hadoop`, `Spark`), a version control system (e.g., `Git`), and querying skills (e.g., SQL).

An emerging role in tech sector is the *machine learning engineer* (ML Engineer)—a position that bridges data science with software engineering. Since data scientists often focus on prototyping an algorithm, they may not be the most skilled with production duties. Most of the effort required to stand-up a production-grade algorithm lies in developing the software and infrastructure that breathe life into the project, then ensuring the algorithm continues to function as designed. ML engineers are responsible for this middle layer that clearly demonstrates the value of a data product.

Data Engineers are keep the data flowing. If data is water, then data engineers build the pipelines to ensure that everyone who needs water has access to it. The work of a data engineer centers of building software that can ingest, transform, and store data so it can be used. In short, they maintain the data platform on which all else depends. Part of the responsibility includes maintaining data infrastructure (e.g., servers, databases, pipelines) and ensuring that data pipelines are continually running. Data engineers automate processing wherever possible to minimize manual work, then schedule the processing to meet operational schedules. This specialized role create the conditions that make data science possible.

Example tasks include:

- Send API request to a data provider, ingest the returned data, then process and store it in a database.
- Develop a program to apply Optical Character Recognition algorithms to extract text from PDFs and store as a dataset.
- Develop a real-time pipeline to transform the data for use in a dashboard.
- Set up data access controls for each column of a highly sensitive dataset.
- Devise a pseudo-anonymization procedure to conceal personally identifiable information in a widely accessed dataset.

Data engineering teams rely on many of the same tools as data scientists, but apply them in different ways. While the specific programming languages vary from one organization to the other, they typically involve a software development languages (e.g., `Python`, `Java`), a large scale data processing framework (e.g., `Hadoop`, `Spark`), a scheduling framework (e.g., `Airflow`, `AWS Glue`, `Luigi`), a version control system (e.g., `Git`), and querying skills (e.g., SQL), among others. Indeed, the skill set is much wider and more complex.

While data engineers are common in tech companies, the role is less common in government. Instead, data engineering tasks tend to be spread across a number of roles in an IT department, such as involving a database administrator to manage the data processing and make available database views (i.e., custom queries) to data users. In small teams, there may not be a difference between data scientist and data engineer roles, requiring all team members to undertake data engineering tasks themselves.

16.3.4   Titles in the civil service system

Chances are that data science roles are not currently defined in government civil service systems. Instead, they are often times classified as an IT analyst role, which has pre-defined salary ranges that might not match market expectations for data scientists. As a manager, being aware of these limitations is important so that you can plan your team structure accordingly. But fear not—the civil service system is not a monolith. Every government agency and jurisdiction has different ways of interpreting civil service law, which will at times give afford more flexibility to how one hires data scientists.

For example, during the Bloomberg Administration in New York City, for example, some data analysts were hired into "Staff Analyst" roles while some data scientists were hired into "Administrative Staff Analyst" roles. Both roles share similar responsibilities as other non-technical roles such as program and project managers. In addition, civil service titles often do not reflect what the data scientist does, but serve as the vocabulary that describe how someone is hired and paid. While it is not expected that every hiring manager will understand civil service nuances, having some familiarity of what is possible in your agency will smoothen the hiring process and identify strategies to easily onboard talent.

16.4   The hiring process

Hiring a data scientist in government is similar to the private sector. There are some practices that can help streamline the hiring process and converge on suitable candidates more quickly. In this section, we provide a few tips on setting up the hiring process and typical steps in an interview process.

16.4.1   Job postings and application review

The hiring process starts with drafting a position description that details responsibilities and meeting with recruiters to align on expectations.

The *position description (PD)* should describe the vision for a data role, what is exciting about the team's mission, the growth opportunities, and the skill requirements. As HR recruiters will need to map the position to civil service requirements, it is also worth conducting some background research on suitable civil service title, salary grade, and permanence.

The civil service cannot hire for job titles that are not in the system, thus recruiters may need to change the title to match government standards. Some recruiters will opt to rename a role to match a current civil service title. For example, a "data scientist" role might be hired as a 2210 in the US Government and be relabeled as "Information Technology Management Specialist"—a title that does not easily translate into everyday vernacular. Instead, we recommend co-locating an "office title" with a more subtle civil service title. For example, "Data Scientist (Job Series 2210)."[6]

In some agencies, recruiters are involved in the initial *application screenings*, relying on a scoring rubric to help identify qualified candidates. The technical nuances, however, might not be apparent—sometimes what differentiates a data scientist from a frontend web developer is not clear. To set expectations and tone, consider introducing recruiters to your team's work by walking through a typical day's work, hot topics in the field, and how roles differ. It can also be helpful to provide keywords and verbiage that signal the right qualities.

16.4.2 Interviews

Data scientists should be assessed on whether they have the skills to deliver on technically demanding tasks, but also if they exhibit the emotional intelligence and the business acumen to work well with others. The precise mix between these two qualities depends on the role, though being trustworthy and personable makes for good work environments. Thus, interviews involve at least two steps: a *fit screening* and a *tech test*.

Fit screening. This initial interview focuses on whether the candidate would fit within the culture of the team. Since the culture of every team is different, there are no absolutely correct answers, but the fit should match those teammates, stakeholders, and managers. This screening also presents the opportunity to check if candidates have had meaningful data science experiences, such as their role on past projects, how do they like to work (e.g., teams, solo), how do they respond to challenging situations.

To calibrate expectations within the team, we recommend that at least two team members (or one team member and one stakeholder) conduct a few screenings together. In a separate session, the interviewers should discuss their impressions of the candidates with one another. This process will not only expose personal preferences but calibrate perspectives.

Tech test. The objective of the technical screening is to determine if the candidate has the right qualities to contribute to the team. While the subject matter of a tech test varies from role to role, the test can be administered in two formats: a *whiteboard assessment* and *take home test*.

Whiteboard assessments are conducted in-person and can be intimidating for the candidate. This format evaluates how the candidate performs when faced with a time-constrained problem and provides up-close data points on their thinking process. The interviewer asks technical questions to the candidate who will then walk through the logic and solution a whiteboard. For example,

- What is the definition of R^2? How would you explain it to a non-technical person?
- In coding language X, write a function that calculates a Fibonacci sequence returning the first k values?.
- Given a non-linear relationship illustrated on a whiteboard, describe three ways to fit a model to the pattern.

For a list of questions, see Appendix B. While whiteboard tests force an answer, they can unnecessarily penalize candidates for wrong answers, especially if the questions are too specific. A data scientist who is interviewing for a computer vision role, for example, may not understand the mechanics of time series models. Asking questions that show their thinking process is a better gage although identifying these questions is challenging.

---

[6]Note that these examples are specific to the United States. The experience in other countries may vary.

Take home tests simulate a simple technical project that the candidate could encounter while on the job. At an agreed time, the interviewer will send a set of tasks and materials to the candidate to complete with a deadline. The candidate submits code, documentation, and outputs for the data team to review. The objective is to evaluate the quality and thoroughness of the work product and determine if it meets a reasonable standard. The specific set of tasks depends on the nature of the role. For example,

- A policy and strategy team may ask the candidate to analyze a CSV of data to answer questions, then write a short presentation and give a short five-minute talk to non-technical stakeholders.
- A transportation infrastructure team that specializes in risk management could provide data about roadway inspections and ask the candidate to predict which bridge (in an unlabeled test set) is likely to have faults and why.
- A clinical research team looking to work on quasi-experimental techniques could ask a candidate to perform a Regression Discontinuity Design on a dataset and write a short research brief.

Take home tests have encountered some controversy in recent years (McEwen 2018). In fact, some tech companies have allegedly abused tech tests as a source of free labor—an ethical violation of the interview process. To avoid this perception, we recommend that the tech test is administered according to the following guidelines:

- Under no circumstances should tests be drawn from *current* or *future* projects.
- If possible, current staff should take the test themselves to see if it is feasible and reasonable.
- The test should not take more than two to three hours.
- Be clear how the tech test was constructed and why it is relevant to the interview process.

At the end of the test, staff should review the test results, then schedule a short follow-up discussion with the candidate to ask questions and give feedback.

Post-Interview. At the end of the interview process, the top candidate will be extended a *preliminary* offer. In the public sector, the "preliminary" status is necessary as civil service jobs are positions of public trust, requiring candidates to successfully pass background and reference checks. If the candidate accepts, then all other candidates should receive feedback. We recommend, however, that other candidates be kept on reserve in case the top candidate withdraws. As governmental hiring processes are lengthy and the data science market is competitive, this unfortunate scenario does happen from time to time. From start to end, hiring a data role can last between three months to five months. Indeed, governmental hiring processes are a weakness that needs to be addressed in order to compete with the private sector for talent.

16.5   Final thoughts

Building a well-functioning data team is a marvelous adventure for any government agency. A well-designed data team can serve as a source of innovation and inspiration. While one can model their team after what has been described in this textbook or any magazine article about management strategy, there is no substitute for being present and thoughtfully engaging people to evolve public service. For anyone aspiring to build a data science team, remember that humans are social creatures. The technical splendors are tools that can enable wonderous use cases. However, building trusted relationships with stakeholders must always come first for trust is the substance that propels data science in public policy.

Appendix A: Planning a Data Product

Developing a prototype of a prediction model is the easiest part of developing a data product. And in a world where rapid development is prized, we can take comfort in this fact. There are many challenges, however. For one, data scientists must have a clear understanding of how the product fits into a policy process and be able to articulate its value. Without this, the product runs the risk of being "a flash in the pan". Furthermore, products must be functional and continue to function for the duration of its useful life, requiring some thought on how to maintain and evolve it as circumstances change. The world, after all, is a chaotic place with unexpected paradigm shifts happening all the time. For these reasons, data products should have a Concept of Product Operations Plan (CPOP)—a document that clearly outlines how the data product will function and all operating considerations.

In many cases, data teams will have a verbal understanding of how products will function. A CPOP can be a helpful tool to formalize processes, identify gaps and risks, and communicate with stakeholders on a product's direction. The questions in the following section can inform the plan, touching upon many of the key issues that should be considered.

If the objective is to deploy a data product that is a web service or decision making engine, then there are likely stakeholders who will need to be briefed. Presenting the CPOP communicates key elements of governance and lends confidence to the initiative.

Key Questions

Motivation. We begin by defining the main motivation for building the data product and articulate why it merits the attention of stakeholders. Furthermore, we identify where does the data product sit in a process, including the users and intended actions:

- *Motivation*: Why was the data product created and what problem does it solve?
- *Theory of change*: How will the data product be used to affect change? (i.e., Is there a clear avenue for success? Does the product prompt a decision such as target something for investigation, or does it serve as contextual information such as an ad hoc report?)

Decisions. Data products produce some set of outputs; However, effective data products have a clear purpose with identified decision points that lead to well-defined actions.

- *Decision point*: Is there a clear decision point at which the algorithm will sit? (i.e., Does the data product influence a current or new decision point? If it is not tied to a clear decision point, why build it?)
- *User*: Who will receive the product's outputs? (i.e., Will the product be received by a human to inform decisions? Or will it trigger a software program to carry out standardized actions like online recommendations or alerts?)
- *Form factor*: What form will the outputs be delivered? (e.g., API, dashboard, graph, etc.)
- *Intended action*: When the user is presented with outputs, what actions will be taken?

© Springer Nature Switzerland AG 2021
J. C. Chen et al., *Data Science for Public Policy*, Springer Series in the Data Sciences,
https://doi.org/10.1007/978-3-030-71352-2

- *Impacts*: Who will be the recipient of these actions? Should the actions and decision process be made explicit to the recipient? (i.e., If a computer is passing judgment on another human, is this fair?)
- *Cadence*: How often does the decision need to be made? Can the algorithm and infrastructure meet that cadence and frequency? (i.e., Beyond the prototype stage, can the data product be continually used to meet regular operational needs? If not, perhaps it is not a sound concept to implement.)
- *Pipeline*: Can the data be delivered reliably given the operational cadence? What are the consequences of delays in delivery? (i.e., If the product goes offline, what processes will be handicapped? Is there a backup process?)
- *Harm*: What is the harm that may be potentially done?

Sample Design and Data Quality. Just having any data is not enough to make a data product. The input data must have good qualities that lend themselves to the use case while some qualities are detractors that need to be carefully considered and managed.

- *Fit for purpose*: How was the data collected and is it appropriate for the intended use case? (e.g., If the data are collected for a research survey, does it hold operational value? If the data is collected on one population, does it generalize onto another?)
- *Data collection protocols*: What were the data collection protocols?

- *Coverage*: Does the data have sufficient detail about all strata that will be impacted? If not, which subpopulations would this model be applicable? (e.g., If data is about income but contains mostly affluent populations, can it be applied to less affluent populations?)
- *Biases*: What biases can appear in this dataset? Were biases actively controlled during data collection? (e.g., historical or echo chamber biases, measurement bias, attitude bias, etc.)
- *Adjustments*: Can biases be adjusted? If not, is this data product safe to use?
- *Missing data*: How are missing values treated? Could imputations or deletions adversely affect the data's characteristics and introduce artifacts to the model? (i.e., If the data has missing values, how do you plan to treat them? If missing values were imputed by the data producer, are those imputations valid?)

Error Evaluation. All quantitative models have errors—the question is how much of each kind. Is it more acceptable to falsely predict something will happen or allow an event to go undetected?

- *Type I Errors*: What are the social and economic costs of a Type I error? (i.e., What if the product mistakenly labels an instance such as a person as a false positive?) = *Type II Errors*: What are the social and economic costs of a Type II error? (i.e., What if the product mistakenly labels an instance such as a person as a false negative? What are the costs of being "blind-sided"?)
- *Disparity in errors*: Do errors affect some substrata more than others? (i.e., Is the level of accuracy consistent across demographic groups or subpopulations? What does this mean for reliability when passing judgment if there are disparities in accuracy?)
- *Acceptable risk*: How much error is acceptable? What is the cost of not deploying the product?

Transparency and Perception. Products need to be acceptable and adopted by the policy process, thus being able to sell the idea and anticipate perceived risks will go a long way to socialize the product and garner support.

- *Transparency*: Which stakeholders should understand the algorithms inner workings? (i.e., Who need to lend support to the initiative and stand by it)
- *Elevator pitch*: How will this model's inner workings be communicated in three sentences or less? (i.e., How can one quickly attract the attention and support of a senior officer?)
- *Technical review*: Is the statistical and algorithmic logic clear and well-documented for experts to review if needed? (i.e., Was the effort developed following a clear methodology or was it hacked together? Will this project stand up to scrutiny?)
- *Explanation*: Is it possible to explain how the data product arrived at a specific prediction? (i.e., If the product produces a prediction, can a user see which specific factors influenced the result)
- *Optics*: Can the product be viewed negatively? If so, what is the source and can it be mitigated? (i.e., can the project be misconstrued or is it already built on dubious grounds)

- *Moral hazards*: Are there unfair or perverse incentives for deploying the data product? (e.g., Is the producer of the product unfairly profiting from the use of the product at the expense of another party?)

Thinking about Automation. Automating some processes will shift some responsibility and power to a machine rather than a human. If the product aims to automatically make decisions, then the downstream consequences should also be explored.

- *In the Loop*: If a process is automated using a data product, how do humans retain control over the process and guide how algorithms evolve?
- *Need for automation*: If the algorithm replaces a human, what responsibilities will displace humans have?
- *Automation failure*: If the data product goes offline unexpectedly, what needs to be done until services come back online?
- *Responsibility*: If automation is the goal, is the organization hosting the product willing to assume the liabilities of biases and error? (i.e., If the algorithm is found to be harmful, will the organization be ready to own the liability.)
- *Model Drift*: Which metrics need to be reviewed on an ongoing basis to ensure the data product is behaving as expected? What are the normal levels of performance?

Privacy and Security. Many data products require sensitive data to give them their edge. Safeguarding the data thus becomes a priority to maintain trust and safety of those who are affected by the products.

- *Security*: What measures and practices will be taken to minimize unnecessary data exposure? (e.g., password protection, access permissions, access logging)
- *Identifiable information*: Is there identifiable information being used? (e.g., social security identifier, phone number, name, e-mail address)
- *Anonymization*: If so, can data be anonymized or withheld without substantially reducing functionality? (i.e., can each record be identified using a pseudo-identifier)
- *Privacy*: What are the consequences of a breach? (i.e., algorithm and data fall into the wrong hands)

Operations and Monitoring. Lastly, the product will be put into production. There will be performance issues when the data product does not behave as intended. What will the data science team do to ensure smooth continuity of operations?

- *Incident reporting*: If anyone identifies an error or issue, how will the issue be reported? (e.g., If data processing is delayed, how will it be reported?)
- *Incident announcement*: Who will be informed of an ongoing issue? (e.g., users of the data product, management and executives, engineering teams)
- *Incident response*: Who will be responsible for addressing the incident and under what timeline?
- *Backstop*: Under what conditions should the algorithm be shut off temporarily? What if permanently?

Appendix B: Interview Questions

Interviews are an opportunity for a candidates and employers to get to know one another and check if there is a good fit. While it would be ideal to cover all aspects about a candidate's experience and skills, the conversation needs to be concise and target the most pertinent issues. In this appendix, we provide a list of sample questions that employers can use to model their own questions, ranging from interpersonal considerations to technical knowledge.

Getting to know the candidate

Like any other interview, get to know the candidate. Below are a few opening questions to help set the stage:

- Tell us about yourself.
- What interested you in this role?

Business acumen

The candidate might give a great first impression, but does he or she have the right demeanor and thinking processes to work in a team in government? The following questions help expose a candidate's business acumen:

- What role does data science play in policy environments? (i.e., do you have a sense of what data science can do for public sector?)
- What factors do you consider when prioritizing projects? (i.e., can you be trusted to figure out order of operations?)
- How do you deal with failure? (i.e., what is your attitude toward hard situations?)
- Do you prefer to work in teams or individually?
- What does co-creation mean to you? (i.e., can you share decision making power?)

Project experience

Let's dive into the candidate's past technical experiences. An easy entry point is to ask:

> Tell us about one of your past data science projects.

Then dig deeper into the details. Keep in mind that any detail that has been mentioned as part of the past project is fair game for further conversation. Consider these follow up questions:

- What kind of user would benefit from the project? How did the project impact the user? (Note: This will only apply to candidates with work experience. Recent graduates will likely recount a class project or thesis.)

© Springer Nature Switzerland AG 2021
J. C. Chen et al., *Data Science for Public Policy*, Springer Series in the Data Sciences,
https://doi.org/10.1007/978-3-030-71352-2

- What techniques did you use? Why was it appropriate for the situation? What were the qualities of the data that informed that decision?
- Describe the core assumptions of that technique.
- Which languages and software did you use?
- Where their limitations to the software that required custom code?
- What ethical implications did you consider before implementing the project?
- How would you summarize this project to a non-technical stakeholder?

Whiteboard questions

In tech organizations, a whiteboard test helps interviewers assess the candidate's core knowledge and her ability to think on her feet. Some require a whiteboard to write or draw a solution, while others use visuals to prompt the question. For each of these questions, we provide some intuition for what they test and why are they appropriate.

Statistics

Q1. Draw the shape of the income distribution of the United States (or any other country). Which is greater: the mean or the median?

Why this question? Working with data requires the ability to understand how to treat different shapes of data. A kernel density plot of income will likely have a peak to the left (closer toward zero) and a long tail to the right. Mean is sensitive to outliers while median preserves the middle point of the distribution – often the peak. Thus, the mean will be larger than the median.

Q2. What is the effect of a logarithm transformation on a continuous variable?

Why this question? Much of a data scientist's job involves transforming data so it is usable. The logarithm transformation, which compresses distributions reducing the skew in the data, is one of the basic tools for improving the usefulness of a variable.

Q3. What are the consequences of multicollinearity in a linear regression? How would you detect it?

Why this question? As regression is one of the most basic modeling techniques, candidates should have a thorough understanding of its assumptions. Multicollinearity is a condition in which two or more variables are highly correlated, which can be diagnosed using a variance inflation factor (VIF). If left unchecked, a coefficients can exhibit large, unusual values due to small changes in a model.

Q4. The COVID-19 outbreak forced all countries to develop testing programs to track the spread of the virus. If a country reports "newly confirmed cases", what is possible to infer from this data? What is not possible to infer?

Why this question? A core problem with COVID-19 statistics is their generalizability. Some countries are able to conduct small random samples to understand full population spread while most reports reflect the number of people who have requested testing (self-selection). Thus, new cases largely reflect self-selected testing and underestimate the true number of infections. The candidate should be able to distinguish between self-selected and random samples.

Q5. Can internet polls (e.g., Facebook poll, Twitter poll) be generalized to a population?

Follow up. How can polls be adjusted to approximate the views of a population?

Why this question? Similar to the COVID-19 question, internet polls also tend to be self-selected. Although self-selected data is more abundant, it will often be biased due to self-selection and coverage.

Causal inference

Q1. Explain why control and treatment groups are necessary in experiments and quasi-experiments.

Follow up. How does one know if a control group is appropriate? What are three conditions that must be observed in order to prove a causal relationship?

Why this question? This question assesses basic knowledge of ignorability (i.e., control group is indistinguishable from the treatment group) and how to test for statistical differences. It is an opportunity for someone who understands the Potential Outcomes Framework to describe the basic tenets of causal inference.

Q2. Suppose you would like to test whether housing subsidies improve health and social outcomes. How would you set up an experiment?

Follow up. How might assignment contamination and sample attrition affect the experiment? If you did not have funding for a randomized control trial, what are some quasi-experimental approaches that could be applied?

Why this question? By framing causal inference in a real world situation, we can assess a candidate's creativity and awareness of experimental design principles.

Estimation versus prediction

Draw one of the reference graphs in Figure B.1, then ask the following questions.

Q1. Suppose some of your policy analysis stakeholders would like to understand the relationship between x and y. What sort of model would you use and why?

Why this question? Policy environments place most emphasis on interpretability and linear models are the *de facto* choice. This question checks if the candidate understands which models lend themselves to estimation of relationships.

Q2. Suppose you would like to build a linear model to fit this graph. What would your model specification look like? What are the limitations of your specification?

Why this question? The goal is to check if the candidate understands estimation strategies and their tradeoffs for polynomial expansions, discretizing the x value to approximate the shape, intercept changes, etc.

Q3. If you were to apply a pruned CART model, draw one predicted line. Explain how the algorithm works and if it is suitable for interpretation.

Why this question? CART and linear methods have different assumptions and outputs. CART models produce jagged or blocky predictions that rely on attribute tests. For continuous problems, CART produces fairly crude approximations.

Q4. Explain the concept of bagging and how it can improve this fit.

Why this question? CART can be improved through bootstrap aggregation and further improved with the random subspace method (e.g., Random Forest).

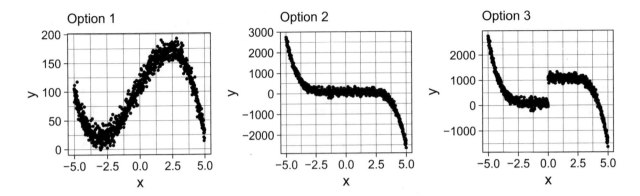

Figure B.1: Example reference graphs for modeling question.

Machine learning

Q1. What makes LASSO and Ridge Regression different from a typical linear regression?

Why this question? Regularized regression relies on penalization of the cost function. For teams that work with high dimensional datasets, regularized regression will be a must.

Q2. What is a hyperparameter? Give an example of a hyperparameter and its relevance.

Why this question? Hyperparameters are a central part of tuning and optimizing machine learning models. Candidates who work with machine learning should know the importance of hyperparameters and their role in model performance. A few examples include the number of variables in Random Forest, the k in k-Nearest Neighbors, the λ parameter in LASSO regression, etc.

Q3. What is an attribute test in tree-based learning?

Follow up. What is an example of an attribute test?

Why this question? This question can screen for candidates who are concerned with the mechanics of algorithms. The attribute test (e.g., information gain, entropy, Gini gain) is the foundation on which tree-based models are built. Any data scientist who has used CART, Random Forests, Gradient Boosting and any other tree model should be familiar with attribute tests.

Q4. What are the steps in a X-algorithm? Provide a step-by-step recipe.

Why this question? If a team has preferred techniques (e.g., neural networks, tree-based learning, etc.), it can be helpful to find candidates who are also familiar with those core methods. This question can be answered with a bullet point answer or a flow diagram that details each step.

Q5. Suppose you are given a $n \times m$ matrix of topic probabilities from a Structural Topic Model. How would you index all documents to one another?

Why this question? A common problem with machine learning projects is to making the outputs useful to an end user. This question asks how to use topic probabilities. One strategy could involve transposing the probability matrix so that documents (n) are positioned as columns, then compute a cosine similarity between each document. The most relevant documents for document d could be identified by sorting cosine similarities in descending order.

Model evaluation

Q1. Let's suppose we have a classifier that has a False Positive Rate that is two times higher than the False Negative Rate. What are your thoughts on this model and its fitness for applied use?

Why this question? This question checks if the candidate understands that False Positives and False Negatives each have different costs in different contexts. There is no right or wrong answer; However, the candidate should point out that the model's fitness depends on context. For example, in the criminal justice system, is it better to have a model that tends to falsely accuse someone or erroneously drop charges? The candidate should reason through these social costs.

Q2. A team member estimated the following specification and developed policy recommendations from the specification $Income = f(Years of Education)$. A number of policy analysts would like to apply the model as a "tool" for income-related policies. What are your thoughts on this?

Why this question? While estimation and data storytelling are typical tasks in policy environments, the bias-variance tradeoff is not typically discussed in public policy schools. This questions helps assess if the candidate sees the potential for the model to be underfit and prone to bias. Thus, the candidate should be able to discuss bias-variance tradeoff and the difference between estimation and prediction.

Q3. Given a table of a binary target and predicted probabilities from a classification model (Table B.1), plot the Receiving Operating Characteristic (ROC) Curve. Explain what do the axes indicate.

Why this question? Many data scientists use functions without understanding how the underlying machinery works. This question tests whether the candidate has the technical knowledge to produce simple ROC curve without software, requiring knowledge of sensitivity and specificity. Consider using this questions for senior technical roles that have mentorship responsibilities.

Table B.1: Table of binary targets and predicted probabilities.

Target	Probability
1	0.9
1	0.7
1	0.4
1	0.3
0	0.4
0	0.2
0	0.1

Communication and visualization

Q1. What is the definition of R^2? How would you explain it to a non-technical person?

Why this question? Explaining regression models will be inevitable task for a data scientist working in policy. Being able to articulate the amount of variability explained is a good indicator if the candidate will be able to liaise with and convince non-technical stakeholders of data science project outputs.

Q2. What is hypothesis testing and why does it matter? How would you explain it to a non-technical person?

Why this question? Non-technical stakeholders will often interpret shifts in indicators as a policy effect without considering statistical significance. There will invariably come a moment when hypothesis testing becomes a central point in debunking faulty assumptions. Similar to the previous question, assesses core explanatory skills.

Q3. How would you explain cross validation? When would you apply it?

Why this question? Many quantitative practitioners who are trained in the social sciences will not be able to articulate the difference between estimation and prediction. In fact, cross validation does not come naturally. This question can screen for those who can work beyond estimation tasks and switch into a prediction mindset.

Q4. How would you improve the graph in Figure B.2? The intended outlet is a policy brief.

Why this question? There will always be poorly designed graphs, yet they are one of the principal modes of communicating with stakeholders. This question focuses on the candidate's attention to detail and clarity of communication. The candidate should ideally identify (1) lack of title, (2) lack of axes, (3) overly dark background, (4) overcrowding of points, (5) lack of a clear takeaway.

Programming

Q1. Given the following string (e.g., "The average household income was $59,039 in the year 2016."), write a pair of regular expressions that extracts each dollar amount and the year.

Why this question? For organizations that work with raw, messy data, data scientists will spend a significant time using regular expressions. Keep in mind, however, that not all qualified candidates will have a command of regular expressions.

Q2. Write and document a function that calculates a Fibonacci sequence returning the first k values? As a hint, a sample Fibonacci sequence is as follows: $0, 1, 1, 2, 3, 5, 8, 13, ...$

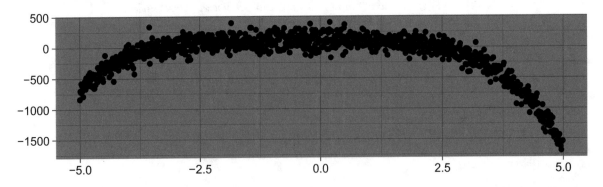

Figure B.2: Example of a poorly designed graph as a point of reference for the visualization question.

Why this question? For teams that need strong programmers, it becomes more important to have clean, functional code. This simple test checks to see programming ability and hygiene. In particular, check if the code returns the requested result, clearly defines the input parameters, and includes documentation. Allot 10 to 15 minutes for this question.

Q3. Write a function to calculate a cosine similarity matrix between two vectors, A and B. Cosine similarity is given as

$$cos(\theta) = \frac{A \cdot B}{\|A\| \|B\|}$$

Why this question? Translating mathematical concepts into code is a useful skill when working adopting new techniques from research articles.

Take-home questions

Take-home tests check if a candidate has a command over key technical skills. Typically, the interviewer will prepare instructions, tasks, and data. The candidate is provided a package of test materials and given a time limit to undertake the test. We recommend choosing a representative task that the candidate would be expected to perform while on the job, but ensure it is either a mock task or an already completed task. In this section, we provide a set of high-level ideas.[1]

The main focus is to assess the quality and thoughtfulness of the answers and delivery. A code review should also be conducted to assess programming hygiene and accuracy. Example ideas:

- A team that focuses on data preparation can provide a dataset of a few thousand sentences of text. The candidate can be asked to extract all dollar amounts, linking each to the sentence from which it was extracted.

- A team focused on politics can prepare a corpus of news articles and ask the candidate to cluster articles, assign a "topic" label (e.g., topic modeling), and write a paragraph about the central themes.

- A clinical research team looking to build capacity in quasi-experimental techniques could provide a dataset and documentation about its context. The candidate can be asked to perform a Regression Discontinuity Design (RDD) at a pre-defined threshold. The output would be a short research brief detailing whether a discontinuity exists, which robustness checks were considered, and the implications.

- A team that regularly develops visualization tools can provide hundreds of time series. The candidate needs to develop an interactive dashboard that allows a user to toggle between time series (e.g., a flexdashboard).

- A policy and strategy team can ask a candidate to analyze a CSV of data to answer questions requiring the use of simple cross tabulations and regression models. The output would be in a form of a presentation along with a short five-minute talk to non-technical stakeholders.

- For a social services agency, provide the candidate with public micro data (e.g., US Census American Community Survey, Eurostat Labour Force Survey) and ask for a recommended set of population segments that reflect distinct economic experiences in the labor market.

- A transportation infrastructure team that specializes in risk management could provide data about roadway inspections and ask the candidate to predict which bridge (in a test set) is likely to have faults and why.

- For a security-focused agency, provide the candidate with multiple time series of server request volumes. The candidate is asked to identify and justify periods when unusual activity likely occurred.

- A GIS-oriented criminology team may need to quickly identify crime hot spots and provides a dataset containing latitude, longitude, and date of crime along with a shapefile of local districts. The candidate is asked to identify dates when districts have unusual spikes in activity (e.g., spatially join data, aggregate, then forecast the time series).

---

[1] Under no circumstances should current projects be given as this would raise some ethical flags.

- For a customer service-focused agency, provide the candidate with a time series of customer requests and staffing levels by time of day. Ask the candidate to forecast requests for the next two months and provide recommendations on staffing levels.

- For a performance monitoring team, provide 10 years of key performance indicators. Ask the candidate to develop labels for distinct types of activity (i.e., clusters of trends).

References

Abowd, John. 2018. Protecting the Confidentiality of America's Statistics: Adopting Modern Disclosure Avoidance Methods at the Census Bureau. https://www.census.gov/newsroom/blogs/research-matters/2018/08/protecting_the_confi.html

Aggarwal, Charu C. 2018a. *Machine Learning for Text*, 1st ed. Springer

Aggarwal, Charu C. 2018b. *Neural Networks and Deep Learning: A Textbook*. Springer. https://doi.org/10.1007/978-3-319-94463-0

Akaike, H.: A New Look at the Statistical Model Identification. IEEE Transactions on Automatic Control **19**(6), 716–23 (1974)

Akaike, Hirotogu. 1973. Information Theory and an Extension of the Maximum Likelihood Principle. In *Proceeding of the Second International Symposium on Information Theory*, 267–81

Alaimo, Katherine, Christine M. Olson, and Edward A. Frongillo. 1999. Importance of Cognitive Testing for Survey Items: An Example from Food Security Questionnaires. *Journal of Nutrition Education* 31 (5): 269–75. https://doi.org/10.1016/S0022-3182(99)70463-2

Alan Turing Institute. 2020. Alan Turing Institute - About Us. https://www.turing.ac.uk/about-us

Amazon. 2018. Amazon Selects New York City and Northern Virginia for New Headquarters, November. https://blog.aboutamazon.com/company-news/amazon-selects-new-york-city-and-northern-virginia-for-new-headquarters

Angrist, Joshua D., Keueger, Alan B.: Does Compulsory School Attendance Affect Schooling and Earnings? The Quarterly Journal of Economics **106**(4), 979–1014 (1991)

Angwin, Julia, Jeff Larson, Surya Mattu, and Lauren Kirchner. 2016. Machine Bias. *ProPublica*, May. https://www.propublica.org/article/machine-bias-risk-assessments-in-criminal-sentencing

Arrington, Michael. 2006. AOL Proudly Releases Massive Amounts of Private Data. *Tech Crunch*. https://techcrunch.com/2006/08/06/aol-proudly-releases-massive-amounts-of-user-search-data/

Arun, R., V. Suresh, C.E. Veni Madhavan, and M.N. Narasimha Murthy. 2010. On Finding the Natural Number of Topics with Latent Dirichlet Allocation: Some Observations. In *Advances in Knowledge Discovery and Data Mining*, ed. Mohammed J. Zaki, Jeffrey Xu Yu, B. Ravindran, and Vikram Pudi, 391–402. Berlin, Heidelberg: Springer

Aschwanden, Andy, Mark A. Fahnestock, Martin Truffer, Douglas J. Brinkerhoff, Regine Hock, Constantine Khroulev, Ruth Mottram, and S. Abbas Khan. 2019. Contribution of the Greenland Ice Sheet to Sea Level over the Next Millennium. *Science Advances* 5 (6). https://doi.org/10.1126/sciadv.aav9396

© Springer Nature Switzerland AG 2021
J. C. Chen et al., *Data Science for Public Policy*, Springer Series in the Data Sciences,
https://doi.org/10.1007/978-3-030-71352-2

Athey, Susan, and Guido W Imbens. 2018. Design-Based Analysis in Difference-in-Differences Settings with Staggered Adoption. National Bureau of Economic Research

Athey, Susan, and Stefan Wager. 2019. Estimating Treatment Effects with Causal Forests: An Application. arXiv:1902.07409

Avirgan, Jody. 2016. A History of Data in American Politics (Part 2): Obama 2008 to the Present. *FiveThirtyEight*. https://fivethirtyeight.com/features/a-history-of-data-in-american-politics-part-2-obama-2008-to-the-present/

Ayalew, Lulseged, and Hiromitsu Yamagishi. 2005. The Application of Gis-Based Logistic Regression for Landslide Susceptibility Mapping in the Kakuda-Yahiko Mountains, Central Japan. *Geomorphology* 65 (1): 15–31. http://www.sciencedirect.com/science/article/pii/S0169555X04001631

Bajari, Pat, Zhihao Cen, Victor Chernozhukov, Ramon Huerta, Junbo Li, Manoj Manukonda, and George Monokroussos. 2020. New Goods, Productivity and the Measurement of Inflation. *AEA Conference*. American Economic Association

BBC. 2017. US Homeless People Numbers Rise for First Time in Seven Years, December. https://www.bbc.com/news/world-us-canada-42248999

Beck, Martin, Florian Dumpert, and Joerg Feuerhake. 2018. Machine Learning in Official Statistics. *CoRR* abs/1812.10422. arXiv:1812.10422

Belloni, Alexandre, Victor Chernozhukov, and Ying Wei. 2016. Post-Selection Inference for Generalized Linear Models with Many Controls. *Journal of Business & Economic Statistics* 34. https://www.tandfonline.com/doi/abs/10.1080/07350015.2016.1166116?journalCode=ubes20

Bertrand, Marianne, Esther Duflo, and Sendhil Mullainathan. 2004. How Much Should We Trust Differences-in-Differences Estimates? *The Quarterly Journal of Economics* 119 (1): 249–75

Bishop, Christopher M. 2006. *Pattern Recognition and Machine Learning*. Berlin, Heidelberg: Springer

Black, Sandra E. 1999. Do Better Schools Matter? Parental Valuation of Elementary Education. *The Quarterly Journal of Economics* 114 (2): 577–99. http://www.jstor.org/stable/2587017

Blei, David M., Ng, Andrew Y., Jordan, Michael I.: Latent Dirichlet Allocation. Journal of Machine Learning Research **3**(4–5), 993–1022 (2003)

Bollen, Johan, Huina Mao, and Xiaojun Zeng. 2011. Twitter Mood Predicts the Stock Market. *Journal of Computational Science* 2 (1): 1–8. https://doi.org/10.1016/j.jocs.2010.12.007

Bolukbasi, Tolga, Kai-Wei Chang, James Zou, Venkatesh Saligrama, and Adam Kalai. 2016. Man Is to Computer Programmer as Woman Is to Homemaker? Debiasing Word Embeddings. In *Proceedings of the 30th International Conference on Neural Information Processing Systems*, 4356–64. NIPS'16. Red Hook, NY, USA: Curran Associates Inc

Breiman, Leo: Random Forests. Machine Learning **56**(1), 5–32 (2001)

Breiman, Leo: Bagging Predictors. Machine Learning **24**(2), 123–40 (1996)

Breiman, Leo, Jerome Friedman, Charles J. Stone, and R.A. Olshen. 1984. *Classification and Regression Trees*. The Wadsworth and Brooks-Cole Statistics-Probability Series. Taylor & Francis

Brennan, Allison. 2012. Microtargeting: How Campaigns Know You Better Than You Know Yourself. *CNN*. https://www.cnn.com/2012/11/05/politics/voters-microtargeting/index.html

Broockman, David, Joshua Kalla, and Peter Aronow. 2015. Irregularities in Lacour (2014), May. https://stanford.edu/~dbroock/broockman_kalla_aronow_lg_irregularities.pdf

Buolamwini, Joy, and Timnit Gebru. 2018. Gender Shades: Intersectional Accuracy Disparities in Commercial Gender Classification. In *Conference on Fairness, Accountability and Transparency*, 77–91

Burgess, Matt. 2019. What Is Gdpr? The Summary Guide to Gdpr Compliance in the Uk. *Wired.* https://www.wired.co.uk/article/what-is-gdpr-uk-eu-legislation-compliance-summary-fines-2018

Burns, Bob. 2012. Individuals on the No Fly List Are Not Issued Boarding Passes. *Transportation Security Administration Blog*, May. https://www.tsa.gov/blog/2012/05/11/individuals-no-fly-list-are-not-issued-boarding-passes

Cain, Aine, Anaele Pelisson, and Shayanne Gal. 2018. 9 Places in the Us Where Job Candidates May Never Have to Answer the Dreaded Salary Question Again. *Business Insider*, April. https://www.businessinsider.com/places-where-salary-question-banned-us-2017-10?r=US&IR=T

Calonico, Sebastian, Cattaneo, Matias D., Titiunik, Rocio: Robust Nonparametric Confidence Intervals for Regression-Discontinuity Designs. Econometrica **82**(6), 2295–2326 (2014)

Cao, Juan, Tian Xia, Jintao Li, Yongdong Zhang, and Sheng Tang. 2009. A Density-Based Method for Adaptive Lda Model Selection. *Neurocomputing* 72 (7): 1775–81. https://doi.org/10.1016/j.neucom.2008.06.011

Card, David, and Alan Krueger. 1994. Minimum Wages and Employment: A Case Study of the Fast-Food Industry in New Jersey and Pennsylvania. *American Economic Review* 84: 772–93

Carlson, Ben. 2019. These Charts Show Just How Awful the Economy Was When Paul Volcker Took over as Fed Chairman. *Fortune*, December. https://fortune.com/2019/12/09/charts-1973-19174-economy-paul-volcker-fed-chair

Centers for Disease Control. 2017. Leading Causes of Death and Numbers of Deaths, by Sex, Race, and Hispanic Origin: United States, 1980 and 2016. https://www.cdc.gov/nchs/data/hus/2017/019.pdf

Chambers, Bill, and Matei Zaharia. 2018. Spark: The Definitive Guide

Chawla, Nitesh V., Kevin W. Bowyer, Lawrence O. Hall, and W. Philip Kegelmeyer. 2002. SMOTE: Synthetic Minority over-Sampling Technique. *Journal of Artificial Intelligence Research*

Chen, Jeffrey C., Abe Dunn, Kyle Hood, Alexander Driessen, and Andrea Batch. 2019. Off to the Races: A Comparison of Machine Learning and Alternative Data for Predicting Economic Indicators. Book. In *Big Data for 21st Century Economic Statistics*, by Katharine G. Abraham, Ron S. Jarmin, Brian Moyer, and Matthew D. Shapiro. National Bureau of Economic Research; University of Chicago Press. http://www.nber.org/chapters/c14268

Chen, Tianqi, and Carlos Guestrin. 2016. XGBoost: A Scalable Tree Boosting System. In *Proceedings of the 22Nd Acm Sigkdd International Conference on Knowledge Discovery and Data Mining*, 785–94. KDD '16. San Francisco, California, USA: ACM. https://doi.org/10.1145/2939672.2939785

Chicago Police Department. 2018. Crimes - 2001 to Present. https://data.cityofchicago.org/Public-Safety/Crimes-2001-to-present/ijzp-q8t2/data

Chollet, Francois, and J.J. Allaire. 2018. *Deep Learning with R*. Shelter Island, NY: Manning Publications

Cillizza, Chris. 2007. Romney's Data Cruncher. *The Washington Post.* http://www.washingtonpost.com/wp-dyn/content/article/2007/07/04/AR2007070401423.html

Cleveland, Robert B, William S Cleveland, and Irma Terpenning. 1990. STL: A Seasonal-Trend Decomposition Procedure Based on Loess. *Journal of Official Statistics* 6 (1)

Cohen, Jacob. 1960. A Coefficient of Agreement for Nominal Scales. *Educational and Psychological Measurement* 20 (1): 37–46. https://doi.org/10.1177/001316446002000104

Cohen, Peter, Robert Hahn, Jonathan Hall, Steven Levitt, and Robert Metcalfe. 2016. Using Big Data to Estimate Consumer Surplus: The Case of Uber. Working Paper 22627. Working Paper Series. National Bureau of Economic Research. https://doi.org/10.3386/w22627

Commission of European Communities. 2002. Commission Regulation (Ec) No 831/2002. https://eur-lex.europa.eu/legal-content/EN/TXT/?uri=CELEX%3A32002R0831

Cook, Thomas D, and Vivian C Wong. 2008. Empirical Tests of the Validity of the Regression Discontinuity Design. *Annales d'Economie et de Statistique*, 127–50

Corbett-Davies, Sam, Emma Pierson, Avi Feller, and Sharad Goel. 2016. A Computer Program Used for Bail and Sentencing Decisions Was Labeled Biased Against Blacks. It's Actually Not That Clear. *The Washington Post*. https://www.washingtonpost.com/news/monkey-cage/wp/2016/10/17/can-an-algorithm-be-racist-our-analysis-is-more-cautious-than-propublicas

Corbett-Davies, Sam, Emma Pierson, Avi Feller, Sharad Goel, and Aziz Huq. 2017. Algorithmic Decision Making and the Cost of Fairness. In *Proceedings of the 23rd Acm Sigkdd International Conference on Knowledge Discovery and Data Mining*, 797–806. KDD '17. New York, NY, USA: Association for Computing Machinery. https://doi.org/10.1145/3097983.3098095

Council of European Union. 2014. Council Regulation (EU) No 269/2016. https://eur-lex.europa.eu/legal-content/EN/TXT/HTML/?uri=CELEX:32016R0679&from=EN#d1e3265-1-1

Cox, Lawrence H. 1980. Suppression Methodology and Statistical Disclosure Control. *Journal of the American Statistical Association* 75 (370): 377–85. http://www.jstor.org/stable/2287463

Crutsinger, Martin. 2019. US Deficit Hits Nearly $1 Trillion. When Will It Matter? *Associated Press*, October. https://apnews.com/caeb6d6c4eff45e4bc5da12db06004bc

Curry, Tom. 2007. The Mechanics of Micro-Targeting. *NBC News*. http://www.nbcnews.com/id/15292903/ns/politics-tom_curry/t/mechanics-micro-targeting/#.XUDzgZNKgb1

Dastin, Jeffrey. 2018. Amazon Scraps Secret Ai Recruiting Tool That Showed Bias Against Women. *Reuters*. https://www.reuters.com/article/us-amazon-com-jobs-automation-insight/amazon-scraps-secret-ai-recruiting-tool-that-showed-bias-against-women-idUSKCN1MK08G

Datta, Amit, Michael Carl Tschantz, and Anupam Datta. 2014. Automated Experiments on Ad Privacy Settings: A Tale of Opacity, Choice, and Discrimination. arXiv:1408.6491

Davenport, Thomas H., and D. J. Patil. 2012. Data Scientist: The Sexiest Job of the 21st Century. *Harvard Business Review*. https://hbr.org/2012/10/data-scientist-the-sexiest-job-of-the-21st-century

DeNisco Rayome, Alison. 2019. How Has Gdpr Actually Affected Businesses? *TechRepublic*, May. https://www.techrepublic.com/article/how-has-gdpr-actually-affected-businesses/

Dillet, Romain. 2019. French Data Protection Watchdog Fines Google $57 Million Under the Gdpr. *Tech Crunch*. https://techcrunch.com/2019/01/21/french-data-protection-watchdog-fines-google-57-million-under-the-gdpr

Dobnik, Verena. 2007. Two Firefighters Dead at 7-Alarm Deutsche Bank Blaze. *New York Daily News*, August. https://www.nydailynews.com/news/firefighters-dead-7-alarm-deutsche-bank-blaze-article-1.238838

Dwork, Cynthia. 2011. Differential Privacy. In *Encyclopedia of Cryptography and Security*, ed. Henk C.A. van Tilborg and Sushil Jajodia, 338–40. Boston, MA: Springer. https://doi.org/10.1007/978-1-4419-5906-5_752

Dwork, Cynthia, Frank McSherry, Kobbi Nissim, and Adam Smith. 2006. Calibrating Noise to Sensitivity in Private Data Analysis. In *Theory of Cryptography*, ed. Shai Halevi and Tal Rabin, 265–84. Berlin, Heidelberg: Springer

Dwork, Cynthia, Aaron Roth, and others. 2014. The Algorithmic Foundations of Differential Privacy. *Foundations and Trends in Theoretical Computer Science* 9 (3–4): 211–407

Edwards, Lilian, and Michael Veale. 2018. Enslaving the Algorithm: From a 'Right to an Explanation' to a 'Right to Better Decisions'? *IEEE Security Privacy* 16 (3): 46–54

Elis, Niv. 2019. 2019 Deficit Nears \$1 Trillion, Highest Since 2012: Treasury. *The Hill*, October. https://thehill.com/policy/finance/467488-2019-deficit-nears-1-trillion-highest-since-2012-treasury

Enamorado, Ted, Benjamin Fifield, and Kosuke Imai. 2019. Using a Probabilistic Model to Assist Merging of Large-Scale Administrative Records. *American Political Science Association* 113 (2): 353–71

Energy Information Administration. 2019. Electric Power Monthly. https://www.eia.gov/electricity/monthly/epm_table___rapher.php?t=epmt_1_01

Energy Information Administration. 2007. Spot Prices for Crude Oil and Petroleum Products. http://www.eia.gov/dnav/pet/pet_pri_spt_s1_d.htm

Eurostat. 2020. EU Labour Force Survey Database. https://ec.europa.eu/eurostat/documents/1978984/6037342/EULFS-Database-UserGuide.pdf

Federal Reserve Bank of Philadelphia. 2019. Real-Time Data Set for Macroeconomists. https://www.philadelphiafed.org/research-and-data/real-time-center/real-time-data

Fellegi, Ivan P., Sunter, Alan B.: A Theory of Record Linkage. Journal of the American Statistical Association **64**, 1183–1210 (1969)

Fitzpatrick, Jen, and Karen DeSalvo. 2020. Helping Public Health Officials Combat Covid-19. *Google - the Keyword Blog.* https://www.blog.google/technology/health/covid-19-community-mobility-reports

Franck, Thomas. 2019. Federal Deficit Increases 26% to \$984 Billion for Fiscal 2019, Highest in 7 Years. *CNBC*, October. https://www.cnbc.com/2019/10/25/federal-deficit-increases-26percent-to-984-billion-for-fiscal-2019.html

Frenkel, Sheera, and Mike Isaac. 2018. Facebook Gives Workers a Chatbot to Appease That Prying Uncle. *The New York Times*, December. https://www.nytimes.com/2019/12/02/technology/facebook-chatbot-workers.html

Friedman, Jerome: Greedy Function Approximation: A Gradient Boosting Machine. Annals of Statistics **29**(5), 1189–1232 (2001)

Gaba, Kwawu Mensan, Brian Min, Anand Thakker, and Chris Elvidge. 2016. Twenty Years of India Lights. http://india.nightlights.io/

Gelman, Andrew, and Guido Imbens. 2018. Why High-Order Polynomials Should Not Be Used in Regression Discontinuity Designs. *Journal of Business and Economic Statistics*, 1–10

Geron, Aurelien. 2017. *Hands-on Machine Learning with Scikit-Learn and Tensorflow: Concepts, Tools, and Techniques to Build Intelligent Systems.* Sebastopol, CA: O'Reilly Media

Gonzalez-Ibanez, Roberto, Smaranda Muresan, and Nina Wacholder. 2011. Identifying Sarcasm in Twitter: A Closer Look. In *Proceedings of the 49th Annual Meeting of the Association for Computational Linguistics: Human Language Technologies*, 581–86. Portland, Oregon, USA: Association for Computational Linguistics. https://www.aclweb.org/anthology/P11-2102

Goo, Sara Kehaulani. 2004. Sen. Kennedy Flagged by No-Fly List. *Washington Post.* http://www.washingtonpost.com/wp-dyn/articles/A17073-2004Aug19.html

Goodman-Bacon, Andrew. 2018. Difference-in-Differences with Variation in Treatment Timing. National Bureau of Economic Research

Google. 2009. Google R Style Guide. https://google.github.io/styleguide/Rguide.xml

Greenberg, Andy. 2017. How One of Apple's Key Privacy Safeguards Falls Short. *Wired*. https://www.wired.com/story/apple-differential-privacy-shortcomings

Greenhouse, Steven. 2002. Labor Lockout at West's Ports Roils Business. *The New York Times*, October. https://www.nytimes.com/2002/10/01/us/labor-lockout-at-west-s-ports-roils-business.html

Hadamard, Jacques: Sur Les Problèmes Aux Dérivées Partielles et Leur Signication Physique. Princeton University Bulletin **13**, 49–52 (1902)

Hamins, Anthony P., Casey Grant, Nelson P. Bryner, Albert W. Jones, and Galen H. Koepke. 2015. Research Roadmap for Smart Fire Fighting. *NIST SP 1191*, June. https://doi.org/10.6028/NIST.SP.1191

Hao, Karen. 2019. We Analyzed 16,625 Papers to Figure Out Where Ai Is Headed Next. *MIT Technology Review*, January. https://www.technologyreview.com/2019/01/25/1436/we-analyzed-16625-papers-to-figure-out-where-ai-is-headed-next

Harrison, David, and Daniel L Rubinfeld. 1978. Hedonic Housing Prices and the Demand for Clean Air. *Journal of Environmental Economics and Management* 5 (1): 81–102. https://doi.org/10.1016/0095-0696(78)90006-2

Hastie, Trevor, Robert Tibshirani, and Jerome Friedman. 2001. *The Elements of Statistical Learning*. Springer Series in Statistics. New York: Springer

Hastie, Trevor, Robert Tibshirani, and Jerome H. Friedman. 2009. *The Elements of Statistical Learning: Data Mining, Inference, and Prediction*. Springer Series in Statistics. Springer

Hazel, James, and Christopher Slobogin. 2018. Who Knows What, and When?: A Survey of the Privacy Policies Proffered by U.s. Direct-to-Consumer Genetic Testing Companies. *Cornell Journal of Law and Public Policy, Vanderbilt Law Research Paper No. 18-18*. https://ssrn.com/abstract=3165765

Hearn, Rose Gill, Nicholas Scoppetta, and Robert LiMadri. 2009. Investigation of Administrative Issues Relating to the Fatal Fire of August 18, 2007 at 130 Liberty Street, June. https://www1.nyc.gov/assets/doi/downloads/pdf/pr_db_61909_final.pdf

Hernandez, Michael J. 2013. *Database Design for Mere Mortals: A Hands-on Guide to Relational Database Design Inc*. USA: Addison-Wesley Longman Publishing Co., Inc

Hess, Abigail. 2019. Women Are Slowly Pursuing More High-Paying Degrees, but the Pay Gap Remains, Says New Research. *CNBC*, October. https://www.cnbc.com/2019/10/04/women-are-pursuing-more-high-paying-degrees-but-the-pay-gap-remains.html

Higgins, Tim. 2016. Sanders Supporters Like Chipotle, While Trump Fans Prefer Sonic. *Bloomberg*. https://www.bloomberg.com/news/articles/2016-02-18/sanders-supporters-like-chipotle-while-trump-fans-prefer-sonic

Highways England. 2020. *Highways England Pavement Management System Network Layer*. https://data.gov.uk/dataset/2b0dd22d-213e-4f5b-99da-8b5ec409112c/highways-england-pavement-management-system-network-layer

HM Land Registry Public Data. 2019. UK Housing Price Index. https://landregistry.data.gov.uk/app/ukhpi

Hoerl, A.E., and R.W. Kennard. 1970. Ridge Regression: Applications to Non-Orthogonal Problems. *Technometrics* 12 (1)

Humphries, Stan. 2019. Introducing a New and Improved Zestimate Algorithm. *Zillow*. https://www.zillow.com/tech/introducing-a-new-and-improved-zestimate-algorithm/

Hundepool, Anco, Josep Domingo-Ferrer, Luisa Franconi, Sarah Giessing, Eric Schulte Nordholt, Keith Spicer, and Peter-Paul de Wolf. 2012. *Statistical Disclosure Control.* John Wiley & Sons, Ltd. https://doi.org/10.1002/9781118348239

Hurvich, Clifford M., and Chih-Ling Tsai. 1993. A Corrected Akaike Information Criterion for Vector Autoregressive Model Selection. Journal of Time Series Analysis **14**(3), 271–79. https://onlinelibrary.wiley.com/doi/abs/10.1111/j.1467-9892.1993.tb00144.x

Imbens, Guido and Karthik Kalyanaraman. 2012. Optimal Bandwidth Choice for the Regression Discontinuity Estimator. *The Review of Economic Studies* 79 (3): 933–59

Imbens, Guido W, and Donald B Rubin. 2015. *Causal Inference in Statistics, Social, and Biomedical Sciences.* Cambridge University Press

Imbert, Fred. 2017. Stocks Close Higher as Wall Street Shakes Off Us-North Korea Tension. *CNBC.* https://www.cnbc.com/2017/08/29/us-stocks-north-korea-missile.html

Isidore, Chris. 2015. West Coast Ports Shut down as Labor Dispute Heats up. *CNN.* https://money.cnn.com/2015/02/12/news/companies/port-shutdown

Jalan, Jyotsna, and Martin Ravallion. 2003. Estimating the Benefit Incidence of an Antipoverty Program by Propensity-Score Matching. *Journal of Business & Economic Statistics* 21 (1): 19–30. https://doi.org/10.1198/073500102288618720

James, Gareth, Daniela Witten, Trevor Hastie, and Robert Tibshirani. 2014. *An Introduction to Statistical Learning: With Applications in R.* New York: Springer

Jensen, Meiko, Cedric Lauradoux, and Konstantinos Limniotis. 2019. *Pseudonymisation Techniques and Best Practices.* European Union Agency for Cybersecurity. https://www.enisa.europa.eu/publications/pseudonymisation-techniques-and-best-practices

Kaye, Kate. 2019. New York Just Set a'Dangerous Precedent' on Algorithms, Experts Warn. *Bloomberg CityLab.* https://www.bloomberg.com/news/articles/2019-12-12/nyc-sets-dangerous-precedent-on-algorithms

Kayser-Bril, Nicolas. 2020. Google Apologizes After Its Vision Ai Produced Racist Results. *Algorithm Watch.* https://algorithmwatch.org/en/story/google-vision-racism/

Kennedy, Courtney, Scott Clement, Mark Blumenthal, and Christopher Wlezien. 2018. An Evaluation of 2016 Election Polls in the U.s. *Public Opinion Quarterly* 82: 1–3. https://doi.org/10.1093/poq/nfx047

Kim, Been, Rajiv Khanna, and Oluwasanmi O Koyejo. 2016. Examples Are Not Enough, Learn to Criticize! Criticism for Interpretability. In *Advances in Neural Information Processing Systems 29*, ed. D.D. Lee, M. Sugiyama, U.V. Luxburg, I. Guyon, and R. Garnett, 2280–8. Curran Associates, Inc. http://papers.nips.cc/paper/6300-examples-are-not-enough-learn-to-criticize-criticism-for-interpretability.pdf

Kleinberg, Jon, Sendhil Mullainathan, and Manish Raghavan. 2017. Inherent Trade-Offs in the Fair Determination of Risk Scores.' *Proceedings of the 8th Conference on Innovations in Theoretical Computer Science.* arXiv:1609.05807v2

Knight, Will. 2019. AI Is Biased. Here's How Scientists Are Trying to Fix It. *Wired.* https://www.wired.com/story/ai-biased-how-scientists-trying-fix

Koman, Richard. 2006. Democrats Discover Database Cavassing in Time for Nov. Elections. *ZDNet.* https://www.zdnet.com/article/democrats-discover-database-cavassing-in-time-for-nov-elections

Krueger, Alan B., Summers, Lawrence H.: Efficiency Wages and the Inter-Industry Wage Structure. Econometrica **56**(2), 259–93 (1988)

Kuhn, Max, and Kjell Johnson. 2013. *Applied Predictive Modeling.* New York: Springer

Kwapisz, J.R., G.M. Weiss, and S.A. Moore. 2010. Cell Phone-Based Biometric Identification. *2010 Fourth IEEE International Conference on Biometrics: Theory, Applications and Systems (BTAS)*, September. https://ieeexplore.ieee.org/document/5634532

LaCour, David, and Joshua Green. 2014. When Contact Changes Minds: An Experiment on Transmission of Support for Gay Equality. *Science* 346 (December): 1366–9. https://science.sciencemag.org/content/346/6215/1366

Laptev, Nikolay, Slawek Smyl, and Santhosh Shanmugam. 2017. Engineering Extreme Event Forecasting at Uber with Recurrent Neural Networks. *Uber Engineering Blog.* https://eng.uber.com/neural-networks/

Lardinois, Frederic. 2019. Google Launches an Open-Source Version of Its Differential Privacy Library. *Tech Crunch.* https://tcrn.ch/2Lvvqfh

Larson, Jeff, Surya Mattu, Lauren Kirchner, and Julia Angwin. 2016. How We Analyzed the Compas Recidivism Algorithm. *ProPublica*, May. https://www.propublica.org/article/how-we-analyzed-the-compas-recidivism-algorithm

Le, Quoc, and Tomas Mikolov. 2014. Distributed Representations of Sentences and Documents. In *Proceedings of the 31st International Conference on International Conference on Machine Learning*, 32:1188–96. ICML14. Beijing, China: JMLR.org

LeCun, Yann, Yoshua Bengio, and Geoffrey Hinton. 2015. Deep Learning. *Nature* 521 (7553): 436–44. https://doi.org/10.1038/nature14539

Lee, David S.: Randomized Experiments from Non-Random Selection in Us House Elections. Journal of Econometrics **142**(2), 675–97 (2008)

Lee, David S, and Thomas Lemieux. 2010. Regression Discontinuity Designs in Economics. *Journal of Economic Literature* 48 (2): 281–355

Li, Xingong, Rowley, Rex J., Kostelnick, John C., Braaten, David, Meisel, Joshua, Hulbutta, Kalonie: GIS Analysis of Global Impacts from Sea Level Rise. Photogrammetric Engineering and Remote Sensing **75**, 807–18 (2009)

Li, Xuecao, Yuyu Zhou, and Xia Zhao. 2020. A Harmonized Global Nighttime Light Dataset 1992–2018. *Scientific Data* 7 (168). https://doi.org/10.1038/s41597-020-0510-y

Lipton, Zachary C. 2018. The Mythos of Model Interpretability: In Machine Learning, the Concept of Interpretability Is Both Important and Slippery. *Queue* 16 (3): 31–57. https://doi.org/10.1145/3236386.3241340

Liu, Bing, Minqing Hu, and Junsheng Cheng. 2005. Opinion Observer: Analyzing and Comparing Opinions on the Web. In *Proceedings of the 14th International Conference on World Wide Web*, 342?351. New York: Association for Computing Machinery. https://doi.org/10.1145/1060745.1060797

Lohr, Sharon. 2019. Sampling: Design and Analysis. https://doi.org/10.1201/9780429296284

Lohr, Steve. 2014. For Big-Data Scientists, 'Janitor Work' Is Key Hurdle to Insights. *The New York Times.* https://www.nytimes.com/2014/08/18/technology/for-big-data-scientists-hurdle-to-insights-is-janitor-work.html

Luraschi, Javier, Kevin Kuo, and Edgar Ruiz. 2019. *Mastering Spark with R.* O'Reilly Media, Inc

Madaio, Michael, Shang-Tse Chen, Oliver L. Haimson, Wenwen Zhang, Xiang Cheng, Matthew Hinds-Aldrich, Duen Horng Chau, and Bistra Dilkina. 2016. Firebird: Predicting Fire Risk and Prioritizing Fire Inspections in Atlanta, KDD '16, 185–94. https://doi.org/10.1145/2939672.2939682

Mantyla, Mika V., Daniel Graziotin, and Miikka Kuutila. 2018. The Evolution of Sentiment Analysis: A Review of Research Topics, Venues, and Top Cited Papers. *Computer Science Review* 27: 16–32. http://www.sciencedirect.com/science/article/pii/S1574013717300606

Mason, Ross. 2017. Have You Had Your Bezos Moment? What You Can Learn from Amazon. *CIO Magazine*, August. https://www.cio.com/article/3218667/have-you-had-your-bezos-moment-what-you-can-learn-from-amazon.html

Mathiowetz, Nancy A., Charlie Brown, and John Bound. 2002. Measurement Error in Surveys of the Low-Income Population. *Studies of Welfare Populations: Data Collection and Research Issues*. https://doi.org/10.17226/10206

McEwen, Melissa. 2018. The Latest Trend for Tech Interviews: Days of Unpaid Homework. *Quartz*. https://qz.com/work/1254663/job-interviews-for-programmers-now-often-come-with-days-of-unpaid-homework

McKinney, Wes. 2017. Python for Data Analysis

Meishausen, Nicolai. 2006. Quantile Regression Forests. *Journal of Machine Learning Research* 7: 983–99. http://www.jmlr.org/papers/v7/meinshausen06a.html

Mervis, Jeffrey. 2019. Can a Set of Equations Keep U.s. Census Data Private? *Science*. https://www.sciencemag.org/news/2019/01/can-set-equations-keep-us-census-data-private

Microsoft. 2018. Computer Generated Building Footprints for the United States. *Github Repository*. https://github.com/microsoft/USBuildingFootprints

Mikolov, Tomas, Kai Chen, Greg Corrado, and Jeffrey Dean. 2013. Efficient Estimation of Word Representations in Vector Space. arXiv:1301.3781

Miller, Tim. 2019. Explanation in Artificial Intelligence: Insights from the Social Sciences. *Artificial Intelligence* 267: 1–38. https://doi.org/10.1016/j.artint.2018.07.007

Mills, Jeffrey, and Kislaya Prasad. 1992. A Comparison of Model Selection Criteria. *Econometric Reviews* 11 (February): 201–34. https://doi.org/10.1080/07474939208800232

Min, Brian. 2014. Monitoring Rural Electrification by Satellite. The World Bank. http://india.nightlights.io/

Min, Brian, Kwawu Mensan Gaba. 2014. Tracking Electrification in Vietnam Using Nighttime Lights. *Remote Sensing* 6 (10): 9511–29

Mohammad, Saif M., Turney, Peter D.: Crowdsourcing a Word-Emotion Association Lexicon. Computational Intelligence **29**(3), 436–65 (2013)

Moskowitz, Eric. 2011. Weapons in the Battle Vs. Potholes. *The Boston Globe*. http://archive.boston.com/news/local/massachusetts/articles/2011/02/09/weapons_in_the_battle_vs_potholes

NASA Earth Science. 2019. *MODIS Collection 6 Nrt Hotspot / Active Fire Detections Mcd14dl*. https://www.nasa.gov/feature/goddard/2019/study-predicts-more-long-term-sea-level-rise-from-greenland-ice

National Institute of Standards and Technology. 2016. NIST Net-Zero Energy Residential Test Facility. https://pages.nist.gov/netzero/research.html

National Oceanic and Atmospheric Administration. 2016. Defense Meteorological Satellite Program (Dmsp). https://www.ngdc.noaa.gov/eog/dmsp.html

Navarro, Gonzalo. 2001. A Guided Tour to Approximate String Matching. *ACM Computing Surveys* 33 (1): 31–88. https://doi.org/10.1145/375360.375365

Nicol Turner Lee, Paul Resnick, and Genie Barton. 2019. Algorithmic Bias Detection and Mitigation: Best Practices and Policies to Reduce Consumer Harms. *Brookings*, May. https://www.brookings.edu/research/algorithmic-bias-detection-and-mitigation-best-practices-and-policies-to-reduce-consumer-harms/

Nielsen, Finn Århup. 2011. A New Anew: Evaluation of a Word List for Sentiment Analysis in Microblogs. *Proceedings of the ESWC2011 Workshop on 'Making Sense of Microposts': Big Things Come in Small Packages*, 93–98

Nigeria National Bureau of Statistics. 2019. General Household Survey, Panel (Ghs-Panel) 2018-2019. https://microdata.worldbank.org/index.php/catalog/3557

NYC Department of Information Technology and Telecommunication. 2020. 311 Service Requests from 2010 to Present. https://data.cityofnewyork.us/Social-Services/311-Service-Requests-from-2010-to-Present/erm2-nwe9/data

NYC Mayor's Office of Operations. 2012. Mayor's Management Report, Fy2012, September. https://www1.nyc.gov/assets/operations/downloads/pdf/mmr0912/0912_mmr.pdf

O'Neil, Cathy: Weapons of Math Destruction: How Big Data Increases Inequality and Threatens Democracy. Crown Publishing Group, USA (2016)

Owano, Nancy. 2013. Accelerometer in Phone Has Tracking Potential, Researchers Find. *Phys.org*, October. https://phys.org/news/2013-10-accelerometer-tracking-potential.html

Page, Carly. 2020. Oracle and Salesforce Hit with $10 Billion Gdpr Class-Action Lawsuit. *Forbes*. https://www.forbes.com/sites/carlypage/2020/08/14/oracle-and-salesforce-hit-with-10-billion-gdpr-class-action-lawsuit

Pak, Alexander, and Patrick Paroubek. 2010. Twitter as a Corpus for Sentiment Analysis and Opinion Mining. In *Proceedings of the Seventh International Conference on Language Resources and Evaluation (LREC'10)*. European Language Resources Association (ELRA)

Patten, Eileen. 2016. Racial, Gender Wage Gaps Persist in U.s. Despite Some Progress. *Pew Research*, July. https://www.pewresearch.org/fact-tank/2016/07/01/racial-gender-wage-gaps-persist-in-u-s-despite-some-progress/

Pearl, Judea. 2009. *Causality*. Cambridge University Press

Pesch, Beate, Benjamin Kendzia, Per Gustavsson, Karl-Heinz Jockel, Georg Johnen Hermann Pohlabeln, Ann Olsson, Wolfgang Ahrens, et al. 2012. Cigarette Smoking and Lung Cancer–Relative Risk Estimates for the Major Histological Types from a Pooled Analysis of Case-Control Studies. *International Journal of Cancer* 131 (5): 1210–9. https://www.ncbi.nlm.nih.gov/pubmed/22052329

Playfair, William. 1801. *The Commercial and Political Atlas, Representing, by Means of Stained Copper-Plate Charts, the Progress of the Commerce, Revenues, Expenditure, and Debts of England*. During the Whole of the Eighteenth Century. London: Wallis

Port of Long Beach. 2019. TEUs Archive Since 1995. http://www.polb.com/economics/stats/teus_archive.asp

Public Law 107-347: Title V - Confidential Information Protection and Statistical Efficiency (116 Stat. 2962). 2002. https://www.congress.gov/bill/107th-congress/house-bill/5215/text

Raftery, Adrian E. 1995. Bayesian Model Selection in Social Research. *Sociological Methodology* 25: 111–63. http://www.jstor.org/stable/271063

Rahimi, Ali. 2017. Reflections on Random Kitchen Sinks. *Argmin Blog*. http://www.argmin.net/2017/12/05/kitchen-sinks/

Rappeport, Alan. 2019. *The New York Times*, October. https://www.nytimes.com/2019/10/25/us/politics/us-federal-budget-deficit.html

Ratcliffe, Caroline, and Steven Brown. 2017. Credit Scores Perpetuate Racial Disparities, Even in America's Most Prosperous Cities. *Urban Wire - the Blog of the Urban Institute*. https://www.urban.org/urban-wire/credit-scores-perpetuate-racial-disparities-even-americas-most-prosperous-cities

Ribeiro, Marco Tulio, Sameer Singh, and Carlos Guestrin. 2016. Why Should I Trust You?: Explaining the Predictions of Any Classifier. arXiv:1602.04938

Rivera, Ray. 2007. 2 Firefighters Are Dead in Deutsche Bank Fire. *New York Times*, August. https://cityroom.blogs.nytimes.com/2007/08/18/2-firefighters-are-dead-in-deutsche-bank-fire/

Roberts, Margaret E., Brandon M. Stewart, and Edoardo M. Airoldi. 2016. A Model of Text for Experimentation in the Social Sciences. *Journal of the American Statistical Association* 111 (515): 988–1003. https://doi.org/10.1080/01621459.2016.1141684

Rocher, Luc, Julien M. Hendrickx, and Yves-Alexandre de Montjoye. 2019. Estimating the Success of Re-Identifications in Incomplete Datasets Using Generative Models. *Nature Communications* 10. https://www.nature.com/articles/s41467-019-10933-3

Roman, Jesse. 2014. In Pursuit of Smart. *NFPA Journal*, November. http://www.nfpa.org/news-and-research/publications/nfpa-journal/2014/november-december-2014/features/in-pursuit-of-smart

Romell, Rick, and Joe Taschler. 2018. Foxconn Purchases Downtown Racine Property as Part of 'Smart City' Plans Across State. *Journal Sentinel*, October. https://www.jsonline.com/story/money/business/2018/10/02/foxconn-develop-downtown-racine-site/1499783002/

Rosenbaum, Paul R., Rubin, Donald B.: The Central Role of the Propensity Score in Observational Studies for Causal Effects. Biometrika **70**(1), 41–55 (1983)

Rowley, Rex J., John C. Kostelnick, David Braaten, Xingong Li, and Joshua Meisel. 2007. Risk of Rising Sea Level to Population and Land Area. *Eos, Transactions American Geophysical Union* 88 (9): 105–7. https://doi.org/10.1029/2007EO090001

Rubin, Donald B. 1976. Inference and Missing Data. *Biometrika* 63: 581–90. https://doi.org/10.1093/biomet/63.3.581

Rudin, Cynthia. 2019. Stop Explaining Black Box Machine Learning Models for High Stakes Decisions and Use Interpretable Models Instead. *Nature Machine Intelligence* 1: 206–15. https://doi.org/10.1038/s42256-019-0048-x

Samuel, Alexandra. 2018. Amazon's Mechanical Turk Has Reinvented Research. *JSTOR Daily*, May. https://daily.jstor.org/amazons-mechanical-turk-has-reinvented-research/

Santos-Lozada, Alexis R., Jeffrey T. Howard, and Ashton M. Verdery. 2020. How Differential Privacy Will Affect Our Understanding of Health Disparities in the United States. *Proceedings of the National Academy of Sciences* 117 (24): 13405–12. https://doi.org/10.1073/pnas.2003714117

Scherer, Michael. 2012. Inside the Secret World of the Data Crunchers Who Helped Obama Win. *Time*. http://swampland.time.com/2012/11/07/inside-the-secret-world-of-quants-and-data-crunchers-who-helped-obama-win/

Schroeder, Robert. 2019. U.S. Federal Budget Deficit Runs at Nearly $1 Trillion in 2019. *Market Watch*, October. https://www.marketwatch.com/story/us-runs-fiscal-2019-budget-deficit-of-nearly-1-trillion-2019-10-25

Schwarz, Gideon: Estimating the Dimension of a Model. The Annals of Statistics **6**(2), 461–64 (1978)

Schwarz, Gideon: Estimating the Dimension of a Model. The Annals of Statistics **6**(2), 461–64 (1978)

SCOUT, NYC. 2020. Street Condition Observations. *NYC Mayor's Office.* https://www1.nyc.gov/site/operations/performance/scout-street-condition-observations.page

Shadish, William R., Thomas D. Cook, and Donald T. Campbell. 2002. Experimental and Quasi-Experimental Designs for Generalized Causal Inference

Shinal, John. 2018. Facebook Has a Case Study Showing How Much Its Ad Targeting Tools Helped Presidential Candidate Gary Johnson. *CNBC.* https://www.cnbc.com/2018/02/21/facebook-targeting-tools-helped-libertarian-candidate-gary-johnson.html

Singh Walia, Bhavkaran, Qianyi Hu, Jeffrey Chen, Fangyan Chen, Jessica Lee, Nathan Kuo, Palak Narang, Jason Batts, Geoffrey Arnold, and Michael Madaio. 2018. A Dynamic Pipeline for Spatio-Temporal Fire Risk Prediction, KDD '18, 764–73. https://doi.org/10.1145/3219819.3219913

Smith, Greg B. 2013. Mayoral Hopeful Bill de Blasio Has Had Three Different Legal Names, Court Records Show. *NY Daily News,* September. http://www.nydailynews.com/news/election/de-blasio-names-de-blasio-article-1.1463591

Social Security Administration. 2017. Popular Baby Names. https://www.ssa.gov/OACT/babynames/index.html

Soundex System. 2007. *The National Archives,* May. https://www.archives.gov/research/census/soundex.html

Staniak, Mateusz, and Przemysaw Biecek. 1988. Explanations of Model Predictions with Live and breakDown Packages. *R Journal* 56 (2): 259–93

Staniak, Mateusz, and Przemyslaw Biecek. 2019. Explanations of Model Predictions with Live and breakDown Packages. *The R Journal* 10 (2): 395. https://doi.org/10.32614/rj-2018-072

Stolberg, Sheryl Gay, and Robert Pear. 2003. Obama Signs Health Care Overhaul Bill, with a Flourish. *The New York Times.* http://www.nytimes.com/2010/03/24/health/policy/24health.html?mcubz=1

The Battle of New Orleans. 2010. *History.com,* February. https://www.history.com/this-day-in-history/the-battle-of-new-orleans

The Economist. 2014. The New Border War, March. https://www.economist.com/united-states/2014/03/22/the-new-border-war

Thind, Jasjeet. 2016. Regression Model for Home Price Growth Using Repeat Sales. *Zillow.* https://www.zillow.com/tech/weighted-repeat-sales/

Thistlethwaite, Donald L., Campbell, Donald T.: Regression-Discontinuity Analysis: An Alternative to the Ex Post Facto Experiment. Journal of Educational Psychology **51**(6), 309 (1960)

Tibshirani, Robert: Regression Shrinkage and Selection via the Lasso. Journal of the Royal Statistical Society, Series B **58**(1), 267–88 (1996)

Toohey, M., and K. Strong. 2007. Estimating Biases and Error Variances Through the Comparison of Coincident Satellite Measurements. *Journal of Geophysical Research: Atmospheres* 112 (D13). https://doi.org/10.1029/2006JD008192

Trentmann, Nina. 2019. Resonate with Humans and Robots. *The Wall Street Journal,* November. http://www.sciencedirect.com/science/article/pii/S187775031100007X

Trump-Pence Campaign. 2020. Official Trump 2020 Campaign Strategy Survey. https://www.donaldjtrump.com/landing/the-official-2020-strategy-survey

Tufte, Edward R. 1983. The Visual Display of Quantitative Information

Umoh, Ruth. 2018. Why Jeff Bezos Makes Amazon Execs Read 6-Page Memos at the Start of Each Meeting. *CNBC,* April. https://www.cnbc.com/2018/04/23/what-jeff-bezos-learned-from-requiring-6-page-memos-at-.html

U.S. Census Bureau. 2017a. Small Area Income and Poverty Estimates (Saipe) Program. https://www.census.gov/programs-surveys/saipe.html

U.S. Census Bureau. 2018a. 2014-2018 ACS PUMS Data Dictionary. https://www2.census.gov/programs-surveys/acs/tech_docs/pums/data_dict/PUMS_Data_Dictionary_2014-2018.pdf

U.S. Census Bureau. 2018b. County Business Patterns. https://www.census.gov/programs-surveys/cbp.html

U.S. Census Bureau. 2017b. Health Insurance Historical Tables, August. https://www.census.gov/data/tables/time-series/demo/health-insurance/historical-series/hic.html

U.S. Equal Employment Opportunity Commission. 1979. Questions and Answers to Clarify and Provide a Common Interpretation of the Uniform Guidelines on Employee Selection Procedures. *Federal Register* 44

U.S. Fire Administration. 2017. U.S. Fire Statistics. https://www.usfa.fema.gov/data/statistics/

U.S. Global Change Research Program. 1990. *Legal Mandate.* https://www.globalchange.gov/about/legal-mandate

U.S. Global Change Research Program. 2018. *Fourth National Climate Assessment.* Washington, DC, USA: U.S. Global Change Research Program. https://doi.org/10.7930/NCA4.2018

Vacca, James, Helen K. Rosenthal, Corey D. Johnson, Rafael Salamanca Jr., Vincent J. Gentile, Robert E. Cornegy Jr., Jumaane D. Williams, Ben Kallos, and Carlos Menchaca. 2017. Automated Decision Systems Used by Agencies.' *New York City Council.* https://legistar.council.nyc.gov/LegislationDetail.aspx?ID=3137815&GUID=437A6A6D-62E1-47E2-9C42-461253F9C6D0

Verma, Sahil, and Julia Rubin. 2018. Fairness Definitions Explained. In *Proceedings of the International Workshop on Software Fairness,* 1–7. FairWare '18. New York: Association for Computing Machinery. https://doi.org/10.1145/3194770.3194776

Viescas, John. 2018. *SQL Queries for Mere Mortals: A Hands-on Guide to Data Manipulation in Sql.* Addison-Wesley Longman Publishing Co., Inc

Wallace, Tim, Derek Watkins, and John Schwartz. 2018. A Map of Every Building in America. *The New York Times,* October. https://www.nytimes.com/interactive/2018/10/12/us/map-of-every-building-in-the-united-states.html

W. Han, L. Di, Z. Yang. 2019. CropScape - Cropland Data Layer. *National Agricultural Statistics Service.* https://nassgeodata.gmu.edu/CropScape/

Wickham, Hadley. 2017. *Advanced R.* CRC Press. https://adv-r.hadley.nz/

Wickham, Hadley, and Garrett Grolemund. 2017. *R for Data Science: Import, Tidy, Transform, Visualize, and Model Data.* 1st ed. O'Reilly Media, Inc

Williams, Henry. 2019. Economic Forecasting Survey. *The Wall Street Journal.* https://www.wsj.com/graphics/econsurvey/

Winkler, William E. 1990. String Comparator Metrics and Enhanced Decision Rules in the Fellegi-Sunter Model of Record Linkage. In *Proceedings of the Section on Survey Research Methods,* 354–59. American Statistical Association

Wolper, Allan. 2012. Bill de Blasio and Allan Wolper. *Conversations with Allan Wolper.* https://beta.prx.org/stories/81520

Zemel, Rich, Yu Wu, Kevin Swersky, Toni Pitassi, and Cynthia Dwork. 2013. Learning Fair Representations, ed. Sanjoy Dasgupta and David McAllester, 28:325–33. In Proceedings of Machine Learning Research 3. Atlanta, Georgia, USA: PMLR. http://proceedings.mlr.press/v28/zemel13.html

Zheng, Qiming, Qihao Weng, and Ke Wang. 2019. Developing a New Cross-Sensor Calibration Model for Dmsp-Ols and Suomi-Npp Viirs Night-Light Imageries. *ISPRS Journal of Photogrammetry and Remote Sensing* 153: 36–47. https://doi.org/10.1016/j.isprsjprs.2019.04.019

Index

A

Accelerometer data, 83
Accuracy
  accuracy metric, 144, 290–292
  error estimate, 203, 204
  error rate, 144
  model accuracy, 155, 182, 205
  model performance, 128, 130, 154, 155, 160,
    174, 197, 304, 338
Affirmative action, 290
Agglomerative clustering, 226
American Community Survey (ACS), 94, 151, 204,
  223, 341
Anonymizing data, 48
Apache hadoop, 314
Apache spark, 58
Arithmetic functions, 96
Artificial intelligence, 12, 210, 283, 315, 319
Autocorrelation structure, 134
Automated variable selection, 158, 196
Average treatment effect, 166, 170

B

Bias
  instrumentation bias, 287
  measurement bias, 287, 289, 332
  prejudicial bias, 289
  recall bias, 287
Bias-variance tradeoff, 137, 175, 176, 339
Bitcoin prices, 176, 177, 183
Bloomberg administration, 299, 321, 328
Boosting
  weak learner, 211
Building footprints, 185, 239

C

Causal inference
  confounding factors, 167, 221
  control group, 113, 150, 163, 164, 166–169,
    173, 337
  control variables, 158, 161

counterfactual, 150, 163, 166, 167, 170, 172,
  174, 182, 183
difference-in-differences, 167, 172, 174
forcing variable, 168
regression discontinuity, 114, 167–169, 330, 341
running variable, 168–171
temporal precedence, 165
treatment effect, 2, 163, 164, 166–171, 173,
  174, 182, 183
Census Bureau, 43, 88, 94, 145, 151, 176, 204, 223,
  224, 297
Chicago police district, 250, 251, 255
Choropleth map, 250, 251
Class
  class function, 16, 21, 22
Classification accuracy, 141–143, 145
  classification error, 145, 194
  classification threshold, 80, 143–146, 149
  concordance statistic, 144
  confusion matrix, 143, 144, 189, 290
  f1-score, 141, 144, 145
  hit rate, 81, 155, 156, 195, 198, 311
  positive predictive value, 144, 290
  ROC curve, 144, 339
  true negative rate, 143, 144
  true positive rate, 143, 144
Classification models
  anatomy of a classifier, 141
  binary outcomes, 3, 141, 147, 151
  classification algorithms, 141, 285
  decision boundary, 145, 146, 186
  minority class, 145, 149, 150, 302
  separability, 142, 223
Climate data, 239, 241
Clustering
  cluster analysis, 217, 226, 232
  clustering techniques, 218, 219, 234, 235, 271
  soft clustering, 234
  total cluster variance, 221, 222, 224
Comma Separated Values (CSV), 26, 30, 34–37,
  59, 241, 244, 245, 247, 248, 264, 318, 330

© Springer Nature Switzerland AG 2021
J. C. Chen et al., *Data Science for Public Policy*, Springer Series in the Data Sciences,
https://doi.org/10.1007/978-3-030-71352-2

Printed in the United States
by Baker & Taylor Publisher Services